Petroleum Exploration Worldwide

Petroleum Exploration Worldwide

A history of advances since 1950
and a look at future targets

John C. McCaslin

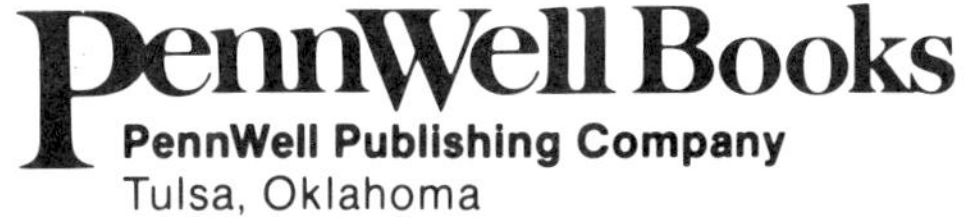

Dedication

Dr. Frank J. Gardner
Former boss, friend, and mentor

PennWell Publishing Company
1421 South Sheridan Road/P.O. Box 1260
Tulsa, Oklahoma 74101

Library of Congress Cataloging in Publication Data

McCaslin, John C.
Petroleum exploration worldwide.

Includes index.
1. Prospecting—History. 2. Petroleum—History.
I. Title.
TN271.P4M38 1983 553.2'8'097 82-22499
ISBN 0-87814-220-7

Printed in the United States of America

1 2 3 4 5 87 86 85 84 83

Contents

Preface

In this book I review the growth and progress of a member of the oil industry whose impact on our global society is unequaled in business history—the explorationist—what he has found since 1950 and what he thinks still lurks in the complex and varied geological basins, structures, sedimentary traps, reefs, arches, anticlines, synclines, and other features of this incredible world. Using the momentous past as a backdrop for the future, we attempt to see what opportunities lie ahead for oil hunters working in one of history's most fascinating industries.

Acknowledgments

The author wishes to thank Bob Lair for his editing of this book and for his advice and counsel in preparing the text and illustrations.

He also wishes to thank the many friends and others who have contributed so much to the literature of the greatest industry on earth and to all of those geologists, alive and passed on, who found and developed the billions of barrels of oil and the trillions of cubic feet of gas that have been discovered around this world since 1950, but not forgetting the thousands who discovered the billions of barrels of hydrocarbons before 1950.

1 History and Outlook

The saga of oil really began in 1854 when George H. Bissel and Jonathon G. Evelth bought the tract containing one of Oil Creek's oil springs in Pennsylvania. Pennsylvania Rock Oil Company was set up, but exploitation of the springs was a failure, and the venture suffered a series of frustrations. Ownership changes then took place with a lease executed to the Seneca Oil Company. "Colonel" Edwin Drake was the man in the driver's seat.

Drake visited the Tarentum salt wells and determined to drill for oil. He hired drillers and bought salt-well drilling equipment and a steam engine. Whether he expected to find oil alone or as a byproduct of brine will always remain a question. However, he built a derrick and began to drill in 1859. On August 27, 1859, the drill had penetrated 30 ft of rock to a total depth of 69½ ft. Oil rose to the surface. A pump was installed, and the well made 30 b/d of oil. Ferris, Kier, and other refiners swarmed into Titusville to buy the crude at $20/bbl.

Drake had shown that oil could be found in big volume by drilling for it through rock.[1] The American oil industry was born then and there. Within days after his well came in, all of the land along Oil Creek was leased. Weeks later, dozens of wells were producing oil. If Drake had not been successful, someone else would have been very soon—in Canada, Poland, Austria, or in other parts of the world—for many were trying at the time of Titusville.

By the spring of 1860, there were 24 producing wells along Oil Creek. One year later there were clusters of producers in other areas of western Pennsylvania, southwestern New York, and Ohio. Producing depths ranged from 60–300 ft. A 2-b/d well was considered successful; a 40-b/d well was phenomenal at $20/bbl.

In 1862, crude production in the area rose to more than 3 million bbl. The price jumped between 10 cents and $2.25/bbl. When the Civil War scarred the land, drilling slowed, demand rose, and the price went to $14/bbl. When the war ended, prices dropped and discoveries again made the news.

As America moved west after the Civil War, so did the oil industry. There were big oil finds reported in Kentucky, Tennessee, Illinois, Kansas, Texas, Colorado, and California. But a big part of the oil industry's early history began in 1901 at the Lucas well on a salt dome 4 miles south of Beaumont, Texas. This historical well at Spindletop gave birth to the modern oil age in America, blowing in for an estimated 75,000 b/d at 1,139 ft. Subsequent drilling proved the field to be a giant. There were 500 wells pumping 17 million bbl/year of oil at that time. The field is still on the active list.

Exploratory work and wildcatting spread throughout the Gulf Coast like a prairie fire. Most of the oil giants that still live today were found between this period and World War II. The years just before the war saw discovery of Conroe gas field in South Texas, the harbinger of a major oil find on the same location.

Old Ocean was found in 1934. This was the forerunner of a wave of geophysical strikes along the Gulf Coast, followed quickly by Hastings and Anahuac fields.[2] Salem field was discovered in 1937 in Illinois. Geophysics was instrumental in locating this giant. This author, as a child returning from a visit to New York and its many sights, will never forget driving through Illinois one summer night during the Salem boom. Light from the gas flares colored the inky nightscape with brilliant orange and red tints. It was my first sight of an active oil field in the making, an experience never forgotten. It was at Williston in North Dakota in 1951 before I saw such a sight again.

Creole was the first offshore field in the U.S., found in 1938 in the Gulf of Mexico. This discovery pioneered revolutionary methods for exploration and development of underwater reserves, creating a new industry within an industry, new technology within technology.

Big Hawkins field was found in East Texas in 1940 as a result of surface and core drill work. This technology developed the field into a multimillion-barrel giant.

The discovery well of the large and very prolific Oficina area in Venezuela was based on a combination of torsion balance and refraction-seismograph data. This 1933 discovery is considered to be one of the world's giant oil fields, with an ultimate recovery of 610 million bbl. In late 1939 came the discovery of big Eocene oil in Lake Maracaibo. The three main operators in this vast lacustrine project were Creole, Shell, and Mene Grande, which developed the Bolivar Coastal field.

Post World War II

Following World War II, many important new oil provinces were discovered in the U.S. In 1943, prior to the end of the conflict, West Edmond field was found as a major oil field and stratigraphic trap in central Oklahoma. In 1946, Rangely, the giant of the Rockies for many years, outlined Colorado's first major oil find.

Western Canada's modern oil history came to life in 1947 at Leduc in an Alberta reef complex. Devonian reef exploration began in western Canada, spawning many large and lucrative oil and gas fields.

In 1947 this author was working as a jug hustler in West Texas, never dreaming that the job was being done on the site of the 1948 Scurry-Snyder reef strike, one of the giants of all time. This discovery had an impact on oil finding throughout the world. Geophysics was notably successful in finding similar producing trends in the vast Permian basin empire as well.

The big news in the Rockies after the war was the birth of the Denver basin as a modern oil province. Ohio Oil opened Gurley field in Nebraska in the late 1940s, heralding the beginning of an oil play that continues today in the area. The big basin had production for many years before the Gurley strike. In fact, Canon City and Florence fields, near Canon City, Colorado, followed closely on the footsteps of Titusville and were considered the second and third significant discoveries in the U.S.

From the 1950s

The Williston basin was opened to the wildcatter in 1951 at Beaver Lodge in North Dakota along the Nesson anticline in an area that "absolutely had no oil prospects." The San Juan basin became a major oil province in 1952 and the home of the world's largest gas deposits, Basin Dakota field. North America's largest field in areal extent to that date (1953) was at Pembina in Alberta, Canada, on a huge stratigraphic trap. And in 1956, the Paradox basin became a major oil province, opening Aneth as a giant Pennsylvanian field.

In Algeria, operators found the Hassi Messaoud field in 1956 with recoverable reserves of 2.5 billion bbl. This was the first discovery in the northern Sahara Desert. Then, in the same year, SN Repal opened the Hassi R'Mel gas field with its vast 42–56 trillion cu ft storehouse of reserves.

Within 14 years after its first discovery, Nigeria became the world's tenth leading producer of oil. The first commercial find was at Oloibiri in 1956. A second field was discovered in the same year at Afam. Shell BP opened Jones Creek, the nation's best, in 1967, when civil war erupted. It was not until 1970 that the oil industry got back on its feet in Nigeria.

In northwestern Europe, Groningen field in the Netherlands was found in 1959. This is the largest gas field in western Europe with 58 trillion cu ft of reserves or more.

After this huge Dutch find, the industry began to look at the nearby North Sea basin as a potential coffer for vast oil and gas deposits. The look was well rewarded. Thirty-seven wells were drilling in British waters in 1965–66, of which four located huge gas fields. However, the big news came in 1970 when the Ekofisk field was found as a billion-barrel-plus pool in Norwegian North Sea waters.

Then a flock of other giant discoveries followed, among them the ultragiant Forties on the British side. Phillips found Ekofisk, and BP opened Forties field.

One of the most significant oil events in American oil history took place in 1957 at Swanson River in the Cook Inlet basin, southwest of Anchorage in Alaska. After this find, drilling moved into Cook Inlet itself, resulting in McArthur River, Granite Point, and other marine discoveries.

However, the discovery well of the century for the U.S., and indeed, one of the most significant strikes in the industry's long and glorious history, was Prudhoe Bay in 1968 on the North Slope of Alaska. This giant of all U.S. giants went on stream in 1977 with reserves set at a conservative 10 billion bbl.

The industry moved into the 1980s with vigor in some areas and hesitancy in others.

Table 1–1 lists the landmark events of the industry's history. If progress continues, the rest of the century will see this list grow, with new giants found in every possible geological province in the country.

Down the road

There are some new provinces to look at, too. Fig. 1–1 shows where the undiscovered resources lie in just one part of the world, the U.S.

The Western Overthrust Belt is being explored and developed cautiously. There is action along the entire length of the Appalachian Mountain system in the eastern U.S. Very little has been found to date, but the interest is there and has never waivered.

Many thought there was not much oil or gas left to be found in Oklahoma, but recent events have disproved this theory. The deep rocks of the Anadarko basin are being probed with vengeance, tapping important new gas reserves at great geologic depths.

Myriad basins of the Rocky Mountains have never been completely explored despite many years of exploration and development. True, some of them seem to be fully explored and are loaded with giant oil fields found many years ago, all of which are still on production. But vast areas of the Powder River, Williston, Paradox, Uinta, Black Mesa, Wind River, and Greater Green River basins still have prospects to drill, explore, and exploit.

In the midcontinent province, many areas of South Dakota, Nebraska, Kansas, and Oklahoma may hold important unknown reserves of oil and gas. The Salina, Forest City, Kennedy, Central Nebraska, Arkoma, Cherokee, and other regional basins and structures have large sections of undrilled townships.

New plays pop up in all of the basins and provinces of Texas from time to time. There are scores of counties in the state that have seen very little drilling. However, this may change in the years approaching 2000.

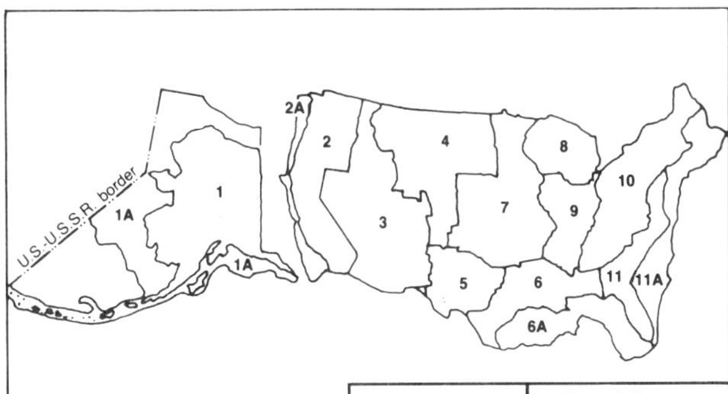

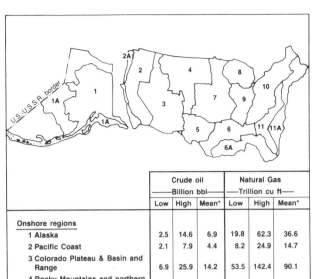

	Crude oil —Billion bbl—			Natural Gas —Trillion cu ft—		
	Low	High	Mean*	Low	High	Mean*
Onshore regions						
1 Alaska	2.5	14.6	6.9	19.8	62.3	36.6
2 Pacific Coast	2.1	7.9	4.4	8.2	24.9	14.7
3 Colorado Plateau & Basin and Range	6.9	25.9	14.2	53.5	142.4	90.1
4 Rocky Mountains and northern Great Plains	6.0	14.0	9.4	29.6	69.0	45.8
5 West Texas and eastern New Mexico	2.7	9.4	5.4	22.4	75.2	42.8
6 Gulf Coast	3.6	12.6	7.1	56.6	249.1	124.4
7 Midcontinent	2.3	7.7	4.4	22.9	80.8	44.5
8 Michigan basin	0.3	2.7	1.1	1.8	10.9	5.1
9 Eastern Interior	0.3	1.9	0.9	1.2	5.0	2.7
10 Appalachians	0.1	1.6	0.6	6.4	45.8	20.1
11 Atlantic Coast	0.1	0.8	0.3	< 0.1	0.4	0.1
Total onshore	41.7	71.0	54.6	322.5	567.9	426.9
Offshore regions (Shelf and Slope)						
1A Alaska†	4.6	24.2	12.3	33.3	109.6	64.6
2A Pacific Coast	1.7	7.9	3.8	3.7	13.6	6.9
6A Gulf of Mexico	3.1	11.1	6.5	41.7	114.2	71.9
11A Atlantic Coast	1.1	12.9	5.4	9.2	42.8	23.6
Total offshore	16.9	43.5	28.0	117.4	230.6	167.0
Total U.S.	64.3	105.1	82.6	474.6	739.3	593.9
Lower48 onshore	36.1	62.0	47.4	288.6	525.9	390.3
Lower 48 offshore	8.7	25.1	15.8	66.1	148.2	102.4

May not add due to rounding. †Includes volumes considered recoverable only if technology permits their exploitation beneath Arctic pack ice, a condition not yet met.

Fig. 1–1 U.S. undiscovered resources (courtesy *OGJ*, 1982)

Along the gulf coastal province, important new discoveries have been made in recent years in the Jurassic Smackover rocks, in the Tuscaloosa sediments of Cretaceous age, the Hosston sands, and in the Cotton Valley sands of the Jurassic. The Tuscaloosa play in South Louisiana has intrigued the industry due to its great depth, expense, and the prolific gas reserves being found.

All of the oil in California has not been found. There are still many oil and gas prospects in this big, old producing province. Much of the state's future oil, of course, lies offshore. It is hoped that the years into 2000 will see

Table 1–1

LANDMARK PETROLEUM HAPPENINGS

Year	Event
1859	Drake discovery at Titusville. First wildcat staked on an oil seep.
1862	Florence-Canon City. First oil field in the Rocky Mountains. Produced from fractured shales.
1894	Corsicana. First of fault-line trend discoveries in Texas.
1896	Coalinga. First of Great San Joaquin Valley fields, followed by McKittrick, Kern River, Midway Sunset, Sunset, and a host of other giants.
1901	Spindletop. First salt-dome discovery. First gusher. Opened the Great Southwest oil industry with a flourish.
	Jennings. First Louisiana salt dome discovery.
1905	Glenn Pool. First important strat-trap discovery in Midcontinent.
1906	Pine Island. North Louisiana's first large oil field. By 1910 field was state's largest at 5 million bbl/year.
1909	Salt Creek. Wyoming's first major oil field set up an active exploration program that found many more productive anticlines.
1910	Electra. First important production on Red River uplift. Sparked North Texas oil boom.
1912	Cushing. This Oklahoma discovery revived anticlinal theory. Re-established geology as a means of finding new oil.
1913	Burkburnett. Town-lot drilling rapidly expanded this North Texas field into giant.
	Healdton. Sparker of southern Oklahoma's first major oil play.
1915	El Dorado. This field made Kansas an important oil state.
1917	Ranger. Surface geology played a major role in this and other similar Texas oil plays of that era.
1918	Panhandle. This became the nation's largest gas field, which covers 5 million productive acres in three states (Hugoton). Panhandle discovery made the area a key oil and gas producer.
1920	Mexia. Started wave of discoveries along fault-line trend, such as Luling, Powel, and Corsicana Deep.
	Burbank. First major oil field in Oklahoma's Osage County. Field produced 20 million bbl/year for first 8 years.
	El Dorado. This Arkansas field opened southern counties and other giant areas.
	Westbrook. The field that opened West Texas to the oil world.
1921	Los Angeles basin. This was the basin's wildcat year with the opening of Signal Hill, Huntington Beach, and Santa Fe Springs.
	Tonkawa. This Oklahoma field became the state's largest by 1922.
1922	Smackover. Arkansas' greatest oil field with 1,000 wells completed in 1 year.
	Kevin-Sunburst. This field opened Montana's oil history.
1923	Seminole. By 1928, new fields in this area had produced 150 million bbl of oil.
	Big Lake. This West Texas field was the region's first major giant strike.
1924	Turner Valley. First important oil production in Canada.
	Atesia. First commercial oil field in New Mexico.
1925	Spindletop. Flank development at 2,588 ft revived this famous oldtimer, setting the stage for an intensive flank exploration on known salt domes.
1926	Yates. Shallow production tipped off a field that is in the billion-bbl-plus category.
1928	Oklahoma City. A large surface anticline became the nation's largest oil field for a time. It still ranks among the giants.
	Kettleman Hills. Deep drilling finally found a major oil reserve on a huge surface structure.
	Hobbs. This field became New Mexico's largest for a time.
1929	Darst Creek. This Texas field was a major fault-line oil discovery.
1930	East Texas. This Dad Joiner discovery put all previous giant discoveries in the U.S. to shame. The 5-billion-bbl-plus field is second only to Prudhoe Bay even today.
1934	Old Ocean. This Gulf Coast discovery was the forerunner of a wave of geophysical successes along the coast. It was followed by Hastings and Anahuac.
1937	Salem. Geophysics revived the Illinois oil search and boom days ensued.
1938	Creole. This was the first Louisiana offshore oil field, heralding an era of intensive offshore work.

1943 West Edmond. This Oklahoma field was a major strat-trap discovery.

1946 Rangely. This Colorado field became the state's first major find.

1947 Leduc. Western Canada's modern oil history began at this Alberta Devonian reef discovery. It was the forerunner of a long period of flush reef finds.

1948 Scurry Reef. Reef production in West Texas had an impact on oil finding worldwide.

1949 Denver basin. The Ohio Oil Company discovery in western Nebraska opened this basin and sired a busy search that lasted well into the 1950s.

1951 Beaver Lodge. The Williston basin got its first oiler at this North Dakota field discovery. Years of intensive wildcatting and development followed.

Oil matches coal as U.S. energy source.

1953 Neutral Zone discovery at Burgan.

Pembina. This huge strat trap became Canada's largest field.

1955 Citronelle. This was and is Alabama's largest oil field.

1956 Aneth. This Pennsylvanian discovery in Utah opened the Paradox basin.

Hassi Messaoud opens in Algeria. Nigerian discovery opens production.

1957 Swanson River. This Cook Inlet onshore discovery opened Alaska's oil era.

1959 Groningen gas field opens in Holland as one of world's largest.

1960 Fairway. This Pettet Cretaceous discovery was one of the most important East Texas discoveries of its time.

1961 Libya begins its oil history. Roma gas field opens in Australia.

1965 Gas found in British North Sea.

1967 Bass Strait, major Australian discovery.

Siberia becomes major oil area.

Mideast War shuts down industry.

1968 Libya becomes 7th largest oil producer.

Prudhoe Bay. The giant of all American giants was opened on Alaska's North Slope with a 10-billion-bbl reserve. Production began in 1977.

Dahomey becomes oil producer.

Syria joins list of world producers.

Mexico's offshore era begins at Atun.

1969 Drake Point gas field opens in Canadian Arctic Islands.

1970 Amazon jungle oil flows from Colombia's Orito to Ecuador coast.

Norway opens North Sea at Ekofisk.

Forties field opened in U.K. North Sea.

Brazil's first offshore discovery flows.

Jay. This Jurassic field established a southeastern anchor to the booming Southeast U.S. Smackover-Norphlet trend.

1971 Romulus. First oil discovery in Arctic Islands.

Taglu field opened in Mackenzie Delta.

1972 Canadian Arctic gas field opened. Chiapas development begins in Mexico.

1973 Arabs cut back production and exports, upsetting western world's industry. Era of cheap energy ends.

Kavala field opens offshore Greece waters.

Iran is 4th largest oil producer in world.

1974 Saudi Arabia reserves hit 138,000 million bbl of oil.

Japan is world's largest importer. Energy crisis hits U.S.

1975 Zaire opens production.

Norway's Ekofisk field tied in to U.K. through pipeline.

1976 Overthrust Belt. This newest of American oil provinces promises to be one of the most important in the Rocky Mountains.

U.S. spacecraft analyzes surface of Mars.

1977 Alyeska. Trans-Alaska pipeline begins operations.

1979 Bohai Gulf.

1980 Anadarko gas era. Overthrust Belt giants.

Williston revival.

Philippines.

1981 Gulf of Thailand.

offshore rules eased, so the industry can really get its feet wet and find the prolific oil and gas stores that lie off California.

The Pacific Northwest has yielded only one commercial field since 1900—Mist in Oregon. However, the interest is there, and the drillers are working diligently, some of them into deep areas. Here, too, many geologists strongly feel, lies an untouched offshore oil province.

Offshore, now and beyond

More than one-fifth of the world's total hydrocarbon reserves are in offshore fields.[3] These reserves total about 200 billion bbl in oil or gas on a BTU equivalent basis. In actual percentages, 23% of the world's crude oil reserves and 14% of the gas reserves are offshore.

Offshore production provides about 20% of the world's crude oil supply and 15% of its gas. That accomplishment required finding solutions to a wide array of engineering problems. However, more problems remain to be solved as global exploration/production moves into deeper waters and increases in hostile Arctic regions.

Nearly half of the oil and gas known to exist offshore has been produced while more than 80 of the offshore resources remains beneath the sea.[4] The reasons for the global industry's accent on land operations are twofold:

1. It has taken time to find and produce oil and gas in deeper and deeper water.
2. The cost of producing a barrel of offshore oil is much greater than for a barrel of onshore oil.

Even so, the offshore industry has developed the technology and hardware to drill and produce just about anywhere, under any conditions of weather and seas.

After reviewing the highlights of the oil industry since Titusville, an in-depth investigation into who found what and where is of merit. Also, the plans for future exploration in the old and new (but undiscovered) provinces of the world will be considered.

REFERENCES

1. *Oil & Gas Journal, Petroleum 2000* (August 1977), p. 99.

2. Ibid., p. 101.

3. H.D. Klemme, *Oil & Gas Journal, Petroleum 2000* (August 1977), p. 108.

4. *International Petroleum Encyclopedia* (Tulsa: PennWell Publishing, 1980), p. 4.

2

Appalachia and the East Coast

The Appalachian Basin, the oldest oil and gas producing region in the U.S., has seen a revival of exploratory interest since 1980 (Fig. 2–1). The drilling emphasis in this multistate area is on the deep rocks. A flurry of wildcats has been probing deep overthrusted features in what has been called the Eastern Overthrust Belt (Fig. 2–2).

Majors and independents alike have been assembling lease blocks and running seismic lines along the entire length of the 1,100-mile belt. Geologists warn, however, that the similarity of names and geology does not warrant comparisons between the eastern and western overthrusts. The older and more established play in the West has found far more reserves than the fledgling effort in the East.

Also, discoveries in the East have been in thrust-related fields in front of the overthrust area, not in thrusted traps such as those found in the Rocky Mountain play in

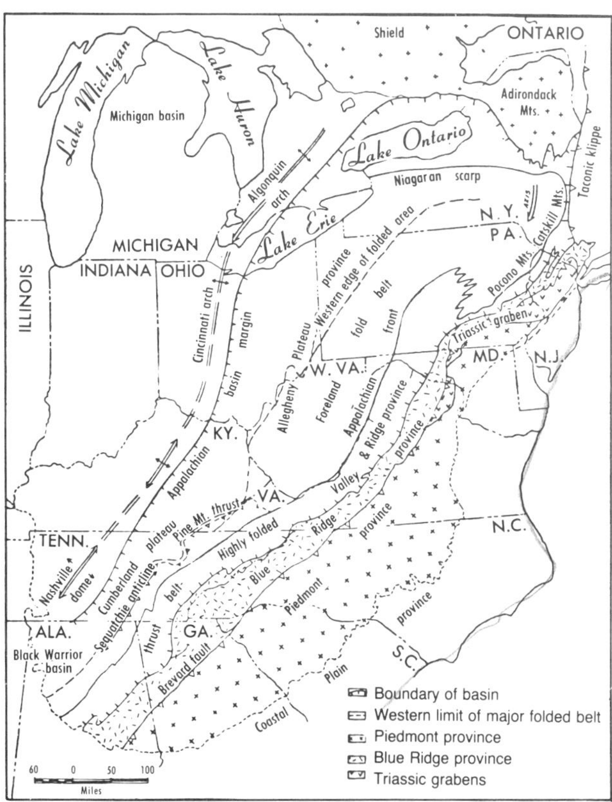

Fig. 2–1 The Appalachian basin (courtesy *OGJ*, October 20, 1980)

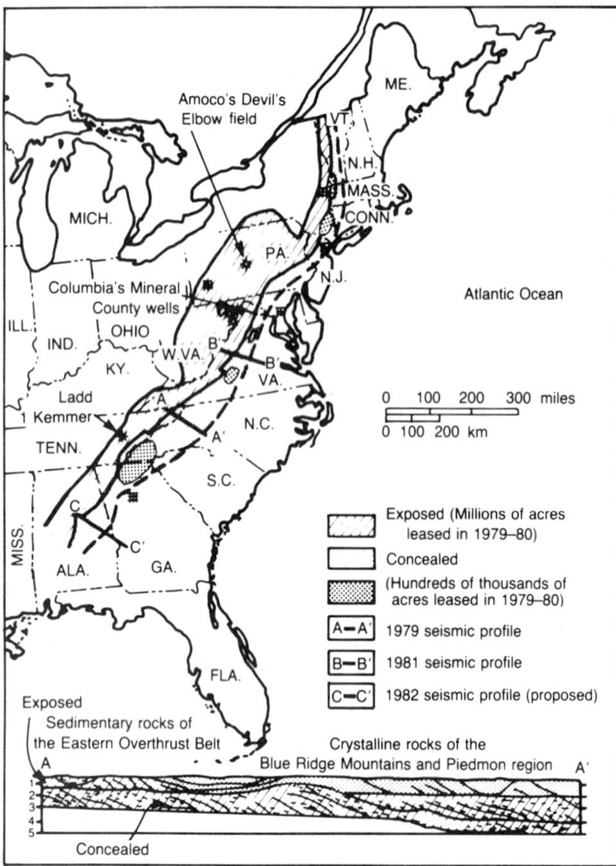

Fig. 2–2 Key finds and tests in Eastern Overthrust (courtesy *OGJ*, after U.S. Geological Survey)

Wyoming and Utah. Nevertheless, the potential exists for the Eastern Overthrust Belt to rival the Western Overthrust, and oil companies have committed millions of dollars to explore that potential along the feather edge of the Appalachian Mountain Belt.

Leasing has been brisk from Vermont to Alabama. Seismic crews have been busy everywhere along the belt, and oil companies are searching vigorously. Operators have been drilling more than 25 holes/year in or near the Eastern Overthrust Belt, up from only a few in previous years. Most of the action in the northern part of the play has been by the major companies, which have amassed huge lease blocks. In the shallower areas of the southern portion of the folded belt, smaller oil companies have led the way.

Geologists feel that the Eastern Overthrust Belt will be gas prone, based on the age of the rocks and the production to date. However, it is still much too early to rule out substantial oil discoveries. Explorationists are now looking for the key to unlock production in the deep Appalachian region, but they are probably 4 years from finding that solution.

1982 activity

Important tests drilling in the region include wells in Tennessee, Virginia, West Virginia, and Pennsylvania. In Buchanan County, Virginia, Columbia Gas (one of the largest operators in the eastern play) is working on two deep Devonian shale tests.

Amoco Production drilled a 9,500-ft Silurian Tuscarora wildcat in Frederick County, Virginia, coming up with a dry hole. Atlantic Richfield drilled to 12,408 ft in a sidetrack hole at 1 Slemp in Lee County, 8 miles south of Big Stone Gap on the Powell Valley anticline. The well was orginally slated for 16,500 ft or basement, but the operator lost pipe in the first hole after reaching 13,252 ft. A whipstock was then set to about 10,510 ft. The well is located in the Yokum Station District of Big Stone Gap Quadrangle.

In Chesterfield County, Virginia, Merrill Natural Resources drilled two 4,000-ft wildcats in the Hallsboro Quadrangle, Clover Hill District. The 1 Georgia Pacific will locate 64 miles south of Hallsboro and 29 miles southwest of Swift Creek Reservoir. The 1 Turner is 29 miles southeast of Hallsboro and 29 miles northwest of the reservoir.

Tennessee action includes a Hancock County wildcat by Amoco and Anschutz at 1 Reed in 10-1s-74e. This is a 17,000-ft Precambrian probe. In Anderson County, Atlantic Richfield also had a 10,500-ft Precambrian hole in 21-4s-64e.

West Virginia tests are in Greenbrier and Hampshire counties. Amoco has a 6,000-ft Tuscarora wildcat at 1 Savoy at Rucker Gap, and Aminoil had a dry hole at 8,489 ft in the Oriskany at 1 McKee in Hampshire County.

Amoco had two tight holes on the Centre County drilling report in Pennsylvania. One is in Devil's Elbow field with a 9,350-ft Tuscarora objective. The other is a wildcat slated to 11,300 ft and the Tuscarora.

Shallow play

Active shallow drilling continues throughout the entire Appalachian basin in and around the old historic hunting grounds for oil and gas (Fig. 2–1). Such plays are widespread from Central Ohio to northwestern and western Pennsylvania and New York state and southward into West Virginia.

Gas flowed at a rate of 7.4 MMcfd at Tennessee Land and Exploration's 1 Bishop, 6-J-71, Clay County, Kentucky. Pay is the Ordovician Trenton, new to the area. Nearest production is Big Lime 3 miles north.

In West Virginia's Cairo-Ritchie field, the Petroleum Development 1 Williams in Murphy District, MacFarlan Quadrangle, flowed 400 b/d of oil on tests, falling to 250 b/d and stabilizing. Production is from six treated Devonian zones at 3,340–5,910 ft. This well was part of an extensive drilling program by the company. It is 1 mile east of a Devonian shale oiler that flowed 500 b/d. It also is 6 miles southeast of 1 Rogers, a 1981 completion that flowed 1,140 bbl of oil in 14 hr, then stabilized at 150 b/d of oil and 3 MMcfd of gas from an open-hole interval in the shale.

In Roane County, Harmony field, the Arabie 1 Greenleaf in Harper District, Walton Quadrangle (4 miles northwest of Walton), flowed 1.86 MMcfd of gas from four Devonian shale zones at 2,906–3,418, 3,503–4,256, 4,325–4,873, and 4,983–5,322 ft. Harmony previously has had only shallower Mississippian production. This new pay discovery is 4 miles southwest of the nearest Devonian shale production.

In Wirt County, Burning Springs South field, West Virginia, Bow Valley 947 McCauley flowed 350 b/d of oil and 500 Mcfd of gas from the Devonian shale at 4,295–4,325 ft. The location is 5 miles north of Devonian shale production in Burning Springs South field and 6 miles southeast of the town of Elizabeth.

Overthrust score

Discoveries by two companies, Amoco and Columbia, fired the interest in the potential of the Eastern Overthrust Belt. In 1977 Amoco found Devil's Elbow field in Centre County, Pennsylvania, with a 6.5-MMcfd gas well. In 1980 the company completed two more wildcats as gas strikes, one 3 miles south of the field and another 3 miles west for 2 MMcfd and 1 MMcfd, respectively.

Columbia in 1979 found Devonian Oriskany gas at a wildcat in Mineral County, West Virginia, flowing 9.8 MMcfd (Fig. 2–2). Step-outs were successful, and drilling continues in this area of the belt. Little else has been found in the continued drilling effort in the Eastern Overthrust Belt. The only oil find was in Tennessee at the Ladd Petroleum 1 Kemmer in Cumberland County. The test was drilled to 10,141 ft but found nothing in the overthrust zones. It was plugged back for small oil and gas shows in the Ordovician.

The potential

Although the Appalachian basin has seen many years of shallow development drilling, the deeper, 60,000-sq-mile thrust belt is one of the least explored regions in the nation. The Appalachian Overthrust is much older in geologic time than the Rocky Mountain Belt, but in many respects it is a mirror image. Both were formed by the thrusting of older rocks up over younger rocks.

A major concern in the Eastern Overthrust has been porosity. Geologists feel their biggest challenge is to find the structural features that have produced enough fracture porosity to contain economic accumulations. Because of the abundance and diversity of structural settings along the 1,100-ft thrusted belt, much drilling will be required before this key is found.

So far, the overthrust-related fields in Pennsylvania and West Virginia have yielded mostly gas, and companies do not expect that pattern to change in the future. No single horizon has come to the forefront as the primary, universal exploratory target. Various geological objectives exist, depending on the well location. Geologists believe that the most favorable outlook for the Eastern Overthrust is discontinuous clusters of 100–150 billion cu ft fields. At present they think the lack of fracture porosity diminishes the chances for fields in the 500-billion-cu-ft class or higher.

The eastern belt is more unexplored than its western partner since it is much newer to the industry, and it is wider and larger. Only about 35 deep holes have been drilled in this vast region. In comparison, companies note that 300 dry holes were drilled in the Rockies belt before the first discovery was made. Even in heavily explored areas of Pennsylvania and West Virginia, only three or four counties have had more than one significant belt wildcat drilled. In the entire state of Alabama, for example, only one deep hole has ever been drilled in the belt.

Leasing plays

Some majors leased broad bands of acreage even before using seismic to define and refine structural trends and prospects. The main operators in the play are Amoco Production, Texaco, Exxon, Columbia Gas Transmission, Atlantic Richfield, Gulf, and Standard Oil of California. Amoco holds more than 5 million acres, at least half in overthrusted areas.

Columbia Gas, which has been in the play since the start, has more than a half million acres and says one of its biggest problems now is infill leasing. Exxon has leased hundreds of thousands of acres in the northern area of the belt.

In many areas, a company's lease is the first ever on the acreage. Land ownership maps are hard to find. Legal descriptions can become nightmarish. Most of the region has been settled for more than 300 years and has been splintered into relatively small tracts.

Seismic surge

Seismic exploration along the Eastern Overthrust Belt has increased five times in as many years. The region has its own particular problems. One is the rugged, densely forested terrain. In the Rockies much of the land is sagebrush; here it is heavy forests. The jumbled topography makes it nearly impossible to set straight seismic lines, so companies often must shoot along existing roads. These roads are few, and they are always crooked and usually short in length. Therefore, the cost of an Eastern Overthrust Belt seismic runs from $5,000–$7,000/mile.

Geology

There are different targets up and down the Appalachian theater of operations. At Devil's Elbow, the shallow Upper Devonian sand has been a conventional target. Below the Oriskany, thrust-related folds in the Silurian Tuscarora sand have yielded high gas flows in this field. In the Pennsylvania-West Virginia section, the Lower Devonian Oriskany sand has been a major target, particularly in thrust-related traps.

In Tennessee, much of the overthrust action has been to the Silurian Clinch. Throughout the region, geologists feel there is yet unproven potential for production from the Cambro-Ordovician, perhaps in the Knox.

The hard Appalachian rocks dull bits and slow drilling progress, but most drilling is not too deep to manage. Many companies use air drilling, since it is cheaper and faster. The firms revert to mud only after hitting water or production or when logging is needed.

Companies say it costs as much to drill to 8,000 ft in the East as it does to drill to 15,000 ft in the Anadarko basin. One said an 11,000–12,000-ft northern Appalachian hole could cost $2.5–3.5 million to complete and an 8,500-ft hole in the southern Appalachians about $1 million.

The stratigraphy of the Eastern Overthrust Belt is essentially continuous with the rest of the Appalachian basin. Therefore, little separation of the belt from the adjacent area of sedimentary rocks exists on a stratigraphic basis. The only major difference in any of the rock units (except the expected predictable facies changes) is the amount of deformation to which they have been subjected. Therefore, adequate well data adjacent to the thrust belt can be extrapolated onto the deformed portion of the basin. Each subsurface unit that is to be interpreted is exposed at the surface somewhere in the belt.

All this vast belt needs in order to come into its own as a potential oil and gas province is another good discovery. The pace seems to be slowing some, but there should be exploration of deep zones for at least another 5 years.

Atlantic offshore

Industry is gearing up to begin exploration drilling in the South Atlantic, which could prove to be the focal point of Atlantic exploration for the near future (Fig. 2–3). Meanwhile, drilling continues in the Georges Bank area of the North Atlantic, and plans have been laid for renewed seismic and evaluation work in the Baltimore Canyon area of the mid-Atlantic (Fig. 2–4).

Although costly, much of the drilling activity in the coming years by major oil companies will be focused on deepwater tracts in the south and mid-Atlantic areas. Also, a proposal to provide incentive pricing for gas produced from water depths 300 ft or greater could provide still more impetus to search for hydrocarbons in deeper waters.

A Jurassic reef, which has yet to be drilled, trends from offshore Mexico through the Gulf of Mexico and up the eastern seaboard off the U.S. This may be a center of much of the future Atlantic offshore action. Choice deepwater tracts offered in OCS Sale 59 lie on each side of the reef trend.[1]

Exploration, however, was dealt a punishing setback when about half of the sale tracts that received bids were rejected.[2] Industry's hopes for economic drilling and development of offshore tracts hinge on changes in govern-

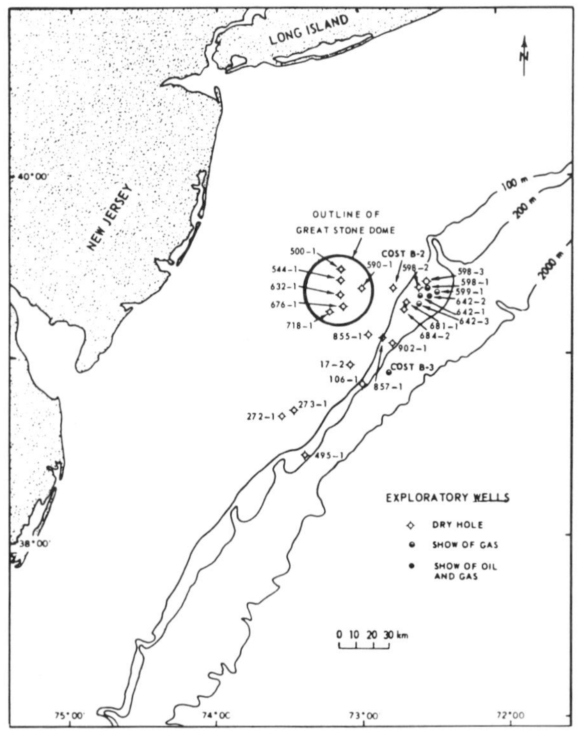

Fig. 2–3 Deep holes drilled in Baltimore Canyon (courtesy *OGJ*)

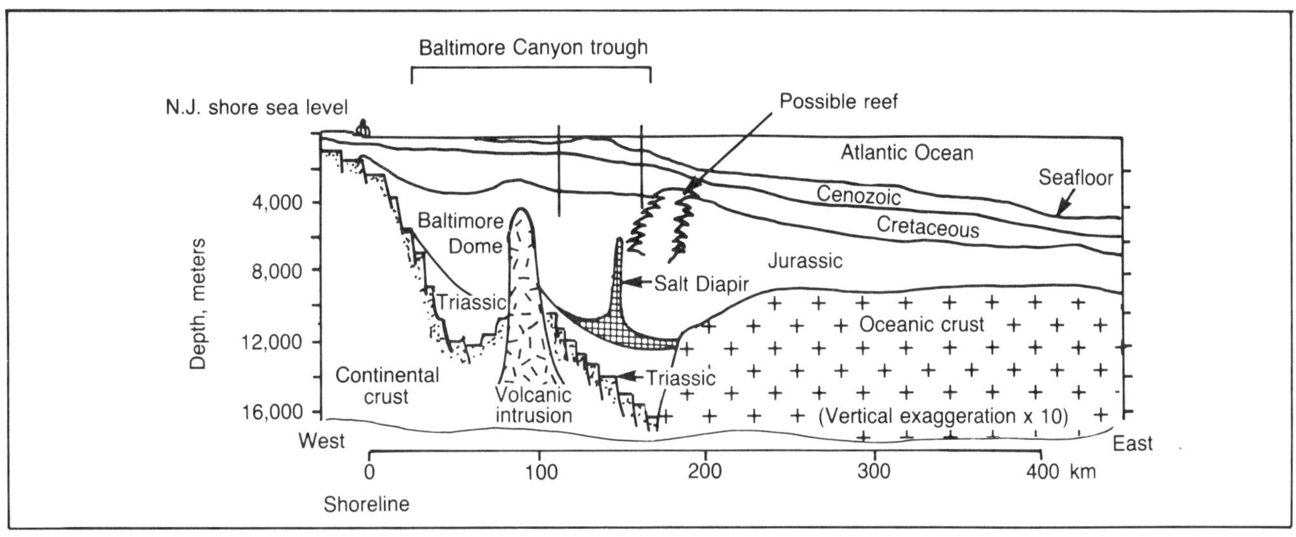

Fig. 2–4 A geological interpretive cross section of the Baltimore Canyon (courtesy *OGJ*)

ment tract evaluation; otherwise, spokesmen say, there is the chance for recurring rejection of bids.[3]

What's being done

Operators are laying the groundwork to begin drilling the South Atlantic off North Carolina in the Manteo region on tracts acquired in OCS Sale 56.[4]

"Offshore North Carolina is going to be the number one area in the mid and South Atlantic" for at least 1983, says Guido DeHoratiis, district drilling engineer, Minerals Management Service (MMS), Northfield, New Jersey. Thus far, the major concern in the area involves currents, which could reduce the amount of drilling activity until more is learned about the gulf stream.

Chevron conducted tests using moored current meters or anchored lines with current meters attached at varying depths. Some previous area measurements indicated a current in excess of 3 knots in the gulf stream. Other data showed that the current runs as high as 5 knots, which would require a lot of energy to keep a rig on location with minimal drillpipe stress.

The current and water depths—ranging from 650–6,500 ft in the Manteo region—limit what types of rigs can work in the area. Both factors also increase the costs of drilling. A rig capable of handling the current and of remaining on station would be needed, such as a drillshop, a world-class semisubmersible, or possibly even a dynamically positioned semi.

DeHoratiis says rig requirements have contributed to the slowness of any activity in the area. The MMS requires an operator planning to drill in the South Atlantic to conduct site-specific current measurements during 6-month periods before drilling, as well as during the entire drilling process. Chevron began measuring before the requirement. ARCO Exploration, which could be the next company to drill after Chevron, also plans current measurement tests.

Leases from OCS Sale 43 in the South Atlantic will expire in April 1983. Observers say no exploratory drilling activity is anticipated in the sale area.

South Atlantic

Chevron has drilled a 20,000-ft wildcat on Manteo region Block 510 in 2,132 ft of water, using the Ben Ocean Lancer drillship. ARCO has submitted a draft exploration plan and environmental report on Blocks 709 and 710 in the Russell Canyon area off North Carolina and south of the Manteo region.

Mobil and Gulf have completed geohazards studies in the South Atlantic. Mobil concentrated on the Manteo region, and Gulf charted the Cape Fear area. Union Oil of California is expected to start similar studies in Currituck Sound and the Manteo region.

Water depths in these areas are fairly deep. Rig availability is limited, and chances are that the companies will share equipment.

Mid-Atlantic

No drilling is under way in the Baltimore Canyon area, but further seismic surveys are planned. Excluding two continental offshore stratigraphic tests (COST), B-2 and B-3, a total of 28 wells has been drilled in the area. Of these, 23 were dry.

According to industry estimates as of December 1981, operators spent an estimated $245.6 million for the two COST wells and geological/geophysical work. Not counting OCS Sale 59 leasing, industry spent about $1.2 billion on sales in 1976 and 1979. Drilling costs for 22 of the dry holes and 5 wells with hydrocarbon shows were about $334.5 million, for a grand total of about $1.749 billion.

The 1978 B-3 COST well holds the U.S. OCS water-depth record. It was drilled in 2,686 ft of water in Wilmington Canyon Block 66 by Chevron using the Ben Ocean Lancer. The well was plugged in January 1979.

A B-4 COST well was proposed initially by E.S.C. Inc., Newhall, California, in August 1980 to be drilled to 20,000 ft in 4,300 ft of water on Hudson Canyon Block 868. It was to be drilled prior to Sale 59. However, the project never materialized due to the timing of the upcoming sale and the lack of a dynamically positioned drillship.

Tenneco Oil, which won approval of a 4-block unit and a 2-year suspension of production, conducted 3-dimensional seismic survey of the blocks which are on the same geological structure. The unit involves Hudson Canyon Blocks 598, 599, 642, and 643. Block 642, owned by a Tenneco-led group, has been certified as producible.

On test, there was flow of hydrocarbons from five of the eight wells that have been drilled on three of the four blocks. Gas has been found on Blocks 598, 599, and 642, as well as oil on 642. No well has been drilled on Block 643, but one well on Block 642 was drilled close to the lease line of Block 643.

According to MMS data, a Texaco group on Block 598 drilled one well resulting in test rates of 7.5 MMcfd and 9.3 MMcfd from two zones. Tenneco, in its 642-1 well, tested gas from three zones of 5.5, 14.2, and 18.9 MMcfd. Its 642-2 well, from three zones, tested 12 MMcfd of gas and 100 b/d of condensate, 1 MMcfd of gas, and 630 b/d of oil. Its 642-3 well tested a total of 9.65 MMcfd from two zones.

Exxon tested a total of 10.3 MMcfd from three zones from its 599-1 well.

Shell plans some high resolution work for a tentative well in the Wilmington Canyon area of Sale 59. This well is to be drilled in the second quarter of 1983, depending on approvals and work progress. Potential drilling sites range between about 3,000–7,000 ft. Gulf also has filed a preliminary application for future drilling on the deepwater Wilmington Canyon Block 328. The water depth averages more than 6,300 ft.

Most of the 39 leases awarded from OCS Sale 49 remain in effect. However, some operators have filed relinquishment notices on some blocks earlier than their scheduled expiration. For example, Exxon asked to relinquish Blocks 639, 946, 989, and 21.

MMS has six exploration plans covering 12 Sale 49 blocks that are still active. Four cover one block each, and one Exxon project covers Blocks 163, 164, 206, 207, and 208 off New Jersey. A plan by Amerada Hess includes Blocks 796, 840, and 839. The other operators and blocks involved are Tenneco, 495; Texaco, 192; Murphy Oil, 107; and Shell, 85.

MMS says all plans are tentative. Most operators intend to begin drilling late in 1983; leases expire in April 1984. The water depths in these areas average 300–500 ft, about 60–90 miles offshore in an area stretching from New Jersey to Maryland.

North Atlantic

A handful of operators continue exploratory drilling in shallow water in the Georges Bank area of the North Atlantic. The industry has spent nearly $1 billion in the area since leasing began, with little success. Three dry holes have been drilled thus far, totaling $58 million in drilling costs alone. Two of the three, drilled by Exxon, cost $27 million, and the Shell dry hole on Block 410 was $31 million. Other expenses for geological/geophysical work have been pegged at about $100 million.

Two COST wells have been drilled in the area. The first was drilled to 16,000 ft about 80 miles east of Nantucket. It was completed in July 1976 and cost $8.2 million. The other, about 120 miles east of Nantucket, went down 21,000 ft and was completed in August 1977. That tab totaled $13.6 million.

Much of the future action could center on deepwater areas. Many of the tracts lie over the ancient Jurassic reef trend.

Conoco, Shell, Mobil, and Tenneco are the only operators drilling. Conoco, the latest to spud, is drilling ahead on its first test on Lydonia Canyon Block 145. Shell and Mobil continue drilling Lydonia Canyon Blocks 357 and 312, respectively. Tenneco's test on Lydonia Canyon Block 187 has been pegged at $35 million.[5]

Meanwhile, more exploration is planned. Mobil filed an exploration plan in April 1982 for six Lydonia Canyon blocks—267, 310, 311, 353, 354, and 397. Water depths average 265–325 ft. It listed as primary drillsite Block 310, and Mobil plans to use the Rowan Midland semi.

The Minerals Management Service has approved an exploration plan for Exxon on Lydonia Canyon Blocks 269, 317, and 359. A Tenneco exploration plan was filed early in May for Hydrographer Canyon Block 172. Water depth averages 230 ft. It listed the Zapata Ugland as a possible rig.

Union Oil also has received approval of an exploration plan for Lydonia Canyon Block 271. Water depths average about 275 ft. A possible rig is the Western Pacesetter II.

REFERENCES

1. *Oil & Gas Journal* (December 14, 1981), p. 61.
2. *Oil & Gas Journal* (January 4, 1982), p. 74.
3. *Oil & Gas Journal* (May 10, 1982), p. 264.
4. *Oil & Gas Journal* (August 10, 1981), p. 38.
5. *Oil & Gas Journal* (March 8, 1982), p. 110.

3

Michigan, Illinois, and Kentucky Basins

Until recently the year of maximum oil production in Michigan was in 1939 when the annual output was 23½ million bbl of oil. This total did not last long. A lower peak was reported after the discovery of rich Ordovician Trenton-Black River oil pools on the Albion-Scipio Trend in 1961. Over 19 million bbl of oil were produced that year. This important discovery was on the southern flank of the Michigan basin. It did much to revive interest in the state's future potential as a good oil and gas producer. Other Trenton-Black River discoveries followed, but the big production remained along the Albion-Scipio Trend.

Then interest waned until the 1970s when the Silurian pinnacle Niagaran reef play swept across the northern counties of the Lower Peninsula. The play shifted to the southern end of the basin. Exploration and development along these trends (which lie on the rim of the Michigan basin) continue today.

Oil and gas have been produced from every Paleozoic system in the Michigan basin except the Cambrian. Through 1967, for example, nearly 552 million bbl of oil and 580 billion cu ft of gas had been produced in the basin. During the past several years, hundreds of pinnacle reefs have been found in the Niagaran reef trend. The results have been fantastic. The state's oil production began to rise rapidly as a result of these flush new strikes and was at record levels for several years. Today's oil output in the state is 92,000 b/d, which is nearly 34 million bbl/year.

The Silurian system makes up one of the thickest and most widespread systems in the basin. Reef and biostrome facies are widespread, especially where Niagaran rocks are thickest. These reefs lie all around the basin edge. The reservoir rocks are the dolomitized remains and debris of reef-building and binding organisms such as algae, stromatoporoids, and corals.

Oil-bearing reefs of Silurian Niagaran age were first discovered at Kingsville in southwestern Ontario at 1 Coste, which was drilled on an anticlinal structure in 1889.[1] This well penetrated a reef in the carbonate bank barrier complex on the southeast edge of the Michigan basin, south of the pinnacle reef trend (Fig. 3–1). Subsequent exploration was by random drilling. The smaller, isolated pinnacle reefs were not discovered until the 1930s.

The main search for pinnacle reefs did not begin until the surface gravimeter method was successfully used in 1947 to locate Kimball reef. The gravimeter detected den-

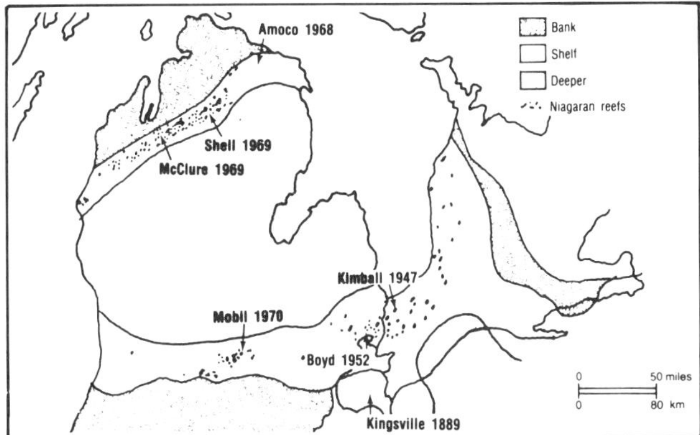

Fig. 3–1 Niagaran depositional environments (courtesy *OGJ*, June 22, 1981)

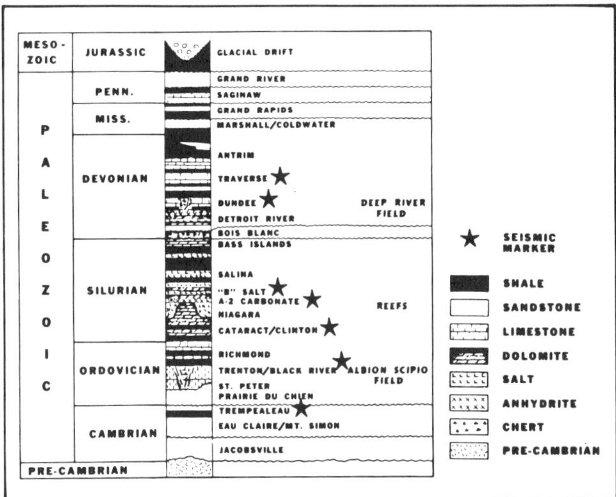

Fig. 3–2 Michigan basin geological column (courtesy *OGJ*, June 22, 1981)

sity contrast between the dolomitic carbonate reef and the surrounding less-dense salt. Continuing into the 1950s, this method was used to locate over 40 reefs—26 containing hydrocarbons with estimated recoverable reserves of 4 million bbl of oil and 170 billion cu ft of gas.[2]

The productive reef trend moved into southeastern Michigan in 1952 with discovery of Boyd reef by Panhandle Eastern Pipeline Co. at 1 Ringle. This well was completed flowing 16.6 MMcfd of gas from the Niagaran. Peters Reef, 4 miles to the northeast of Boyd, was discovered in 1955, also by Panhandle Eastern, and is the largest reef in the area covering some 1,780 acres.

The first phase of the southeast Michigan reef play declined in the late 1960s after discovery of 50 productive reefs with recoverable reserves of 12 million bbl of oil and 187 billion cu ft of gas. Subsurface geology predicted the reef containing shelf environment was circular and was centered on the lower peninsula of Michigan. Anticipating exploration for reefs in the onshore northern and southern parts of the reef trend, Shell, Amoco, and Mobil began leasing large acreage blocks and began to conduct CDP seismic programs (Fig. 3–2).

In 1968, Amoco drilled 1 Draysey in Presque Isle County. This well was a localized reefal buildup on the barrier complex. It only produced 4,000 bbl of oil and could not be considered worthy of offsetting commercially. However, it did prove CDP seismic is a valuable exploratory tool and also confirmed oil-bearing reefs existed in other parts of the basin.

Seismic crews recorded lines on every available road in northeastern Michigan, looking for more reefs. The search led them into the shelf area of Otsego, Kalkaska, and Grand Traverse counties.

The oil potential of the northern trend was finally tapped when the Shell Oil 1-11 Horicon, the third of a 7-well program, was completed for 135 b/d of oil and 179 Mcfd of gas. This discovery was quickly followed by McClure Oil at 1 State-Union in Kalkaska County, flowing 18 MMcfd and 10 bbl of condensate.

The northern reef trend, or fairway, now extends for more than 160 miles from Manistee on Lake Michigan's shore, to just past Gaylord. It averages 6 miles in width. As of 1981 more than 200 hydrocarbon-bearing reefs had been found. Later successes in northeastern Presque Isle County indicate that the trend indeed crosses the state completely from lake to lake. The cumulative for the trend's small features is several hundred million barrels of oil. Mobil opened the modern-day southern end of the trend at 1-A Ellsworth Brown, the fifth well in a 5-well program.

The basin

The Michigan basin is an intracratonic basin that subsided from late Cambrian through Mississippian time. Silurian time began with conformable deposition of the Cataract-Clinton shales and carbonates. Middle and Late Silurian times were characterized by a greatly accelerated rate of subsidence and formation of an extremely thick Salina evaporite section in the basin interior with a belt of Niagaran reefs on the shelf edge. These reefs are thought to have grown in a broad, open, shallow sea relatively far from land on shoals surrounding the rapidly sinking Michigan basin.

In the south central area, Niagaran reef growth shows several distinct stages, though not all are necessarily present (Fig. 3–3). In 1981 a deep drilling wave swept through the Michigan basin as a result of a discovery in the 10,000-ft class in Missaukee County. The Dart Oil 7-36 Edwards in

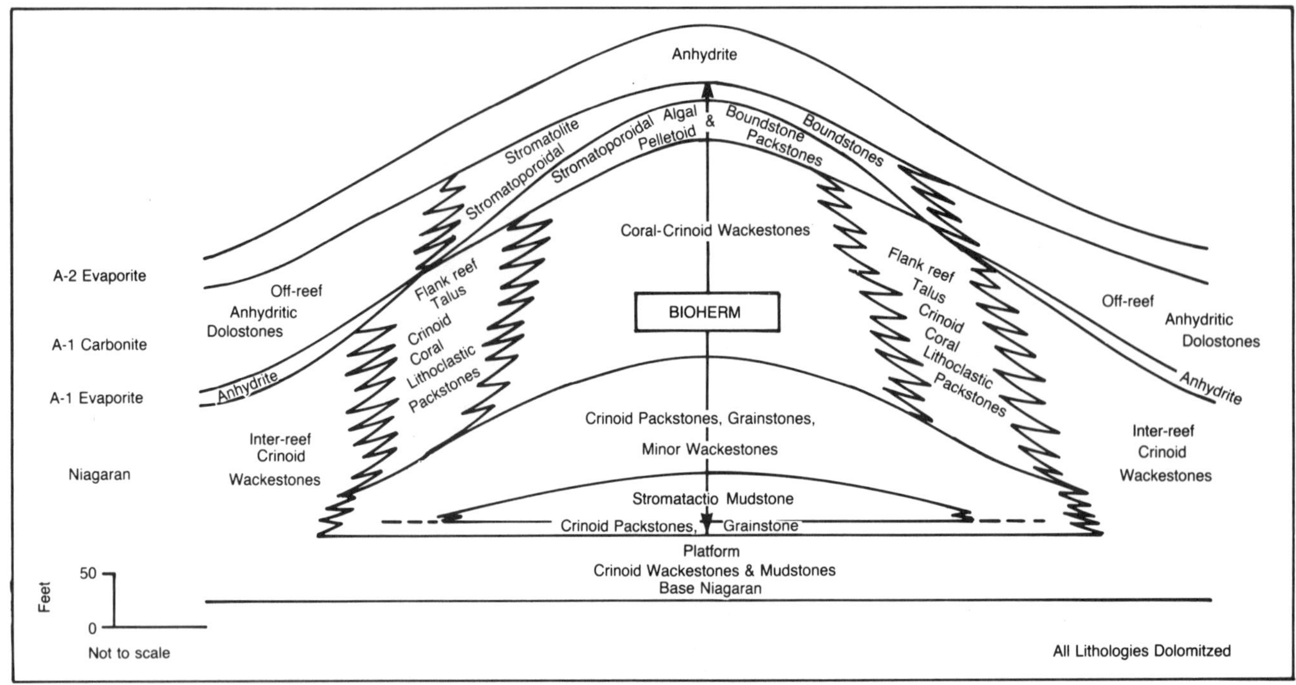

Fig. 3–3 Reef facies relationships—a schematic vertical section, Ingham County, Michigan (courtesy *OGJ*, June 22, 1981)

36-22n-7w, Reeder Township, tested gas in the Prairie du Chien of Ordovician age, coming in just below the Trenton-Black River. Gas was reported near the 8,000-ft level at this interesting well, which set off a flurry of lease activity in 1980 and 1981. Several subsequent dry holes, however, dampened the new enthusiasm for the prospects of deep production in the center of this classic basin.

Some geologists suspect that the Michigan deep basin involved much more complex features, including major faulting, than concepts previously held. Features not unlike the Overthurst area in the West may be found. The so-called basement may be as deep as 25,000 ft in the Clare County area.

Future Michigan exploration will no doubt probe these deep prospects once again. There also will be new reefs found but at a slower rate. There will definitely be a trend to look at the older and shallower horizons in the basin because not all of the oil has been found in Michigan.

THE ILLINOIS BASIN

Illinois has produced gas since 1885 and oil since 1904. In 1908, the first peak of oil production was attained with 34 million bbl. The peak fell into bad times, and it wasn't until 1936 when new oil discoveries in the Illinois basin sent the industry back into this region with furor. In 1940 production reached 148 million bbl of oil. Today's output is 68,000 b/d of oil. This means that new exploration is needed in this three-state basin, covering portions of Illinois, Indiana, and western Kentucky (Fig. 3–4).

Cumulative production from the Paleozoic rocks in this area is more than 4.3 billion bbl of oil since production began. Waterflood projects are prolific throughout the basin.

There are several factors that make the area inviting for future exploration and development because it contains a variety of marine sediments. Only about half of this section has been adequately explored. Most depths to objectives are less than 10,000 ft. The bottom is probably at 15,000 ft in the basin. Markets and transportation are in extremely good supply. And the states are known as oil producers, welcoming industry exploration.

Many geologists feel Illinois and its neighbors need to look for Cambro-Ordovician oil in the Knox formation. Only a smattering of basin wells has penetrated the Knox. Less than a dozen or so have drilled through the Mount Simon sand of early Cambrian age; six of those hit the basement. Pre-Knox control in the deep part of the basin is sparse. There is some evidence the basement may have high ridges and probable faults where the fault expressions are not necessarily brought to surface. These structural features are ideal traps for hydrocarbons.[3]

The Mount Simon sand has had two prominent depositional basins:

Fig. 3–4 Possible exploration areas in Great Lakes region

1. The present area of the Reelfoot basin in Tennessee.
2. The basin in northeastern Illinois. Here the sand reaches 3,000 ft in thickness and bears fresh water.

The Knox group should be considered a highly potential source of oil. At the close of Early Ordovician time, the basin was tilted down to the south. The Knox reached a thickness of 6,000–7,000 ft in extreme southern Illinois. A hinge line trending northeast-southwest appears to have developed just south of a large, broad, shelf-like area that probably reflects the Sangamon arch. There should be some good Knox production found in the future in the southern one-quarter of Illinois.

Production from the St. Peter sand in the lower part of Illinois is also possible. The sand is from 0–425 ft in thickness here. Trapping in the St. Peter would most likely come from one of the following areas:

1. Rapid wedgeouts of the sand against high Knox surface, with a thick cover of Glenwood shale.
2. Where the sand is cut by the upside of a normal fault with the dip of the beds being basinward so as to have open mitigation from basinward side.
3. Regions of similar-type faulting but where upward migration of hydrocarbons has taken place from the fractured Prairie du Chien.

Since the early 1970s, there has been renewed interest in exploring for oil in the Salem limestone of Mississippian age. This interest was fired by the discovery of Salem Oil at Zenith North field in Wayne County in 1972. There were 27 Salem oil discoveries during a 6-year period ending in 1977, involving action in 11 Illinois counties.

Good Salem production has been found under structural noses that have no apparent closure. More drilling may provide data for further study of the deposition and diagenetic history of the Salem limestone. Better understanding of this history should aid geologists in predicting the location of reservoir rocks in unexplored areas.

A good example of Salem interest is that in 1977 over half of all the test holes drilled in Illinois went to this limestone or deeper. Continued interest will certainly find new oil.

Further exploration is expected also in Indiana with a look at deeper formations. There will certainly be new Mississippian fields found in the Indiana portion of the Illinois basin.

KENTUCKY BASINS AND FAULT ZONES

One of the most geologically interesting and unproductive areas of the lower Eastern Interior province of the U.S. is the Jackson Purchase region of Kentucky. This entire section of the state is virtually nondrilled. No more than 35 wells are known to have been drilled into the Paleozoic rocks, but there have been reports of oil shows through the years. In central and eastern Kentucky oil and gas have been produced for many years from Devonian and Silurian formations and from Ordovician formations as deep as the Knox. The Jackson Purchase covers a 2,356-sq-mile, 8-county area in western Kentucky. This interesting region lies in the northern end of the Mississippi embayment of the Gulf Coastal Plain, where Mesozoic (Late Cretaceous) and Cenozoic sediments have filled a structural trough that was developed in Late Cretaceous and Tertiary time on the eroded Paleozoic rocks.[4] The axis of the trough coincides roughly with the present course of the Mississippi River and plunges to the south.

Regional dip of the Paleozoic section is northward into the Eastern Interior basin, away from the Ozark uplift and the now-buried Pascola arch. The Mesozoic and Cenozoic sediments were deposited on the Paleozoic beds ranging in age from Ordovician in the southwest to Mississippian in the southeast.

Geologists believe the Devonian would only be prospective for oil reservoirs in the Purchase area where it is covered by the New Albany shale. South of this subcrop, any reservoir beds would be exposed to flushing from ground water penetrating the Tertiary and Cretaceous formations.

Silurian formations offer little in the way of porous reservoirs except at the pre-Cretaceous unconformity, but the Cretaceous-Tertiary formations are not lithofied and likely would provide a seal for oil entrapment.[5] Oil shows have been reported in the Trenton or Kimmswick formation. However, this formation would only be prospective as a reservoir to the north of the Maquoketa subcrop where a seal might be provided by the Maquoketa shale. The very thick Plattin-Joachim-Dutchtown sequence might provide carbonate porosities for oil accumulation, but the sands lack porosity in this part of the section.[6]

The possibility of fault traps or hydrodynamic traps in the St. Peter sandstone should not be overlooked. There is at least one good porosity zone in the Knox, possibly just below the base of the Everton or Smithville, where an apparent unconformity exists. The early Knox and older Ordovician strata no doubt have porous zones with possible structural and stratigraphic traps. Most of the tests drilled to these rocks in this region have encountered porosity zones, almost exclusively in the secondary dolomites.

A stratigraphic section of Paleozoic beds 8,000–10,000 ft thick offers an opportunity for explorers in a province that certainly has the right ingredients to become an oil province.

Rough Creek fault system

By the mid-1970s, more than 50 million bbl of oil had been produced from the Rough Creek fault zone of Kentucky, and more is available. The deeper pays in the area have scarcely been touched. Numerous shows of oil lend encouragement for exploring the Devonian-Silurian rocks and the Trenton-Black-River formations of the Ordovician with more vigor. The Knox formation is practically virgin in the area, and the underlying Cambrian section is completely unscratched.

This fault system is an east-trending belt of normal and reverse faults located in the Illinois basin near its southeastern margin, mostly in western Kentucky. Regionally, this fault zone, together with the Cottage Grove fault of southern Illinois, divides the Illinois basin into two lobes: a northern basinal area, the deepest part of which is called the Fairfield basin, and a southern basinal area, commonly called the Moorman syncline.[7]

Some of the large oil fields in the area are Morganfield South and Guffey, having produced more than 9 and 6 million bbl of oil, respectively. Most of the production has come from the more basinward, western part of the fault zone. This partly is because the pays of the western counties are absent, thin, and too shallow to produce in the eastern part of the area.

Existing production is almost entirely from the upper 1,000 ft of the Mississippian Chester and Ste. Genevieve pays, but some come from the overlying Pennsylvanian rocks. There is some minor oil production from the Middle and Early Mississippian carbonates. Pre-Mississippian formations produce from just five pools along the fault zone. The deepest production is in Webster County, both stratigraphically and in feet.

Geologists who know this area say the probable future reserves of oil and gas in and near the Rough Creek fault zone are centered in those formations already producing and in those types of traps already found. However, these places have not been sufficiently tested. The best chances appear to be in the fault blocks that have had little drilling. The entire faulted belt is composed of numerous blocks and slices of varied sizes and shapes. Some are grabens, some horsts, and many are parts of a step fault system. But all are isolated by faulting and thus may contain separate potential reservoirs.

Early Mississippian, Devonian, and Silurian formations should be the objectives for more drilling ventures in this area. If pre-Knox hydrocarbons are found in the area, then gas (or gas and condensate) can be expected where the reservoir rocks are 12,000–15,000 ft deep or deeper.

Pine Mountain thrust

Back in 1978 the Cumberland thrust block in southeastern Kentucky yielded an important discovery southeast of the Pine Mountain thrust. The first three wells in Canada Mountain field in Bell County each flowed in excess of 10 MMcfd of gas from overthrust Mississippian Big Lime above 3,100 ft. Numerous shows of oil and gas have been recorded in scattered wells drilled on the Cumberland thrust block. Potential reservoirs range in age from Pennsylvanian to Early Ordovician.

The Canada Mountain field discovery was highly significant. It was the first commercial production on the Cumberland thrust block of southeast Kentucky. The importance of the thrusting to the field also highlights the potential within these and other reservoirs along the leading edge of the Pine Mountain thrust. The coincidence of overthrust structural traps, true source beds, and high deliverability reservoirs makes the Cumberland thrust block an ideal objective for more exploration in this region.

Rome trough

Studies of the Rome trough in central and eastern Kentucky delineated several large, untested areas containing thick, pre-Knox, Cambrian sandstone reservoirs at moderate (5,500–10,000 ft) depths (Fig. 3–5).[8]

Numerous shows of oil and gas from the Cambrian and Ordovician age have been logged in the relatively few test wells drilled to date in the trough. It is believed that

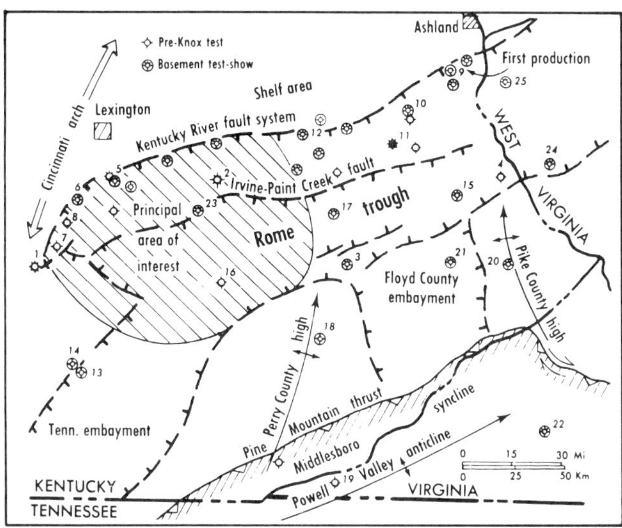

Fig. 3–5 Rome trough of eastern Kentucky (courtesy *OGJ*, November 14, 1977)

there are more attractive, mostly clastic, reservoirs in numerous undrilled traps at lesser depths than those in West Virginia, Pennsylvania, and Virginia.

REFERENCES

1. *Oil & Gas Journal* (June 22, 1981), p. 93.
2. J. Labo, W.G. Werner, and P.H. Pan, *Oil & Gas Journal* (June 22, 1981), p. 93.
3. John Avila, *Oil & Gas Journal* (January 28, 1974), p. 174.
4. Report of Investigations 10, Kentucky Geological Survey, Howard R. Schwalb (1969), p. 7.
5. Ibid., p. 17.
6. Ibid., p. 18.
7. *Oil & Gas Journal* (July 8, 1974), p. 133.
8. O.D. Weaver, *Oil & Gas Journal* (November 14, 1977), p. 250.

4

The Anadarko Basin

The Anadarko basin is one of the largest U.S. basins. It covers parts of western and central Oklahoma, the Texas Panhandle, southwestern Kansas, and extreme southeastern Colorado (Fig. 4–1). This vast basinal area contains more than 40,000 ft of Paleozoic rocks, all of which are prime objectives for the wildcatter. Through the years, most of the basin's production has come from its vast shelves. However, recently the explorationists have forged into the axial trough of the basin, coming up with the deepest hole in the world as well as the deepest producer.

Hundreds of deep and near-deep fields have been discovered with production coming from both structural and stratigraphic traps. Sediments in this famous basin are thick and widespread. Late Cambrian, Early Ordovician Arbuckle, and Devonian Hunton are mostly carbonate reservoirs. The Pennsylvanian Morrow-Springer series, high on the basin's exploration list, is made up of mostly clastics. The basin's axis runs northwest-southeast through the Fletcher-Cyril area and becomes progressively shallower to the north and east (Fig 4–2).

The pays

The deepest part of the Anadarko basin abuts a buried mountain range slicing southeastward from the Texas

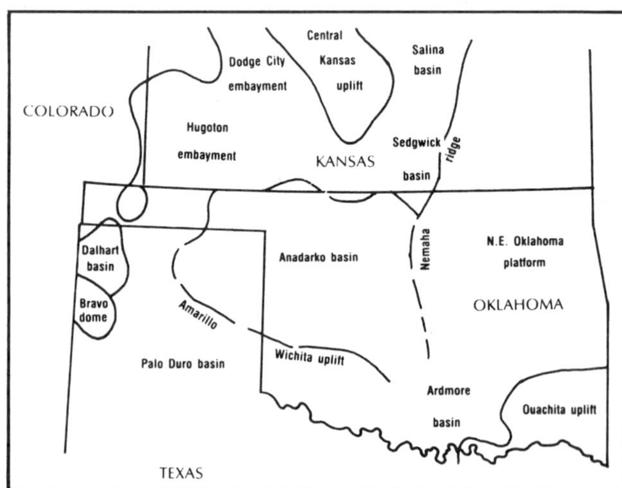

Fig. 4–1 The Anadarko basin (courtesy *OGJ,* January 26, 1981)

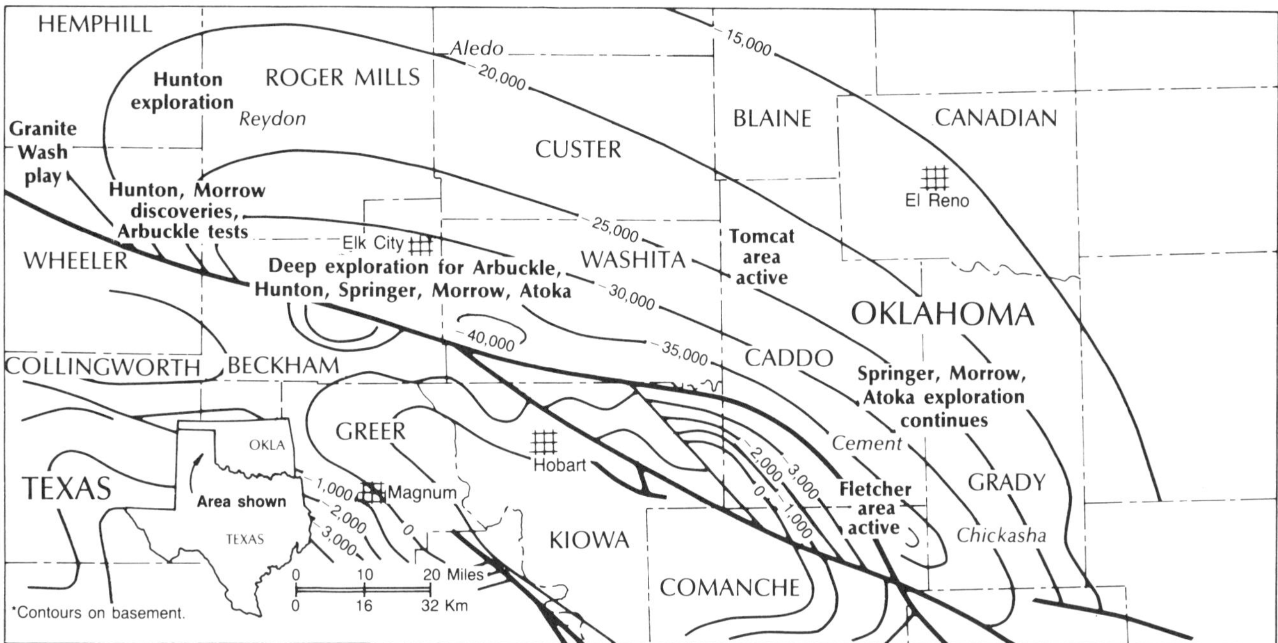

Fig. 4–2 Activity in the deep Anadarko basin (courtesy *OGJ*, 1982, from Petroleum Information Corp.)

Panhandle across western Oklahoma. The geology generally is simple and is often described as layer cake. Operators drill through productive zones that offer good uphole potential in case deeper targets cannot be reached or proven unproductive.

The deepest formations, the Cambro-Ordovician Arbuckle and the Siluro-Devonian Hunton, are structural. Most other pays are stratigraphic with faulted anticlines.

The famous Arbuckle zone is the deepest target in the basin, occurring at more than 40,000 ft near the axis of the basin, geologists believe.

In the western part of the basin, the Granite Wash offers a widespread pay package. There, and elsewhere, the Pennsylvanian Cherokee and Red Fork sands also provide great uphole potential. Other well-known and prolific objectives are the Pennsylvanian Springer, Atoka, Morrow, Puryear, and the Goddard.

Most of the ultradeep tests to the Arbuckle and Hunton are being drilled in the western part of the basin around the town of Elk City and westward into Texas. Red Fork is the common sand target east of Elk City in southern Roger Mills County and central Custer County.

In the eastern deep basin, Springer, Morrow, and Goddard are the main objectives. Drilling surged in this area in 1981 and 1982 around Fletcher field, described by area geologists as a subbasin east of a southward turn of the Mountain View fault. It bifurcates, although the fault continuations are different in nature. As a left-lateral-moving fault where rocks on the basin side were shoved westward, the fault continues as a system along the north edge of Cement and Chickasha and on into Golden Trend field.

As a fault of major vertical throw, it turns southeast and runs down the east side of Apache. The result is an unusual subbasin bordered by the Chickasha horst, the Cement horst, and the big fault on the southwest.

The Hunton has been considered the most prolific of Anadarko formations in the deep basin. In Aledo field (Custer and Dewey counties, Oklahoma) some wells produced 70–80 billion cu ft of gas each since going on stream (Fig. 4–3).

Good Arbuckle wells have reserves of nearly 30 billion cu ft/well, and good upper Morrow producers boast reserves of more than that. Deep Morrow, Atoka, and Springer wells around Elk City are believed to have reserves of more than 75 billion cu ft each.

History

The exploratory history of the Anadarko basin is quite short. Little exploratory action occurred until the early 1950s. The big Panhandle oil and gas field and the Hugoton gas fields, however, had been on production for many years. Carter-Knox was opened in 1922 in the shallow Permian sandstone. Later, gas and oil were found in the Springer series of the Early Pennsylvanian. The field grew rapidly,

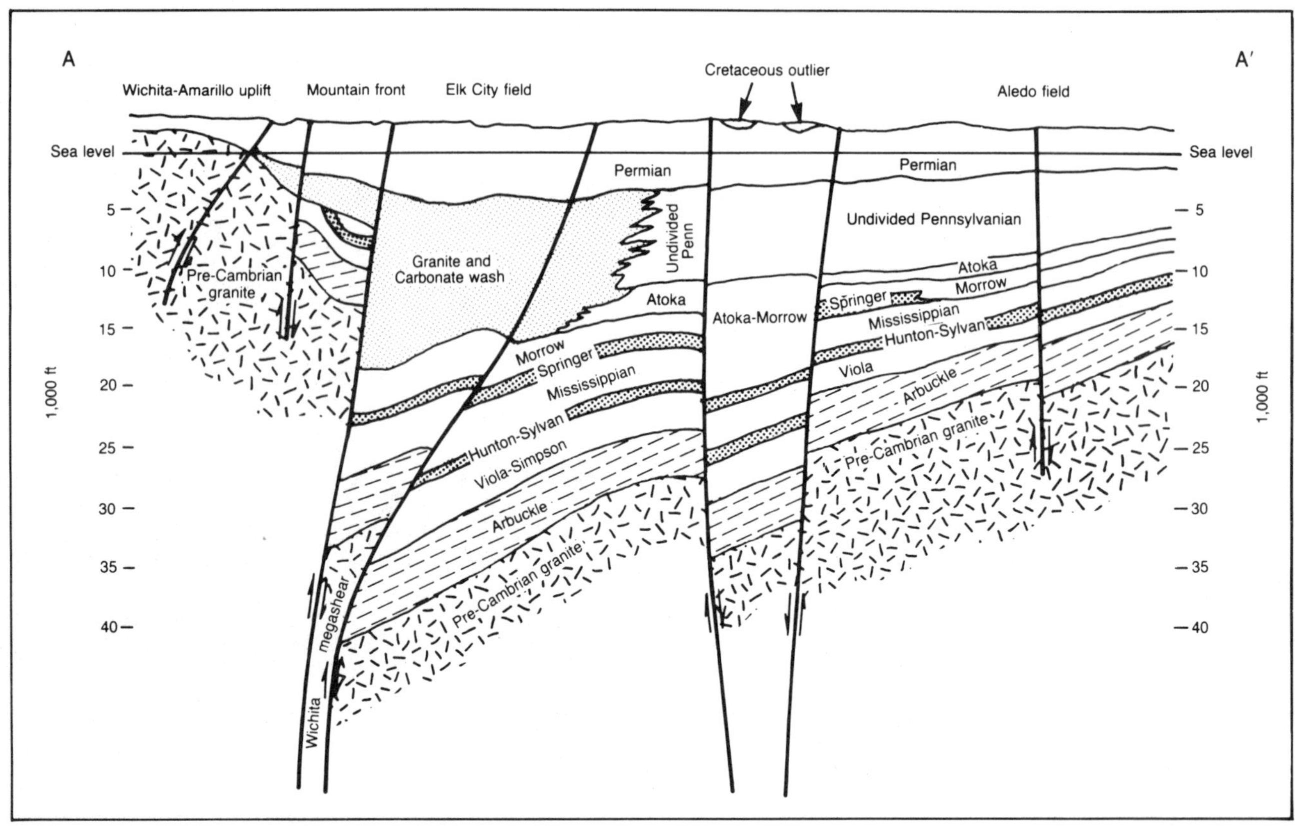

Fig. 4–3 Anadarko basin cross section (courtesy *OGJ*, from GHK Co.)

becoming 10 miles long and nearly 1 mile in width.[1] In 1956 a wildcat in this area found deep gas and condensate below 15,000 ft.

The Chickasha gas field is another oldtimer in the Anadarko basin. It was discovered in 1922 with production from the Permian and Pennsylvanian. Oil came along in 1954, also in the shallow Permian sands. In the last few years, Marchand and other Pennsylvanian zones have been found in wide areas around this field, which has now grown into a multigiant area of pools called the Watonga-Chickasha Trend.

Landes notes one of the oldest oil fields in the basin is Cement, found in 1916 in southeastern Caddo County in the southeastern portion of the big basin. Additional pays through the years in deeper zones caused this field to grow into a prolific pool. Another important basin field was opened years ago at Elk City, now the scene of considerable ultradeep drilling in all directions.

The first production in the Texas Panhandle part of the basin was in Potter County in 1918. This wildcat was on a surface anticline mapped by C.N. Gould.[2] Additional drilling activity confirmed this huge gas field, which crosses the central part of the Panhandle and is one of the largest gas accumulations in the world.

The oil boom began in 1925 when more wildcatting uncovered the big oil pools along the northern edge of the gas field. Following this was the Hugoton gas field in 1922 in Kansas, which lopped over into Oklahoma's Panhandle. There was a long period of quiet exploration in the basin after the Hugoton spread. A discovery at Lips in 1949 opened the modern Anadarko basin chapter. Many oil and gas fields in the Pennsylvanian and other Paleozoic zones were found, and the map of the basin began to look like a big oil and gas field. There was Gageby Creek in Wheeler County, Texas. A well in this field flowed 1.74 billion cu ft of gas/day from the Hunton and the Simpson at 15,000–16,000 ft, startling the industry and causing another brisk exploration and development program to sweep across the prairies.

One of these fields was Washita Creek, an Anadarko classic in southeastern Hemphill County, Texas, about 20 miles southeast of Canadian. The field lies in the deep western sector of the Anadarko basin and north of the synclinal axis (Fig. 4–4). The producing area is on a dome-shaped anticlinal feature slightly elongate in a northwest-southeast direction. Hunton dry gas was found in 1966. Gas from the Morrow went on production in 1969 and from the Hunton in 1970. The Pennsylvanian upper Morrow sand lies at the

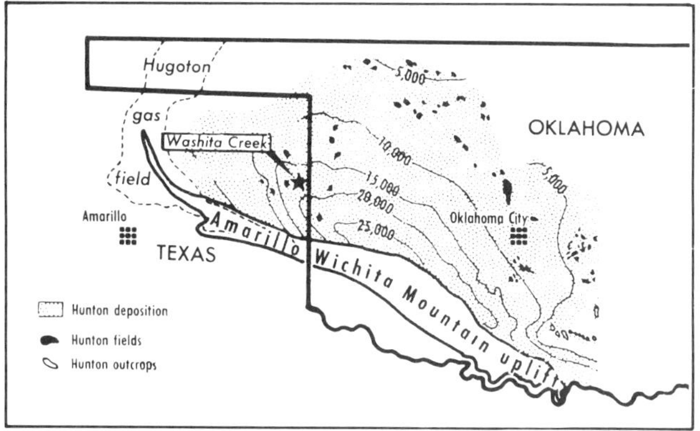

Fig. 4–4 The Hunton structure (courtesy *OGJ*, October 29, 1973)

13,600-ft level in the subsurface. Five lower Morrow members are between 15,200–16,000 ft. Primary gas production is from the Siluro-Devonian Hunton reservoir at depths between 19,300–20,400 ft. The Hunton group of limestone and dolomite rock averages 500 ft in thickness. Porosity developed as a result of dolomitization in the Hunton.[3]

Tutten notes that 200 miles southeast, the Hunton rocks are found on the surface in the Arbuckle Mountains. The Lone Star 1 Baden, at that time, the world's deepest hole at 30,050 ft, 40 miles southeast toward the basin's center, encountered the Hunton some 9,000 ft deeper than at Washita.

The deepest hole in the western world is still in the Anadarko basin at Lone Star 1 Rogers in Washita County, Oklahoma, 27-10n-19w. It went to total depth of 31,441 ft (finding gas at 13,100 ft) and was completed in 1974. There are other tests drilling in the basin that may well sink this depth record.

Hunton rocks are quite irregular in character due to high erosion during and after deposition in the Siluro-Devonian period. Pronounced trends of Hunton production are not indicated by the position of the producing fields in the Hunton depositional basin. Deposition of the Hunton took place in shallow widespread seas; its original areal extent was much greater than it is today. During a major hiatus, the Hunton was completely truncated from the northernmost areas of the shelf and partially or completely truncated from other structural features. Periods of submergence were interrupted at numerous times by uplift, during which erosion caused the Hunton to vary in thickness.[4]

There was also complex faulting in this area of the Anadarko basin. The faults are usually high-angle normal or reverse with step and block faulting being evident. Tutten says evidence of faulting rarely is seen above the Morrow.

Gageby Creek field is a horst block with nearly 500 ft of closure above water. Buffalow Wallow is an anticlinal fold with about 900 ft of closure and is bounded by a down-to-the-east fault. A down-to-the-north fault closes the north end of the fold. Washita Creek is a dip-closed anticline.[5]

Current progress

There had been nothing comparable to today's activity in the industry. In mid-April 1982, there were 650 rigs operating in the Anadarko basin, or 18.5% of the 3,509 rigs working in the U.S. at that time. But of most significance was that 349 of the active Anadarko rigs were drilling toward objectives below 18,000 ft (Fig. 4–5).

While drilling in the basin increased tremendously in 1981, some in the industry wondered how long the activity could continue. The play, the doubters pointed out, was fueled by decontrol of natural gas produced from below 15,000 ft under the Natural Gas Policy Act (NGPA) 1978. Deep gas packages brought prices approaching $10/Mcf at their 1981 peak, twice the level thought possible under total decontrol.

By late May 1982, active rigs in Oklahoma had fallen to 723. An exact count for the Anadarko basin was not attainable but was believed to have slipped to about 550, as both superdeep and smaller rigs were stacked. Operators in the area generally expected activity to continue at lower levels, until the air cleared on federal tax legislation and until crude prices rebounded. High success ratios, higher gas prices, lengthy reserve life of successful wells, ready access to natural gas markets, and steadily improving seismic and drilling technology contributed to the great surge over the past 2 years in Anadarko basin activity.

Elk City in Beckham County, Oklahoma, is the focal point of the deep play. It is the gas capital of the basin. The play spreads all over that county, into Roger Mills through Custer and Washita counties and through the northeastern part of Comanche County. In 1982 61 rigs were busy at one time in the Fletcher area, as well as in Wheeler and Hemphill counties, Texas. The gas hunt also surged into Caddo County where the Ports of Call 14-1 Tomcat well blew out at tremendous rates of gas.

The Fletcher chapter

Saxon Oil completed the 1 Quanah Parker in the "L" sand of the Springer at 18,070–18,117 ft. The well flowed during tests at the rate of 9.6 MMcfd through choke, with flowing tubing pressure of 10,100 psi. Open flow was more than 150 MMcfd.

This well and another successful completion in the area, the Hadson Petroleum 1-7 Livingston, set the stage for a flurry of drilling in Comanche and southern Caddo counties.

Fig. 4–5 An Oklahoma wildcat, the Lear Petroleum 1 Rice in Washita County in the Anadarko basin, was completed dry at 17,200 ft

The Livingston well flowed 3.7 MMcfd during tests of the Goddard at 19,837–19,898 ft with flowing pressure of 4,785 psi. This well is 4½ miles west of the 1 Parker, opening East Midway field.

The Fletcher field sits directly above one of the deepest parts of the Anadarko basin. It appears that a massive Springer sandstone section has been found at 17,000–18,000 ft, extending down to about 22,000 ft. There are multiple sands to be found in this section.

The Fletcher area is called the Cyril subbasin, lying east of a southward turn of the Mountain View fault. Geologists say it bifurcates, although the fault continuations are different in nature. As a left-lateral-moving fault where rocks on the basin side were shoved westward, that fault continues as a system along the north edge of Cement and Chickasha and on into the Golden Trend.

As a fault of major vertical throw, it turns southeast and runs several miles east of Apache. This leaves a subbasin that is bounded by the Chickasha horst, the Cement horst, and the big fault on the southwest.[6] The basin could cover a gross area of three to four townships, but of course, it is too early to assume that the Springer production will cover the entire area.

Solutions

The big splurge of drilling activity in 1981 and 1982 around the Quanah Parker discovery, like the increase in other parts of the basin, results in part from solutions developed for the technical problems of deep drilling. Changing pressure regimes, like those in the area around the discovery well at Fletcher, are common throughout the basin.

Estimated drilling expenditures in the deep Anadarko increased from about $150 million in 1977 to $1 billion in 1980, or roughly one-quarter of the U.S. total money output for deep drilling. The basin ranked second among U.S. geologic provinces in 1981 by expenditures for drilling and by completions, with an outlay of $3.79 billion.

There were 5,456 wells drilled in the Anadarko basin and its shelves in 1981, including 634 new field wildcats, the third largest number in the U.S. well count for 1981. *Petroleum Information* notes that 81 new oil fields were found and 72 new gas fields, for a success rate of 24.1%. The national average for that year was 18.4%.

The Anadarko basin ranks second in North America only to the Gulf Coast province in the amount of gas found to date. Elk City is indeed a giant gas field. It may contain as much as 1 trillion cu ft of gas in the upper Morrow and lower Morrow-Springer formations alone, according to El Paso Natural Gas. The company still feels that the Anadarko basin is the hottest exploration spot in the U.S. today—and will remain so for years to come.

Ouachita Overthrust

Not all of the Midcontinent exploratory interest will be focused on the Anadarko basin in the coming years. A fascinating exploratory play has sprung up in the Ouachita Overthrust Belt that extends from southwestern Arkansas through southeastern and southern Oklahoma, across the Red River into North Texas, and then southwestward to the Marathon Mountains of West Texas.

The Arkansas Novaculite and the Texas equivalent Caballo formation have been studied for over 100 years in outcrops in the Marathon and Ouachita Mountains. Oil has been noted in outcrops, and some of the wells drilled into the Novaculite have had shows of oil and gas. However, until the discovery by Westheimer-Neustadt Corp. at Isom Springs in southern Oklahoma, the Novaculite was not considered to be a potential oil reservoir.[7]

In 1977 the operator in a joint venture with George Ramsey drilled the 1 Wallace discovery well in 2-8s-5e. The zones 3,426–3,440, 3,415–3,420, 3,403–3,409, and 3,390–3,395 ft were perforated. Flow was 250 b/d of oil and 159 Mcfd of gas. In 1978 the laminated zone above the fault was opened and the well was dually completed. Both zones flowed at a combined rate of 350 b/d of oil. By September 1980 this well had produced 180,000 bbl of oil.

Many wells have been completed in Isom Springs, and the cumulative production is well past the 2-million-bbl mark. Active operators are Marshall Oil and Westheimer-Neustadt.

Explorationists began to view the potential of the Ouachitas more seriously after the Isom Springs discovery. This field, with reserve estimates of 10,000–400,000 bbl/well of oil plus associated gas, proved that geologically bothersome Ouachita rocks can contain enough fracture porosity and permeability. The subsurface at Isom Springs is so faulted and folded that wells on 40-acre spacing, about 1,320 ft apart, produce from separate fault blocks. Infill drilling, which will cut well distances to only 660 ft, is expected to tap production from new fault blocks.

More interest was aroused with the opening of McKay Creek field in Terrell County, West Texas, with oil production from the Novaculite equivalent Caballos formation. The play has fanned out into other areas of southeast Oklahoma, through Atoka and Pushmataha counties and into Bryan County near the Texas border. There is new exploratory interest in neighboring Choctaw County as well as several deep wildcats on the way. This deep play also has crossed the Red River into Grayson County, North Texas. Vast areas have been leased throughout the region, all the way from Arkansas southwestward through southeast Oklahoma and across Texas to the Marathon uplift country.

Northern frontal action

Three gas discoveries have been tested along the Arkoma/Ouachita boundary in Pittsburg County, Oklahoma, in what the industry is calling a significant development. Hamilton Brothers Oil 1-30 Indian Nations gauged 1.7 MMcfd of gas with 2,400 psi surface pressure during drilling in the folded transitional Wapanucka at 6,000 ft. The well tested 3.5 MMcfd of gas with 3,600 psi surface pressure after penetrating a normal Wapanucka zone at 9,000 ft.

About 6 miles southeast, 1-8 Blue Creek was completed for 950 Mcfd from the Woodford and Hunton at 12,618–12,758 ft. But the producing formations were siliceous, not the Ordovician Hunton lime and Woodford shale usually associated with the Arbuckle facies. A west offset got 13 MMcfd of gas from subthrust (normal) Wapanucka on a drillstem test. Below the Wapanucka is the pre-Pennsylvanian series: Caney, Sycamore, Woodford, and Hunton with indications of Ouachita facies, Viola, Simpson, First, Second, and Third Bromide, McLish, Oil Creek, Joins, and Arbuckle.

Central Ouachitas

Several million cubic feet of Stanley, Arkansas Novaculite, and Big Fork chert gas production are shut in for pipeline connection in Pushmataha and Atoka counties, Oklahoma. There is gas at Southwest Moyers field, Daisy field, and South Jumbo field.

Patient explorationists are slowly unlocking commercial oil and gas production in the Ouachita Overthrust Belt. Discoveries since 1977 have shattered the Ouachita's undeserved reputation as a nonproducing schist belt. Fourteen fields are producing or capable of production from pays that are part of the Ouachitas or thrust beneath them.

All but one of the fields are in Oklahoma. Except for two, the fields have resulted from discoveries completed since 1977, and several remain one-well fields. This Ouachita Thrust Belt remains one of the least drilled oil and gas provinces in the Midcontinent. Part of the mystery lies in the fact that Ouachita rocks (mostly buried 1,300 miles from western Alabama across Mississippi, Arkansas, Oklahoma, to Southwest Texas) outcrop only in Arkansas, Oklahoma, and

Texas. There were less than a dozen rigs working this vast area in 1982, nearly all of these in Oklahoma. Operators look for action to pick up considerably in 1984–85, as leases begin to expire.

Bulk of the current action is centered around the Oklahoma Frontal Ouachitas to the north and west of the thick main belt. Commercial oil and gas have been found in Ouachita facies and the Arbuckle facies that lie beneath. Prospects are good for new capacity within a few years.

Seismic surveys may cost $20,000/mile. One mountainous 2-acre well location required 1.5 miles of road and cost $150,000. Subsurface hallmarks of Ouachita formations are hardness, 30–90° dip, and low permeabilities that complicate drilling and completion.

Against the backdrop of discoveries, leasing, and additional drilling plans, two operators size up the Ouachitas like this:

"A province this size and the potential that can be indicated and multiplied by the price of tight sand gas and multiplied by only a modest rate of inflation get you into some pretty serious numbers."

"It's a natural for big companies looking for elephants," says another.

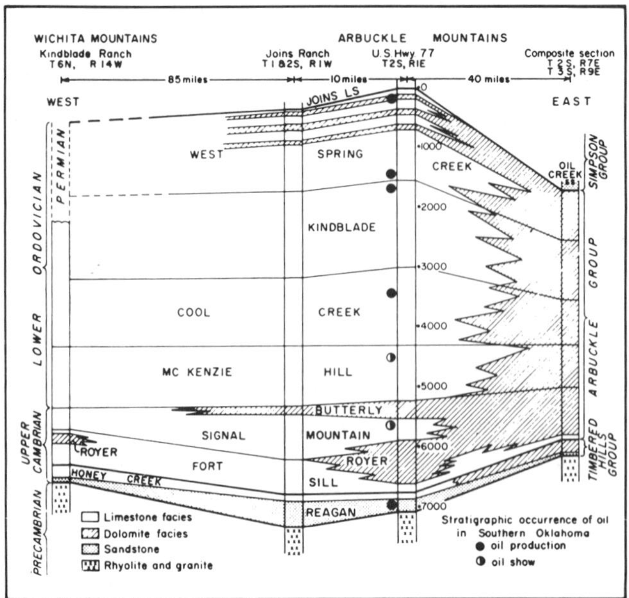

Fig. 4–6 Classification and regional facies of the pre-Simpson strata in the Arbuckle and Wichita mountains (courtesy *OGJ*, December 15, 1958)

Southern Oklahoma

This is a geologically complex and peculiarly oil-rich province that covers about 13,500 sq miles between the Red River on the south, the Anadarko basin on the north, and the Ouachita Mountains on the east. The province, long rich in oil production, is still quite active, and much of the area, which is about 225 miles long and 60 miles wide, remains untouched.

This relatively small field contains two major uplifts, the Arbuckle and Wichita mountains, with a section of Paleozoic rocks nearly 30,000 ft thick.[8] The area consistently produces millions of barrels of oil yearly from about 300 named pay sands of Permian, Pennsylvanian, Mississippian, Devonian, Silurian, and Ordovician age. It locally has more unconformities, faults, and stratigraphic pinchouts per square mile than any other part of Oklahoma. Oil is produced from 500–15,500 ft and deeper.

Near the base of the stratigraphic succession is the Arbuckle group of limestones and dolomites (Fig. 4–6). The Arbuckle limestone is one of the outstanding examples of uninterrupted thick carbonate rocks of the Paleozoic type. Despite prolific production from the Arbuckle in northern Oklahoma and Kansas through the years and from the partly equivalent Ellenburger group in Texas, no oil was found in the Arbuckle in southern Oklahoma until 1940. West Frederick in Tillman County was the discovery well for Arbuckle oil in this province. However, the find did not stir up too much interest. Then, in 1955, Arbuckle oil was found at Southeast Hoover field on a steeply folded anticline at the northwest edge of the mountains. This revived the Arbuckle search in a big way. Other Arbuckle fields were found at Southeast Frederick, Butterly, Royal, Southwest Lone Grove, Loco, and Woodrow.

Geologists feel there is much more oil in this zone to be found in southern Oklahoma. Meanwhile, the industry continues its strong search for Ordovician and other Paleozoic oils throughout the southern part of the state. The search is concentrated in and around older producing areas, but important discoveries also are being made in between these older fields. All of the Paleozoic oil in this province certainly has not been tapped.

Palo Duro basin

A recent Marathon discovery in Briscoe County in the extreme southern part of the Texas Panhandle and an Anadarko Production strike in Potter County have combined to give new exploratory hope to the Palo Duro basin, an area of very few oil fields (Fig. 4–7).

Marathon's rank wildcat pumped 136 b/d of oil from Pennsylvanian rocks at 7,906 ft. This strike set off a brisk leasing play throughout the basin, which runs from eastern New Mexico through the southern Panhandle region of Texas north of the Matador arch, into Hardeman County in northwestern North Texas. It comes to an end in the Hollis

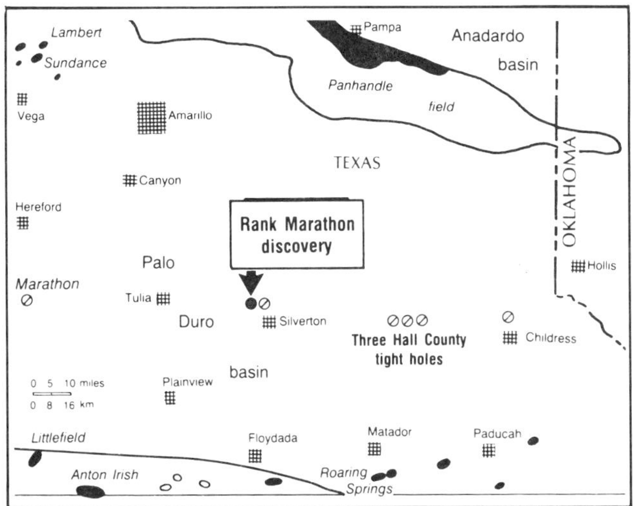

Fig. 4–7 Palo Duro exploration (courtesy *OGJ*, March 22, 1982)

basin of extreme southwestern Oklahoma. There are very few oil fields in this big area, and most of them are found snug against the Matador arch or in the eastern lobe in the Hardeman basin.

Anadarko's Potter County discovery flowed 145 b/d of oil, setting off a Granite Wash play in the Lambert area.

The Palo Duro basin is virtually surrounded by some of the largest producing areas in North America, yet it has not had the success that the industry enjoyed in the nearby Anadarko and Permian basins (Fig. 4–8). Geologists for years have proclaimed the Palo Duro basin as a good place to look for hidden hydrocarbons. But for some reason the industry just never got moving in the basin. There were some

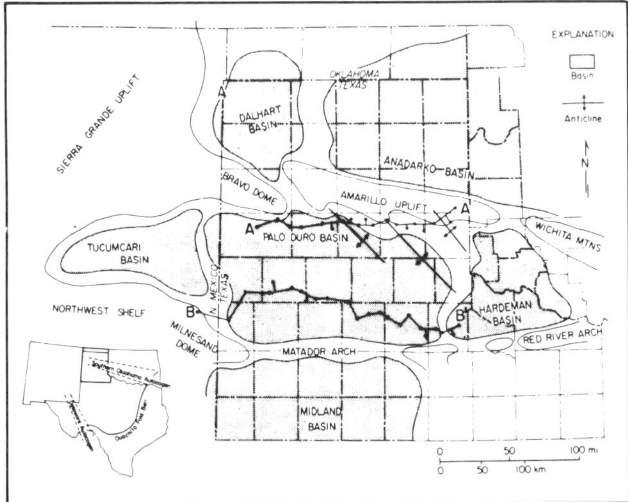

Fig. 4–8 Palo Duro basin—regional location (courtesy *OGJ*, August 20, 1979)

good oil discoveries in the eastern lobe, or Hardeman basin. There was also some wildcatting in the Hollis basin but with little luck. Wildcatting is active in the extreme western end of the basin in New Mexico and on into the Tucumcari basin. New fields also opened along the Matador arch through the years. This structure separates the nearly dry Palo Duro country from the lush oil of the Permian and Midland basins to the immediate south.

C. Robertson Handford of the Bureau of Economic Geology at Texas University in Austin notes Pennsylvanian and Permian strata comprise most of the basin fill and together reach about 10,000 ft in thickness (Fig. 4–9).[9] Lower Permian Wolfcampian strata consist of carbonate and terrigenous clastic sedimentary rocks deposited in fan delta, high constructive delta, carbonate shelf and shelf margin, and slope and basinal environments.

Handford notes three principal fairways that may contain stratigraphically entrapped hydrocarbons in the basin. These fairways are dolomitized-shelf margin carbonates, delta-front sandstones, and fan-delta arkoses.

System	Series	Group	General lithology and depositional setting
Quaternary			Fluvial and lacustrine clastics
Tertiary			
Cretaceous			Nearshore marine clastics
Triassic		Dockum	Fluvial-deltaic and lacustrine clastics
Permian	Ochoa		Sabkha salt, anhydrite, red beds, and peritidal dolomite
	Guadalupe	Artesia	
		Pease River	
	Leonard	Clear Fork	
		Wichita	
	Wolfcamp		Shelf margin carbonates, basin shale, and deltaic sandstones
Pennsylvanian			
Mississippian			Shelf limestone and chert
Ordovician		Ellenburger	Shelf dolomite
Cambrian			Shallow marine(?) sandstone
Precambrian			Igneous and metamorphic

Fig. 4–9 Generalized stratigraphic column and depositional facies, Palo Duro basin (after C. Robertson Handford, courtesy University of Texas at Austin)

Each fairway is proximal to potential source beds, and each consists of porous strata contiguous with relatively nonporous sealing beds. Carbonate-shelf margin stata may be sealed by slope and basinal facies, prodelta shale, and nonporous shelf limestone. Similar producing shelf margin configurations occur at Empire-Abo and Kemnitz fields in New Mexico, according to the Texas geologist.

Handford also says lenticular fan delta and high constructional delta sandstones are up to 200 ft thick in the Palo Duro basin and are similar to producing deltaic sandstones of Mobeetie field in the Texas Panhandle and Morris Buie-Blaco fields in North Texas. In some places arkose seems to lie in contact with uplifted fault blocks along the Amarillo uplift, thus forming possible seals to prevent updip migration of hydrocarbons.

Although lower Permian porous reservoir facies are abundant in the Palo Duro basin, questions about the presence and quality of potential petroleum source beds must be resolved before the true oil potential of the basin can be fully evaluated. There will be more geologic studies in the basin as well as more wildcatting. No doubt some decent oil discoveries will be made in the future in this three-state area of the Southwest.

Arkoma basin

Two geologic provinces in southeastern Oklahoma, the Arkoma basin and the Ouachita Mountains, have been considered of minor interest to the oil industry. But this feeling has definitely changed in the last few years. Exploratory programs are under way in both regions, and development drilling is being conducted in both Arkansas and Oklahoma.

The Choctaw fault forms the boundary between the basin and the Ouachita Mountains in Oklahoma. North of the fault lies a series of northeastward and eastward-trending, gentle synclines and narrower, more sharply folded anticlines. Called uplifts, some of these are structurally more complex than simple anticlines.[10] South of the Choctaw fault is the area of complexly folded and thrust-faulted rocks of the Ouachita Mountains.

Up to 1960, gas had been produced for local consumption in the Oklahoma portion of the Arkoma basin for more than half a century. Production was from the Hartshorne of early Des Moines Pennsylvanian age at 500–3,500 ft. The wells had initial rock pressures between 150–550 psi. Yet some were still producing in 1960.

The first deep gas production of significance in southeastern Oklahoma was discovered in 1930 in the basal Atokan sandstone in LeFlore County on the crest of the Milton anticline. It was called Spiro field, now Cartersville. LeFlore County Gas and Electric and Carter Oil drilled on the Milton anticline in 1958, 7 miles southwest of Cartersville field. Spiro gas at 5,515 ft was found in 17-8n-23e and opened the West Milton field.

Superior Oil completed a wildcat gas discovery in Haskell County in 1952 on the North Kinta anticline with Spiro and Cromwell Morrowan pay at 5,609–6,100 ft. Combined open-flow potential at the 1 Allred well in 18-8n-20e was 30 MMcfd. This was the second indication of big gas reserves in Arkoma basin. Other big gas reserves were discovered during the remainder of the 1950s in both Oklahoma and Arkansas, in new areas, and in and around older and shallower producing areas.

Drilling in the basin continued sporadically through the years. There were a few discoveries, but nothing of real importance until 1982. Regional operators are calling it the first commercial Spiro discovery in the northern part of the Arkoma basin in more than 30 years. Earlsboro Oil and Gas had a rank Spiro gas strike in Sequoyah County, eastern Oklahoma, in 23-11n-23e. The gas flow was a mere 340 Mcfd on choke, but open flow was 1,782 Mcfd. The significance of the well cannot be denied. The pay is Spiro at 4,063–4,090 ft, located 4¼ miles northwest of Cromwell gas at the Brent field (Fig. 4–10).

This gas discovery is the first important wildcat event in this part of the basin in many years. Most of the basin action has been in the trough areas, in the synclines, and on

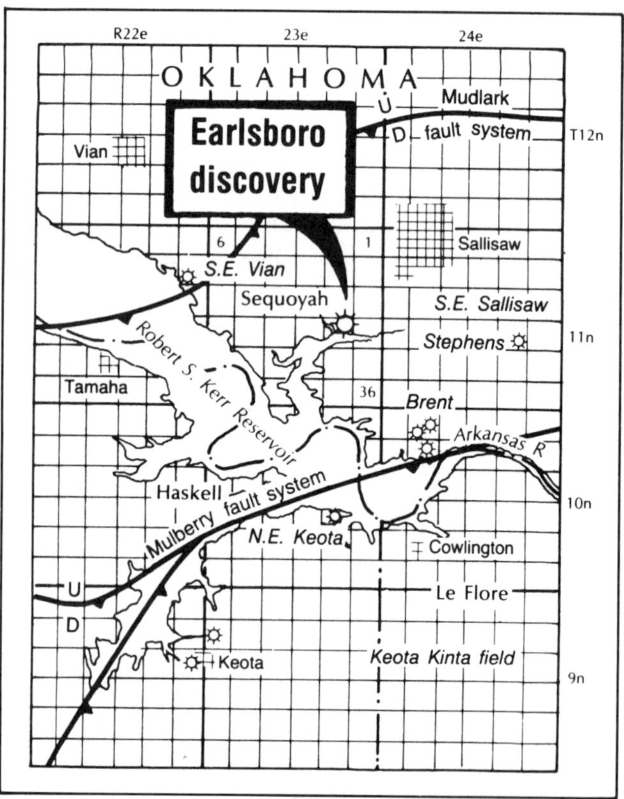

Fig. 4–10 Arkoma basin rank-gas find (courtesy *OGJ*)

the many anticlines that run east-west through the basin. It is the first commercial Spiro strike north of the Mulberry fault in more than 30 years, and is located on the shelf of the basin characterized by a high frequency of faulting. As seen on Fig. 4–10, the Mulberry fault system is a significant structural feature, showing 4,000–5,000 ft of displacement in the area. The Mudlark fault system is similar with throws of 1,500–4,000 ft. Faulting in the magnitude of 100–500 ft is common place.

The Sallisaw area appears to have productive potential in at least five different zones:

1. Upper Atoka sands
2. Middle Atoka sands (cumulative production = 1.7 billion cu ft of gas)
3. Basal Atoka (Spiro) sand
4. Cromwell
5. Hunton

However, to date the commerciality of all but the Spiro is unknown. Generally, the Hunton can be penetrated above 5,500 ft in the Sallisaw area, and an exploratory play for the Spiro in the area north of the Mulberry fault will likely encounter the reservoir in the 1,500–5,000 ft range.

Geologists working the area point out that although there is a marked lack of well control in the Sallisaw area, it can be shown that the Spiro channels are widespread. They have been found in more than 60% of the wells drilled north of the Mulberry fault.

Another Arkoma basin discovery of 1982 which should open some new exploratory paths in the basin was made by Moran Exploration at 1 Reaper in 12-8n-8w, 18 miles southeast of production in Quitman field, White County, Arkansas. The well flowed 1,300 Mcfd of gas from the Cambro-Ordovician Arbuckle pay through perforations at 6,634–6,644 ft on a ¼-in. choke. The nearest production from this zone or its equivalent is in Mississippi's Black Warrior basin and 200 miles to the southeast in Marshall County, Oklahoma, at Cumberland field.

In Arkoma basin is an east-west trending Paleozoic basin that merges with the Black Warrior basin under the Coastal Plain of the Mississippi embayment (Fig. 4–11).

The entire region can be considered a nonoil-associated dry gas province. Important reservoirs are the Early Pennsylvanian sandstones, which contain most of the known gas reserves in the two-state basin. Some pays also exist in the Chester of Mississippian age and the Early Devonian-Silurian rocks. Explorationists have looked for

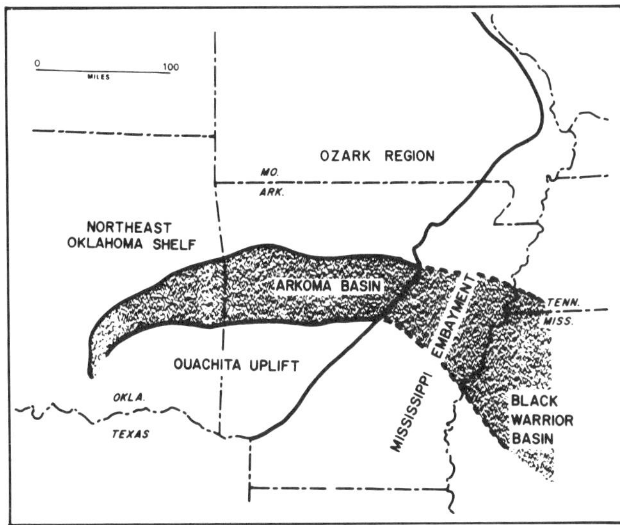

Fig. 4–11 Arkoma basin tectonic features (courtesy *OGJ*, February 12, 1979)

production in the Ordovician Simpson group in the Oklahoma portion of the basin but not in Arkansas.

There will definitely be more interest in deeper horizons in the Arkoma basin, especially in the eastern segment of the basin where it dives under the embayment to join up with the Black Warrior basin. Recent discoveries in Alabama and Mississippi may have stimulated the Arkansas interest kindled by Moran's discovery.

REFERENCES

1. Landes, *Petroleum Geology of the U.S.* (New York: Wiley, 1970), p. 141.

2. Ibid., p. 313.

3. Tutten, *Oil & Gas Journal* (October 29, 1973), p. 130.

4. Ibid.

5. Ibid.

6. *Oil & Gas Journal* (August 10, 1981), p. 35.

7. Lawrence Morrison, *Oil & Gas Journal* (May 11, 1981), p. 170.

8. William E. Ham, *Oil & Gas Journal* (December 15, 1958), p. 152.

9. *Oil & Gas Journal* (August 20, 1979), p. 190.

10. Branan and Jordan, *Oil & Gas Journal* (August 8, 1960), p. 120.

5 Northern Midcontinent

Millions of barrels of oil are yet to be discovered in the northern Midcontinent region of the U.S. This vast area covers several states from Kansas and Missouri, northward through Nebraska, and into North and South Dakota. Many basins are in this part of the country, the largest of which is the Williston basin. Other large basins in the area are the Salina, the Kennedy, the upper reaches of the Anadarko, the Cherokee, and the Forest City basins.

Kansas is the largest producer of oil in this sector, ranking eighth among U.S. oil-producing states. Its current daily output is 191,000 b/d of oil. Nebraska and Missouri are listed among the small oil producers. North Dakota's output is 125,000 b/d, up considerably over previous years.

KANSAS

The oil industry in Kansas has enjoyed success for many years. Important exploration has occurred in every basin and on most arches and uplifts in Kansas. Most drilling has been on the Central uplift. The major oil and gas provinces in this area are the Central Kansas uplift and the Hugoton embayment, which is the northwestern extension of the Anadarko basin. More than 2 billion bbl of oil has been produced on the central uplift. It is still the location of all of the state's giant oil fields, and most of the state's daily output is from this area. More than 70% of the oil found on the uplift has been from the Cambro-Ordovician Arbuckle formation.[1]

The structural geology of Kansas is quite simple.[2] Although at no place in Kansas is there a complete, representative geologic section, portions of all systems younger than Precambrian are represented.

The Paleozoic rocks thin to 2,500 ft or less over Nemaha ridge and the Central Kansas uplift. They thicken toward the southwest into the Anadarko basin where at the Oklahoma-Kansas line they exceed 7,500 ft in thickness. Thicknesses in eastern Kansas vary from 4,000 ft along the east flank of the Nemaha ridge to 1,500 ft on the flanks of the Ozark uplift. These rocks have produced great amounts of oil, both in eastern and western Kansas. All commercial oil and gas in Kansas comes from these rocks, except for small amounts of Precambrian and Cretaceous Niobrara oils in northwestern counties. Permian rocks are in the western four-fifths of the state, while the Pennsylvanian has almost statewide distribution.

Structural provinces

The largest structural feature in Kansas is the southeastward extension of the Cambridge arch, known as the Central Kansas uplift (Fig. 5–1). This feature covers about 5,700 sq miles, with sediments reaching a maximum thickness of less than 5,000 ft. On the crest of the structure, Precambrian rocks are overlain by Pennsylvanian sediments. On the flanks of the uplift, pre-Pennsylvanian strata are uplifted, truncated, and onlapped by Pennsylvanian beds. A section of Permian rocks extends across the structure.[3]

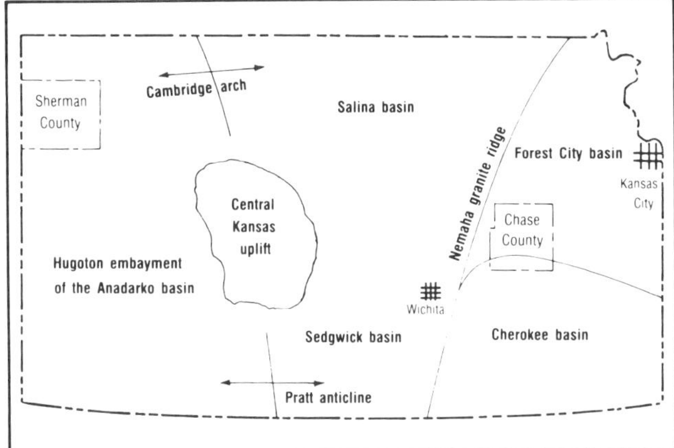

Fig. 5–1 Far-flung Kansas fronts (courtesy *OGJ*, December 29, 1980)

Hugoton embayment

The Hugoton embayment is a large shelf-like extension into western Kansas from the Anadarko basin to the southeast. It covers about 28,600 sq miles, or one-third of Kansas, and is a major structural unit. The Central Kansas uplift borders it on the east, and the Las Animas arch in Colorado on the west. The embayment plunges and sediments thicken southward into the Anadarko basin. It is the deepest structural basin in Kansas, since Precambrian rocks are overlain by as much as 9,500 ft of younger sediments.

Las Animas arch

Only the northeastern end of this large anticline extends into northwestern Kansas. Most of the arch is in Colorado (*see* Chapter 10). It plunges northeastward and separates the Denver basin from the Hugoton embayment. The Kansas portion of the arch is of recent particular interest, due to a successful Niobrara gas exploratory program in the northwestern part of Kansas, western Nebraska, and northeastern Colorado.

Nemaha ridge

The textbook structure of Kansas is the Nemaha granite ridge. This is a major Pennsylvanian feature that crosses Kansas from north to south and extends into Nebraska and Oklahoma. The anticline figures highly in the geologic history of all three states and is well-known by all geologists. This ridge divides the Forest City and Cherokee basins on the east from the Salina and Sedgwick basins on the west. Precambrian rocks along the uplift's crest lie within 600 ft of the surface near the Nebraska line. However, it plunges southward. At the Oklahoma line, the granite is about 4,000 ft below the surface.[4]

Salina basin

The Salina is the state's second largest basin, covering more than 12,700 sq miles. This section is quite thin, and little oil activity has occurred in most of the area. However, the industry feels there are still pools to be found. Rocks of Cambro-Ordovician, Silurian, Devonian, Mississippian, Pennsylvanian-Permian, Jurassic, Cretaceous, and Tertiary vintage exist in the basin. The thickest section encountered is about 4,500 ft thick.

Other Kansas features include the Sedgwick basin, the Bourbon arch, the Cherokee and Forest City basins. There have been hundreds of oil fields found in these areas, some of them dating back to post-Civil War days. Some of the fields have produced considerable amounts of oil.

Forest City basin

New exploratory interest exists in the Forest City basin, one of the nation's oldest but least-tested petroleum provinces. Cities Service completed two dual-zone producers in Nemaha County, confirming the discovery of Ordovician Viola and Simpson production in McClain field. The field was opened by Pendleton Land and Exploration and Petro-Lewis 1 McClain, C SE NE 18-4s-14e. Pump had 65 b/d of oil at 3,453–3,457 ft in the Viola and 45 b/d of oil in the Simpson at 3,664–3,667 ft at 1 McClain A in C NE NE 18-4s-14e. Other Cities wells will add to this field's importance.

Only Iowa lacks production in the Forest City basin. Most of the area's best fields are in southeastern Nebraska, Richardson County, but recent activity there has been slim.

The first oil and gas production in the basin dates back to Civil War days in Kansas City. After the war, drilling spread from Kansas City into western Missouri along the Missouri River into Ray, Clay, Carroll, and Clinton counties. Kansas' role in the basin history began much later.

Over 80 million bbl of oil have been produced in this area. Most of it comes from rocks of Pennsylvanian age and the remainder from the Devonian and Middle Ordovician sediments. Geologists think future Forest City basin activity will uncover more oil from the Ordovician and Devonian.

This basin covers some 32,000 sq miles. It is strictly a Paleozoic basin and is both structural and depositional. This basin contains some of the most famous reservoir sands in oil history. It is a challenge to the thinking oil hunter. Wildcat density after all these years is still sparse. The subsurface is replete with unconformities, disconformities, faults, folds, and stratigraphic traps. Source rocks and reservoir rocks abound. This basin should not be written off. Somewhere in the files of some oil companies, the key to more Forest City oil lies hidden away. A new look at early maps, old seismic records, or even a yellowed magnetometer survey might well provide the tool that's needed by some discovery thinker. The Forest City basin is indeed a sleeper in the Midcontinent province.

Production and outlook

Up to 1979, Kansas had produced 4.6 billion bbl of oil and almost half of this had come from the Central Kansas uplift.[5] By comparison, Oklahoma had produced almost three times this or 11.2 billion bbl of oil.

Six fields in Kansas have produced more than 100 million bbl of oil. Five of these are on the Central Kansas uplift. The state's biggest giant is old El Dorado on the Nemaha ridge. It has produced more than 287 million bbl of oil from a large structural closure, containing Pennsylvanian and Ordovician reservoirs at 630–2,600 ft. The field, found in 1931, produced 1,109,000 bbl of oil in 1981.

Central Kansas uplift operators still find this area attractive at depths to the Arbuckle of 3,000–3,500 ft.[6] Bemis-Shutts is a giant in this area, found in 1928. It has produced more than 229 million bbl of oil from the Arbuckle at 3,700 ft. And there are still 6.5 million bbl or more to be produced.

There is Ordovician Viola and Simpson production on the east flank of the Central Kansas uplift and the Sedgwick embayment on structures aligned with and west of the Nemaha granite ridge. There is little production in Kansas from the Hunton, but some has been found and may be found yet in the Forest City basin. Mississippian oil and gas production from all horizons throughout the column from Chester through and including the Kinderhook is found in Kansas. These rocks produce in many areas of Kansas, except over the crests of major uplifts. Large accumulations of oil and gas have formed by the regional truncation of the Mississippian units around the Central Kansas uplift. This area is a good place to look for more oil. Some of the larger Mississippian fields in Kansas are Pleasant Prairie, a St. Louis producer discovered in 1954; Aldrich, a Spergen-Warsaw producer found in 1929; Bindley field, a Warsaw producer found in 1972; Hanston-Oppy, a 1961 discovery in the Osage chert; Glick, an Osage producer of note; and Spivey-Grabs field, found in 1949.

There is considerable new interest shaping up in northwestern Kansas where Niobrara chalk oil and gas reservoirs of Cretaceous age have been found in recent years. Target depths are about 2,000 ft in a very porous chalk with low permeability and low deliverability. Good gas prices will keep this play on fire (Fig. 5–2).

The following points have been learned by studying the Niobrara play:

1. New shallow gas reservoirs will be nontraditional, such as chalks and shaly sands.
2. The technology to find new reservoirs exists now, as evidenced by the scanning electron microscope and improved logging techniques.
3. Shows on mud logs, previously described as shale gas, may be the shallow gas plays of the future.
4. New drilling and completion techniques will be necessary for exploring these gas reserves.[7]

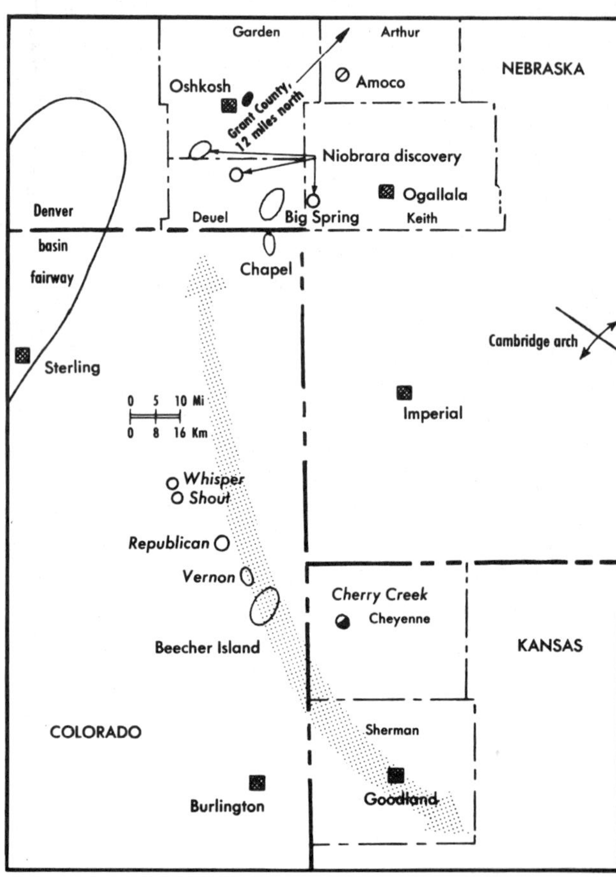

Fig. 5–2 Niobrara gas play (courtesy *OGJ*, July 31, 1978)

Large areas of Kansas can still be tested for accumulations of petroleum in Arbuckle beds, especially deeper parts of the basins. In these areas (such as the Hugoton embayment) possibilities for future oil discoveries are best.

A 1981 discovery by Texaco in Logan County points up the need for further exploratory work in this part of the state. The 1 Howard Dirks, 15 miles northeast of abandoned Carwood field, pumped 219 b/d of oil, 36.6° gravity, from the Pennsylvanian Marmaton at 4,412–4,424 ft. This opened Dirks field and will certainly be a guide for further exploratory work in the western counties where gas has been the big reward for years. This oil discovery prods explorationists to look again at the shelves of the Hugoton embayment in the western basin. There will be some more Pennsylvanian oil found there.

No doubt, exploration for oil and gas in Kansas may have reached the mature level, but drilling continues at a high level in the state. There were 2,257 wells drilled in the state during the first quarter of 1982; 1,003 of these were for oil, 227 for gas. Drilling should spur exploratory enthusiasm in the basins that have yet to be explored completely—Forest City, Salina, and the Hugoton embayment.

NEBRASKA

"What Nebraska needs is wildcats—hundreds of them." This lead appeared in an article in the *Oil & Gas Journal* in October 1956, and the prophetic title is as true today as when it was written. The only thing new is that the wildcats of today can be based on better subsurface control.

Also, Nebraska is geologically and economically more attractive today than in 1956. Most of the state's wildcatting through the years has been in the Denver basin counties in the southwest. However, there are other areas of Nebraska in which to search for oil and gas.

The state's first oil production was in 1939 at Falls City in the far southeastern niche of Nebraska in Richardson County in the Forest City basin. This was also the basin's first producer of hydrocarbons. But the real historymaker in Nebraska was the discovery of Gurley field in Cheyenne County, birthplace of the modern Denver basin play in the state. After Gurley, wildcatters swarmed over the southwestern counties, picking up a large number of Cretaceous fields. This play continues today, with no end seen in the near future.

The first discovery for the Cambridge-arch part of the state was made in Harlan County in June 1950. Success followed at Sleepy Hollow and other south central Nebraska pools (Fig. 5–3).

However, despite years of wildcatting, a large area of this state remains unexplored. Explorers have sought production up and down the Chadron arch, in the Salina and

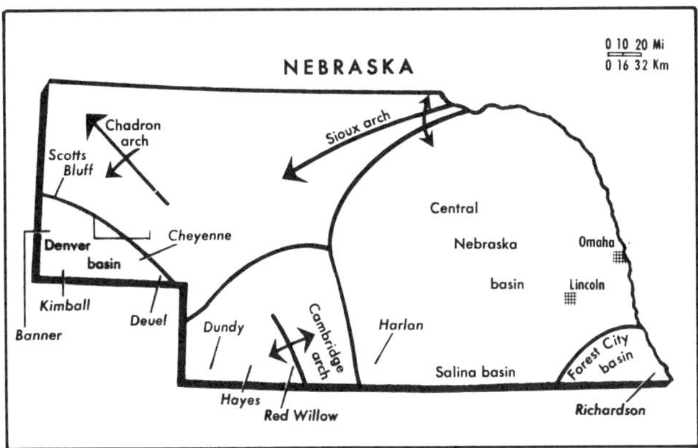

Fig. 5–3 Nebraska exploration areas (courtesy *OGJ*)

Central Nebraska basins, and along the Cambridge arch. The results have been disappointing to say the least. But the Central Nebraska basin alone covers 25,000 sq miles. Many more wildcats are needed throughout Nebraska.

One area of the state that can definitely be tagged as virgin wildcat land is the Chadron arch where only a few deep wells have been drilled on a structure covering 50 townships. Chadron is a large, partly exposed uplift in the northern part of the Panhandle region of Nebraska (Fig. 5–4).[8] The arch effectively separates the Denver basin on the southwest from the Williston basin on the northeast. The arch extends along an axial strike for about 100 miles northwest-southeast and is about 40 miles in width with a structural relief of about 3,000 ft.

Cretaceous rocks are exposed only on the north end of the arch, and probably the rather steep west dips (up to about 20°) in these rocks were noticed by geologists and

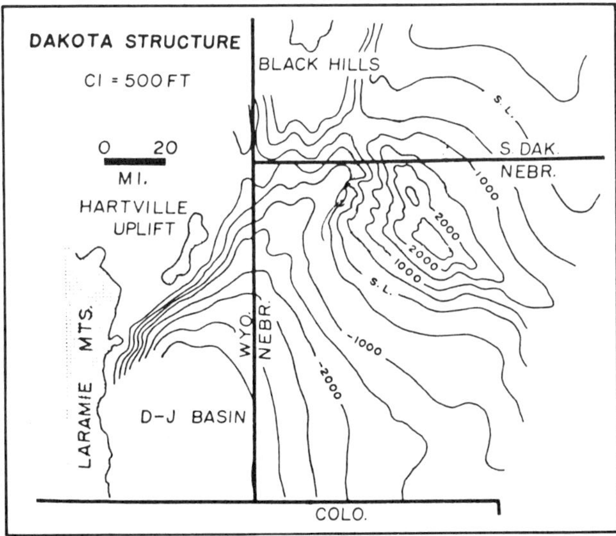

Fig. 5–4 Index map showing Chadron arch (courtesy *OGJ*)

prospectors before 1900. The first oil test was the Miller well drilled in 1902 on a local structural high along the crest of the anticline in 34n-46w. It went 1,000 ft into the Dakota sandstone of Cretaceous age. After some surface work, Midwest Oil drilled a Jurassic Morrison wildcat in 1917 about 3 miles northwest of the Miller hole. There were no shows, but during the 1919–23 period, extensive surface mapping was done on the north end by several companies. Several small closures were located, the best in 35n-47w, and several wells were drilled. A surface closure on the north end of Chadron arch in South Dakota was drilled prior to 1931. Many of the wells found oil and gas shows in various horizons.

Pre-Tertiary rocks have a total thickness of about 2,000 ft on the crest of the Chadron arch and thicken in all directions away from the structure. Shows have turned up in the Minnelusa sandstones, in the Lakota. But the best shows have been in the Dakota D and J sands of the near-shore facies of the Mowry shale along the crest and west flanks of the Chadron arch. The Lakota channel sandstones also are highly prospective. The Converse sandstone at the top of the Minnelusa on the north end of the arch offers traps similar to the prolific sandstones of the same age in the Powder River basin which is nearby in Wyoming.

SOUTH DAKOTA

Probably one of the most undrilled states in the interior portion of the continental U.S. is South Dakota (Fig. 5–5). The state has been among the producing family of states since 1954 when Buffalo field was discovered in the

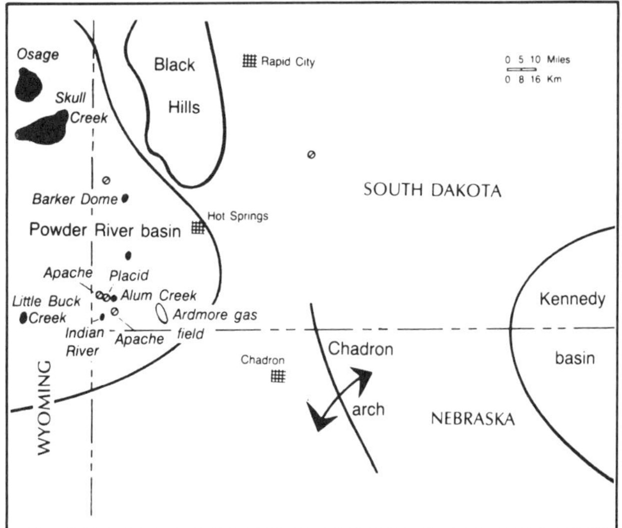

Fig. 5–5 The Leo play (courtesy *OGJ*, March 29, 1982)

northwestern corner of Harding County. Production there is from the Ordovician Red River. Several other similar pools have been found nearby in the intervening years. Each new discovery set off another flurry of exploratory excitement in this corner of the Williston basin.

There have been very few wells drilled in South Dakota to depths older than the Pennsylvanian. The Cretaceous section is similar to that of nearby Wyoming equivalents. Dakota and Muddy sandstones are important producers of oil and gas in Wyoming, and their counterparts should be looked at with great interest in South Dakota. Only a very small part of South Dakota can really be written off as an unlikely spot to find hydrocarbons. However, due to the lack of drilling throughout large areas of this state, no area can really be knocked off the list of prospects. The Williston basin covers a large part of the state, and all of this region can be considered a good spot to look for oil.

The eastern third of the state can be labeled as more or less unfavorable. Precambrian outcrops scratch any prospects in the Black Hills and in parts of extreme eastern South Dakota near Minnesota. Regional geologists feel that the best hopes for new production in South Dakota center on the area from Pierre to the western borders—the Williston and Powder River basins. There could well be some prospects on the extreme northern lobe of the Chadron arch that juts up from Nebraska. Also the industry might be able to count on the southern part of the Kennedy basin. None of these areas mentioned has had sufficient well density to make any rational decisions about how much oil and gas might be discovered. This is another area that badly needs wildcats.

An almost unheard of corner of the Great Plains, southwestern South Dakota, has been having a revival of oil and gas development. Following the discovery and development of the prolific Buck Creek and Little Buck Creek fields and the initial Indian Creek discovery in South Dakota (which flowed 360 b/d of oil), there has been a flurry of leasing and drilling activity.

Action is now centered on Alum Creek field in Fall River County. Apache recently completed a new producer pumping 371 b/d of oil from the Second Leo of Pennsylvanian age at 3,326–3,336 ft. This was the fourth oil well in the field and is ½ mile east of the field discovery well that flowed 1,600 Mcfd of gas from the Leo in 3-11s-1e. Placid Oil then got 199 b/d on pump at 3-16 Federal in SE SE 3-11s-1e at 3,429–3,432 ft in the Leo. Alum Creek field is 7 miles north-northwest of the Indian Creek field and 32 miles southwest of the spa at Hot Springs. Edgemont field in the same area is also a Leo producer.

Geologically, the Leo play is in an interbasin between the Powder River, Denver, and Kennedy basins. Geographically, the Leo play is in the tristate area where South Dakota, Nebraska, and Wyoming meet.

Potential for the generation of substantial amounts of hydrocarbons has been supported by regional source studies in this South Dakota area. The source is local, as highly organic and radioactive shales are interbedded with the Leo sands.

The shales can be correlated throughout the entire area with an average, total organic carbon content of up to 20%. Equally important as the presence of source rock is the fact that reservoir rock has been established in at least eight zones of the Permo-Pennsylvanian section.

Having established the source and reservoir but with an average penetration rate of only five wells per township, operators must drill the wells necessary to discover and develop the reservoirs in this highly potential play.

WILLISTON BASIN

Since the discovery of oil by an Amerada deep test in the heart of the Williston basin in April 1982, the volume and intensity of exploratory activity throughout this broad sedimentary unit have been almost miraculous. A short time ago this area was not considered for development because it was too far from markets, its reservoir rocks were too deep, and it was too difficult to explore.

These words were spoken by Philip C. Ingalls, then of *Oil & Gas Journal* in November 1951. He noted that just a week earlier Wallace E. Pratt had told the American Association of Petroleum Geologists in Austin, Texas, that the Williston basin is an excellent example of his "philosophy of oil finding." He said that oil finders must not let what they know keep them from being constantly alert to what they do not know about the occurrences of oil.

Here is a basin that was rated by some as having poor prospects for discoveries. It was said that Williston was not conducive to the generation of oil (though no one has yet proved just what is conducive to the generation of oil) because it was too big and inactive and because sedimentation was too gradual.

Fortunately for the oil industry, not everyone believed the Williston basin was a bad place to look for hydrocarbons. The discovery of oil in 1951 set off one of the great oil booms of our time. A single pint of oil recovered on a test in the Amerada Petroleum 1 Inverson electrified the industry and launched an unprecedented amount of activity. In terms of acreage involved, it was the greatest lease play in history to that time, with some 30 million acres leased during the first year of excitement. Geologists, operators, mapmakers, leasehounds, news services, seismic crews, and this author swarmed into North Dakota and whipped up exploratory whirlwinds unlike anything since East Texas.

Here was the biggest oil basin yet laid before the U.S. oilman. It covered an area of 118,500 sq miles and offered nearly 76 million acres of hunting grounds for new fields. A sedimentary column varying in thickness from 2,000 ft in Manitoba to more than 12,000 ft at the basin center offered myriad multiple pay prospects unrivaled anywhere else on the oil scene. Fig. 5–6 shows how the basin looked in 1953.

The North Dakota oil discovery in April 1951 marked the first production for the U.S. portion of the international basin. A few months earlier, a small oil well had been completed on the basin edge in southwestern Manitoba. And, a few months after the Iverson well, oil was found in the Montana portion of the huge basin.

Though considered to be a Rocky Mountain basin, Williston lies wholly within the Great Plains physiographic province. It has no obvious limiting features except for the Black Hills and Bowdoin dome. A rather generally accepted outline for the basin is as follows:

On the northeast side, the basin is edged by the feather edge of the Mississippian rocks as they wedge out toward the Canadian shield. On the east and south, the basin outline conforms rather closely to the zero contour on the Dakota sandstone of Cretaceous age. The outline skirts the Black Hills and takes in the Sheep Mountain and Blood Creek synclines of eastern Montana. It swings eastward around the Bowdoin dome. In Saskatchewan, there is no definite geologic or topographic break between the Williston basin and the southern portion of the Alberta monocline. Here the Williston basin outline includes the southern portion of the southeastward plunging Moosejaw syncline and then swings southeastward to tie in with the approximate limit of the Mississippian wedgeout in Manitoba.

The important deep-seated structures in the basin are the Nesson anticline in North Dakota and the Poplar and Cedar Creek anticlines in eastern Montana. There is a rather complete section of Paleozoic and Mesozoic rocks in the Williston basin.

Production in the basin is from rocks of Cambrian, Ordovician, Silurian, Devonian, and Mississippian. Also, some Cretaceous Eagle gas production exists on the Cedar Creek anticline and some Jurassic Spearfish oil in northwestern North Dakota (Fig. 5–7).

The total number of wells drilled in the basin since the 1951–52 initial discovery period (Figs. 5–8 and 5–9) rose between 1951–1958. It dropped off rapidly to 1963 and then vacillated until 1977. Since 1977, an extremely rapid increase in total number of wells drilled has occurred. Wildcat drilling during the initial discovery period peaked in 1954; the stratigraphic Mississippian play peaked in 1958; the Bowman County Red River Ordovician play peaked in 1964; and the shallow Muddy sandstone play of 1967–69 resulted in the highest number of wildcats during any year (1968) until 1980, even though no discoveries were made as a result of it. Since 1972, wildcat action has generally been up.

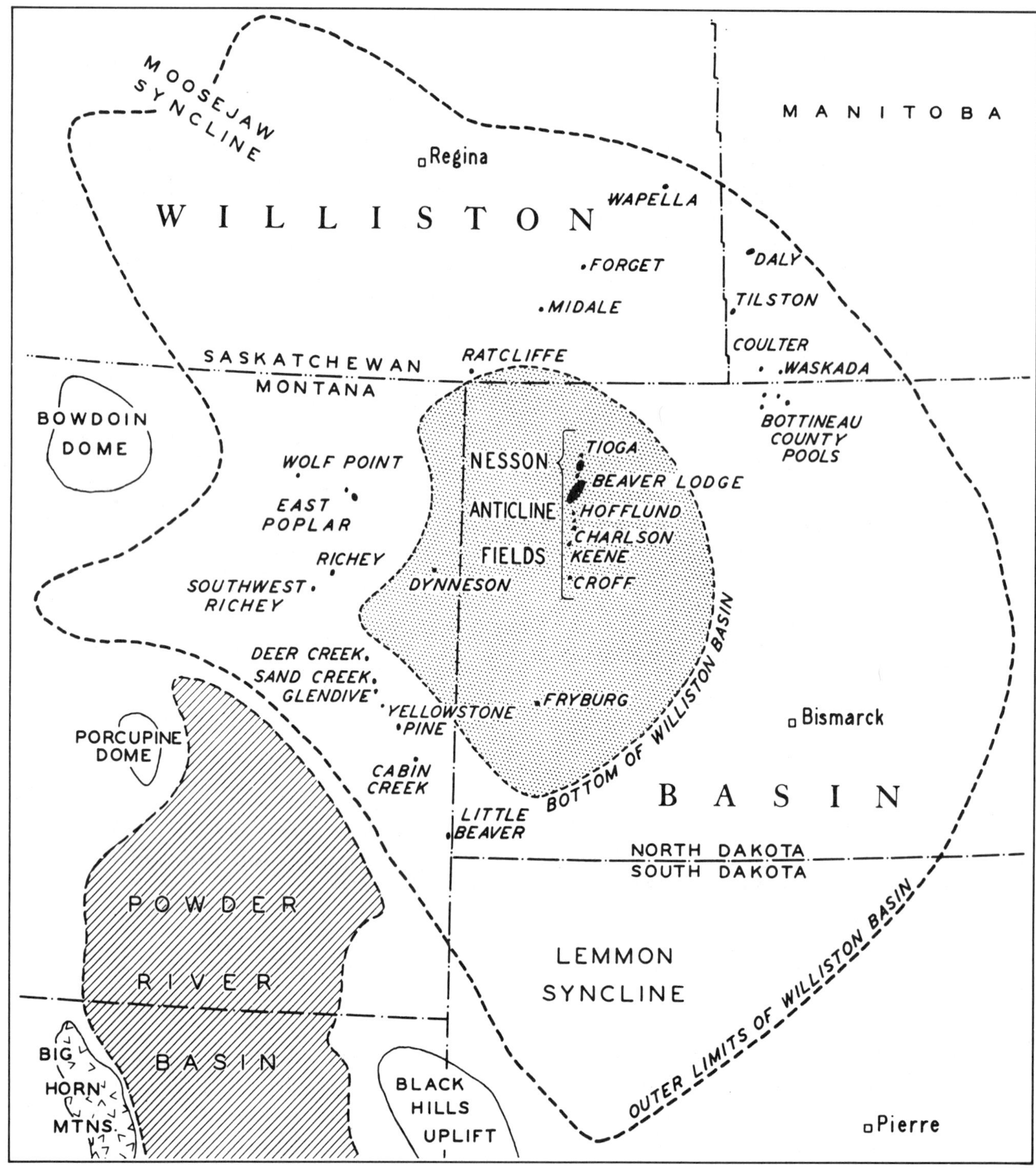

Fig. 5–6 The Williston basin in 1953 (courtesy *OGJ*, August 31, 1953)

Annual production levels also reflect drilling activity (Fig. 5–10). Oil production from the original Nesson anticline discoveries already had begun to level off when the Madison stratigraphic traps north and northeast of the anticline began to produce in 1955.[9] This additional production resulted in overall increases in production until 1962. The slight decrease in 1963 was to be expected in view of the small number of wells drilled and the lack of wildcats being

Systems	Groups	Rock Units
Quaternary		Glacial
Tertiary	White River	
		Golden Valley
	Fort Union Group	Sentinel Butte
		Bullion Creek
		Slope
		Cannonball
		Ludlow
Cretaceous		Hell Creek
		Fox Hills
	Montana Group	Pierre
		Judith River
		Eagle ☼
	Colorado Group	Niobrara
		Carlile
		Greenhorn
		Belle Fourche
		Mowry
	Dakota Group	Newcastle
		Skull Creek
		Inyan Kara
Jurassic		Morrison
		Swift
		Rierdon
		Piper
Triassic		Spearfish ●
Permian		Minnekahta
		Opeche
	Minnelusa Group	Broom Creek
Pennsylvanian		Amsden
		Tyler ●

Systems	Groups	Rock Units
Mississippian	Big Snowy Group	Otter
		Kibbey ●
	Madison Group (Informal intervals)	Poplar interval
		Ratcliffe interval ●
		Frobisher Alida interval ●
		Tilston interval ●
		Bottineau interval ●
		Bakken ●
Devonian		Three Forks ●
		Birdbear ●
		Duperow ●
		Souris River
		Dawson Bay ●
		Prairie
		Winnipegosis ●
Silurian		Interlake ●
		Stonewall ●
Ordovician		Stony Mtn
	Big Horn Group	Red River ● ☼
	Winnipeg Group	☼
Cambrian		Deadwood ☼ ●
Precambrian		●

● Oil production
☼ Gas production

Fig. 5–7 North Dakota stratigraphic column (courtesy *OGJ*, 1982)

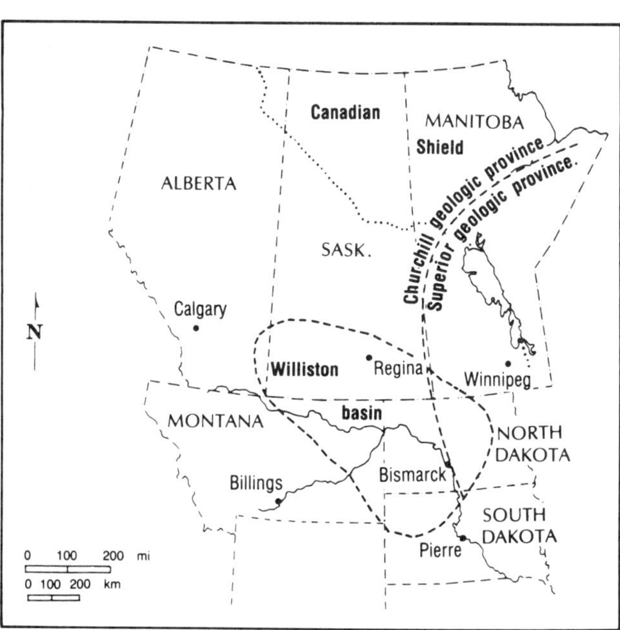

Fig. 5–8 The Williston basin (courtesy *OGJ*, July 27, 1981)

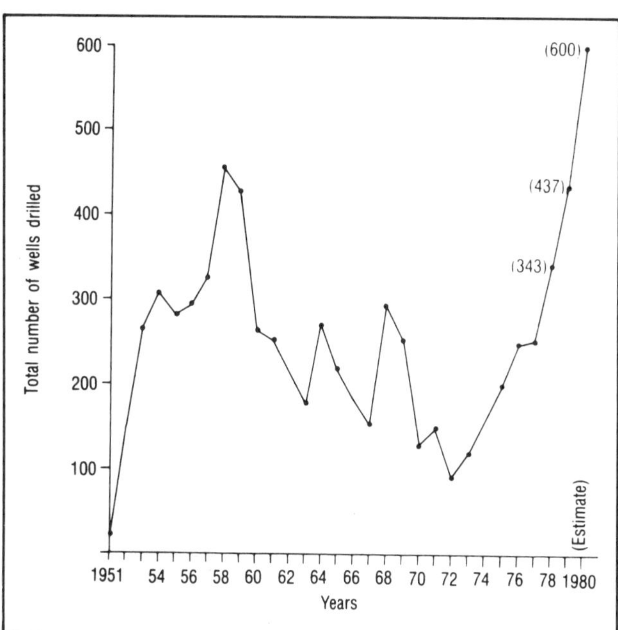

Fig. 5–9 Number of wells drilled in North Dakota for oil and gas since 1951 (courtesy *OGJ*)

completed successfully. Discovery, however, of the Red River oil in Bowman County gave a boost to production from 1964 to the 1966 peak. Fig. 5–11 illustrates the rapid developments during the 1970s in the booming Williston basin.

OPEC's actions resulted in the first substantial increase in the price of oil in 1973. Exploratory drilling increased in 1974 and 1975. As a result of this renewed interest in exploration, two giant fields were discovered: the Mississippian Mondak and Little Knife.

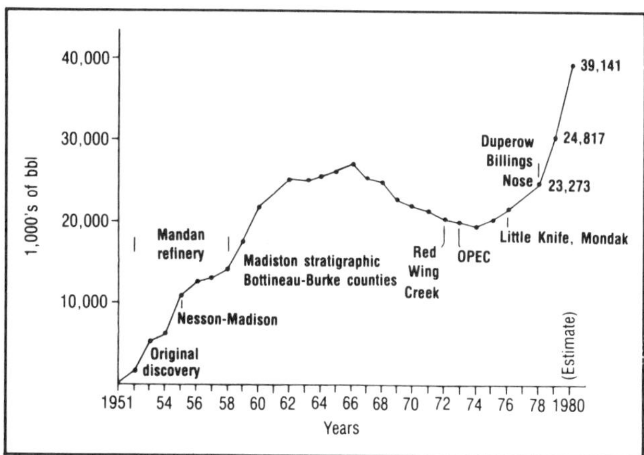

Fig. 5–10 North Dakota crude oil production (courtesy *OGJ*, July 27, 1981)

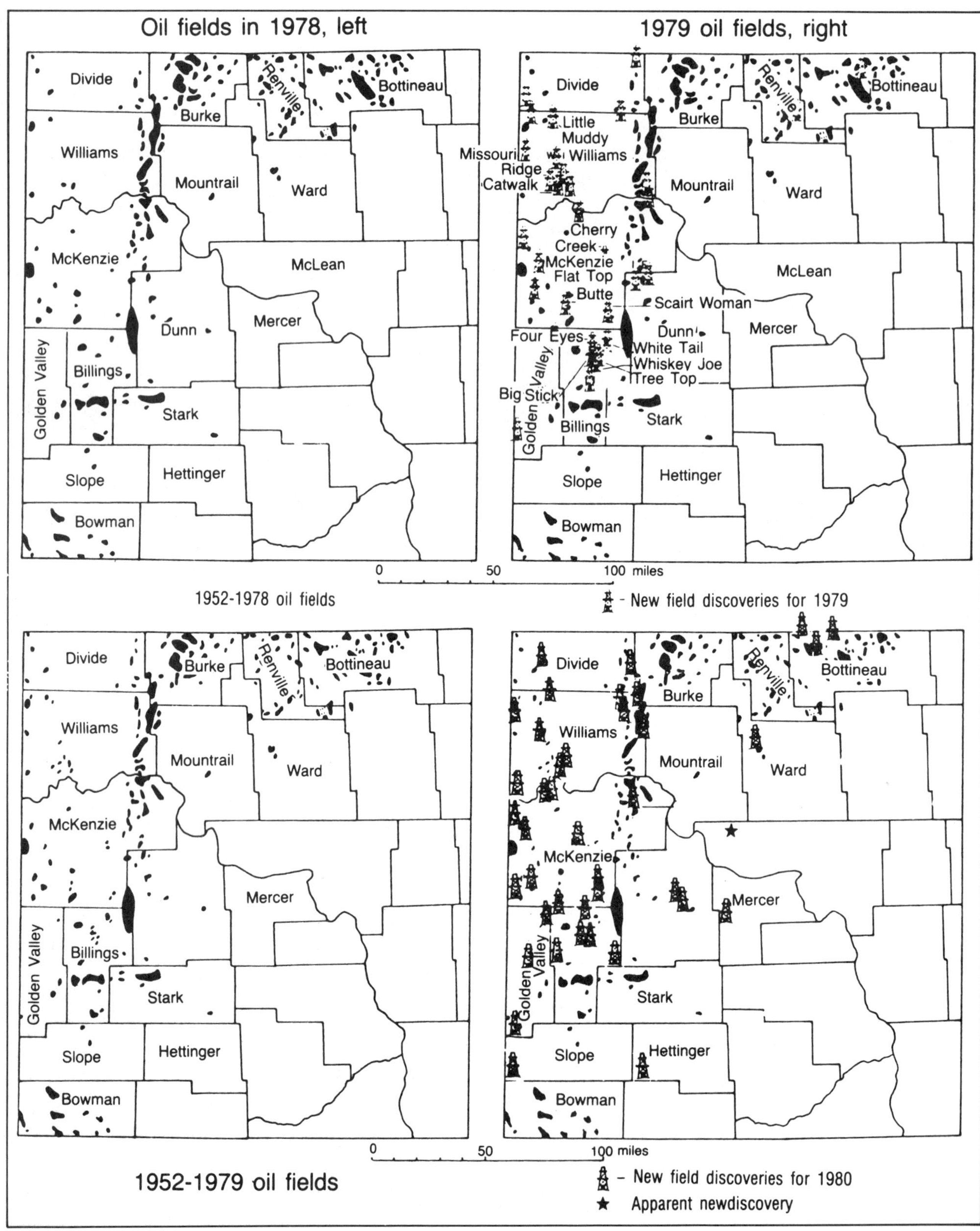

Fig. 5–11 Williston basin development (courtesy *OGJ*)

The slump

The dramatic slide in the Williston basin rig count in 1982 was expected to level off, promising further exploratory work in this basin where only the top has been scratched. Declining lease prices will cause drilling to rebound. However, that may take 2–3 years. Remaining action will continue to focus on the deep part of the basin along the North Dakota-Montana border. The area is appealing because it could have upside reserve potential. This area has more pays, greater extent, and higher success ratios than elsewhere in the basin.

Pennzoil, among the most active drillers in the deeper part of the basin, recently reported two discoveries that typify results in this area. The firm's 35-24 Bowline BN, McKenzie County, North Dakota, during tests flowed 659 b/d of 40° gravity oil and 1.1 MMcfd of gas through a ½-in. choke with 365-psi tubing pressure from the Ordovician Red River at 13,292–13,310 ft.

The well, 3 miles southeast of nearest production at Boxcar Butte field, has production potential in the Mississippian Mission Canyon, Devonian Nisku and Duperow, and two other Red River zones.

Pennzoil's 25-32 Riverside BN wildcat flowed 866 b/d of 50° gravity oil from the Red River during tests of two zones at 13,104–13,136 ft. Gas flow was 1.3 MMcfd. The McKenzie County discovery is 4½ miles southwest of nearest production at Flat Top Butte field. Both wells are on Burlington Northern land for which Pennzoil holds exploration rights under a 5-year exploration program that will expire in 1982.

Also in the deep basin, drilling is active around Lone Butte field, discovered in 1981 by Gulf 1-13 Bob Creek, a Mission Canyon oil and gas producer. The field is north of Little Knife field and is thought to be separated from that giant by a saddle structure or porosity change. Operators continue to drill wildcats between the fields that have been found since 1978 in McKenzie, Dunn, Golden Valley, and Billings counties.

Other basin action is concentrated around traditional producing areas such as the Cedar Creek anticline, Mondak complex, Nesson anticline, and producing areas along the Canada border.

New finds and the outlook

Some discoveries have been made on the basin flanks. Sun Exploration and Production made the basin's westernmost discovery late in 1981 with its 1 Gaunche in Valley County, Montana, a Mission Canyon discovery. Several more wells are planned in this new area, but details have been sketchy.

Several areas of North Dakota are still virtually untouched. The central basin area east of the Nesson anticline is now proved by the Stanley field, but most of that region is untouched by the wildcat drill. Structure mapping by the North Dakota Geological Survey suggests that several major structural lineations and elements present in the eastern basin have not been tested.

All of the area overlying the boundary of the Churchill and Superior provinces of the Canadian shield is potentially

Table 5–1

PAY ZONE BREAKDOWN

Formation	Number of penetrations	Percentage untested
Established		
Charles	4,200	64
Mission Canyon	3,756	71
Duperow	1,701	82
Interlake	1,415	81
Red River	1,314	39
Frontier		
Kibby	4,238	98
Lodepole	2,092	95
Bakken	1,785	95
Winnipegosis	1,431	91
Stony Mountain	1,317	92
Deadwood	289	94

productive. East of that line little drilling has been done, although several large drilling programs were once planned. Perhaps these will return to the drawing boards.

Table 5–1 shows the breakdown of the pay zones in the Williston basin into two categories: established and frontier horizons.[10] Note the clastic pay horizons in the Mississippian and Ordovician show fairly low values in the percent untested column. The Devonian and Silurian only recently have become more important pay zones in the basin, and their values for percent untested are significantly higher than the other formations.

The formations in the frontier pay category are not well established as major producing intervals and, in general, have encountered fewer drilling and testing shows. Geologists should begin looking for conventional pay zones in unconventional traps and unconventional pays in both conventional and unconventional traps. Drillers also should consider both established and frontier pay zones in unconventional areas.

Geologists feel that the industry needs to look for pays in the established areas of the Williston basin. It should also head for the deeper part of the basin and stay there, as well as drill on the eastern flank. The South Dakota portion of the Williston basin is far from being completely scoured by the wildcatter. Some hidden traps might exist on the southeastern flanks of Williston, too.

REFERENCES

1. American Association of Petroleum Geologists, *Memoir 15,* volume 2 (1971), 1046.
2. Harris and Larsh, *Oil & Gas Journal* (April 30, 1979), p. 323.
3. Ibid., p. 328.
4. Ibid., p. 331.
5. Ibid., p. 336.
6. Ibid., p. 336.
7. Mary Beth Cooper, *Oil & Gas Journal* (March 9, 1981), p. 218.
8. Wulf, *Oil & Gas Journal* (February 24, 1964), p. 148.
9. Lee Gerhard and Sidney Anderson, *Oil & Gas Journal* (July 27, 1981), p. 331.
10. D. Hoffman, *Oil & Gas Journal* (February 2, 1981), p. 140.

6
Southeast U.S.

Some of the world's largest oil and gas fields lie in the southeastern portion of the U.S. This area extends from East Texas through South Arkansas, North Louisiana, Mississippi, Alabama, and southern Georgia to the tip of Florida (Fig. 6–1).

The giant East Texas oil field lies in this region—the big daddy of U.S. oil fields until Prudhoe Bay in Alaska. Also included in this area are the Monroe gas field (a world giant), a host of Tuscaloosa, Hosston, and Jurassic oil fields, plus a significant number of Cretaceous oil and gas accumulations.

East Texas

For two decades after its discovery, the East Texas field was regarded as the largest oil accumulation known to man. This huge field has an estimated 6 billion bbl of recoverable oil reserves. It is still on production after many prolific years. This field is the ideal stratigraphic trap, and since its discovery, it has been a guide for similar traps elsewhere in the world (Fig. 6–2).

In 1930, eastern Texas was a poor area bothered by poor times. It was an unfriendly land whose chief crops were cotton, peanuts, pigs, and timber. Kilgore was the major city in the immediate area. The main streets were paved with crushed seashells from the Gulf of Mexico to keep the dust down. Then along came Dad Joiner and his discovery at Daisy Bradford 3. He found the biggest field in the world to that date. The countryside was transformed into a prosperous land with hundreds of oil rigs. Shacks were replaced with neat towns and cities. Prosperity came to a land that had known nothing but poverty.

The East Texas field added vast amounts to the U.S. oil reserve and played a major role in World War II when the "Allies floated to victory on a sea of oil." It was many years before anyone came close to the truth about the astonishing East Texas field. After the discovery well gushed, a group of geologists and engineers estimated the total ultimate production of the field would be 2.1 billion bbl. This figure was raised in 1935 to 2.8 billion bbl and to 4.4 billion bbl in 1940. Today's estimate is 6 billion bbl of recoverable oil.

This field grew by leaps and bounds and now stretches across five Texas counties: Gregg, Rusk, Smith, Upshur, and Cherokee. It covers 9 miles at its widest point and stretches

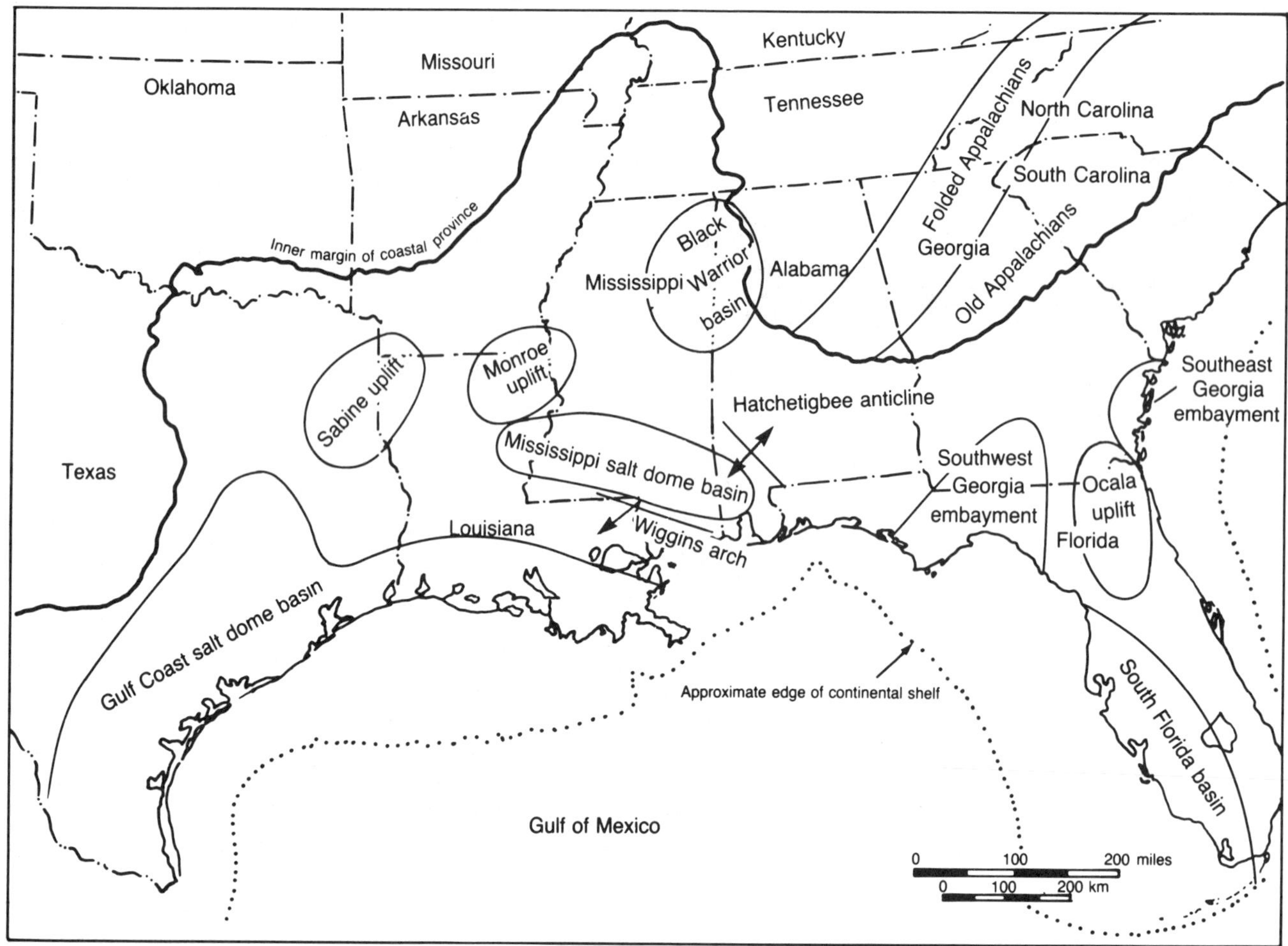

Fig. 6–1 Southeast U.S. geologic features

42 miles northeast-southwest. There were five producers in the field at the end of 1930, making 556 b/d of oil. In 1931 3,612 wells produced over 370,000 b/d of oil. At one time the East Texas field had 26,000 wells producing simultaneously from the Cretaceous Woodbine sand, and the daily output was about 1 million b/d of oil.

Today's output is 55.3 million bbl/year from 11,685 wells. The field has produced more than 4.6 billion bbl of oil and has 1.3 billion bbl left to flow. Of more than 29,000 wells, only 575 dry holes have been drilled, an astonishing success ratio.[1]

After East Texas

When the boom days died in East Texas the industry began looking for more giants but found none. It was said there would never be another East Texas. Prudhoe Bay finally squashed that myth.

Drillers did find more Woodbine fields, and Jurassic rocks also became attractive. As the hunt continues, better reservoirs have been found in the transitional facies belt, which runs northeastward into the Arkansas-Louisiana (Ark-La) region. Pays here are the Cotton Valley and Smackover.

There has been considerable exploration along the Angelina-Caldwell flexure in eastern Texas. The most attractive targets along the flexure are the upper Woodbine and lower Eagle Ford. The postulated preserved transitional sediments deposited along the southeast flank of the giant upper Woodbine delta offer a substantial reservoir.[2]

The industry has been exploring in the Jurassic rocks of northeast Texas for many years. Drilling has been done along the porous belt, which includes the Upper Jurassic Smackover formation. This porous unit is the oldest favorable Mesozoic reservoir. The Cotton Valley group formation is also a good drilling target. This exploration, of course, involves deeper drilling on proven structural closures. Depths range from 10,000–15,000 ft and deeper.

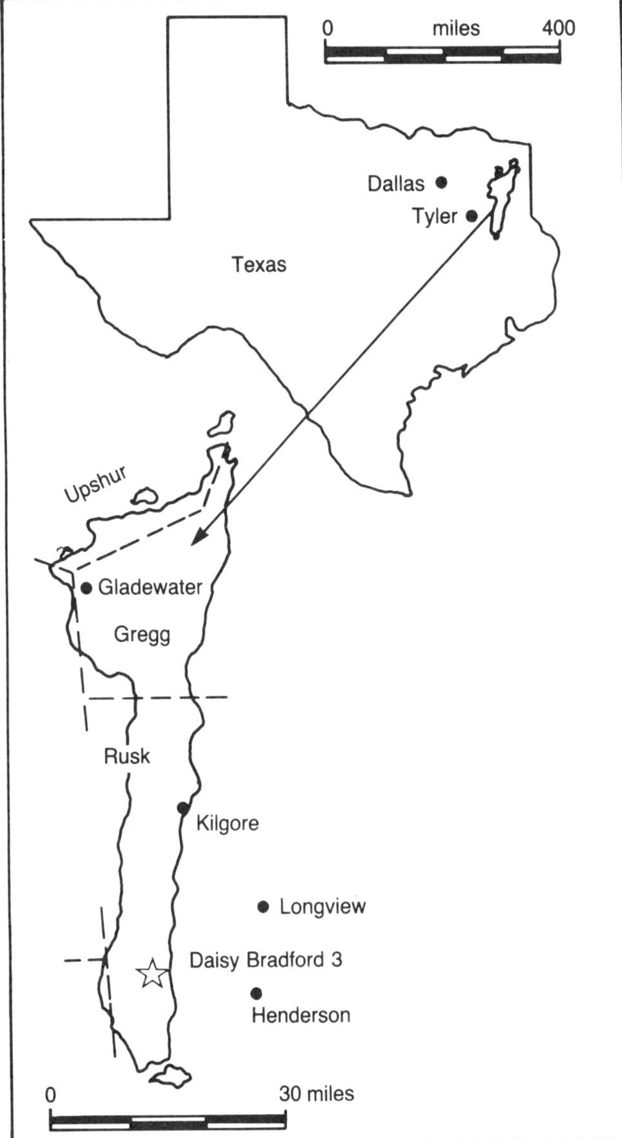

Fig. 6–2 East Texas field (courtesy *Our Sun*, 1955)

More structures should be found on the Angelina-Caldwell flexure. Both stratigraphic and stratigraphic-structural traps of good size can be expected from future exploratory work. There are pays to be found in the Louann salt equivalent, the Pettet limestone, the massive anhydrite, the upper Glen Rose and the Fredericksburg and Washita equivalents (Fig. 6–3).

Fairway

In 1960, the most significant field in this area since the East Texas field was Fairway in Anderson and Henderson counties (Fig. 6–4). At the time of its discovery, geologists asked themselves how a field of this size remained undiscovered for so long, particularly in northeast Texas.

The discovery well was Fairway Operating 1 Wofford. It flowed 420 b/d of oil from the James Reef of Cretaceous age and was completed on August 5, 1960. By March 1963 there were 142 producers and only five dry holes. Today the field has 90 producing wells, making 4.8 million bbl/year of oil. Estimated remaining reserves are 41 million bbl and the cumulative production is 169 million bbl.

The first well in the area to penetrate the James Reef was Humble 1 Milner in 1948 (Fig 6–5). The reef section was 128 ft thick with no reported shows.[3]

Texaco completed 1 Hanks in September 1948 from 15 ft of James Reef as a gas-condensate discovery, opening the Isaac Lindsey field. Cities Service completed the discovery well of the Frankston field in 1955 flowing 220 b/d of oil from 40 ft of saturated James Reef over water.

Before finding the Fairway field, a porous James Reef development had been found to the south and east of the present field outline. Production from the James had been found in two gas-condensate wells at the Isaac Lindsey field and in one oil well at Frankston.

Prior to the Fairway field, many geologists did not consider the James porosity development to be a reef. Fairway Operating became interested in the area because of the production performance at Cities E-1 White in the Frankston field. The first 5 years of production indicated that the well was producing from a large reservoir. The well had produced about 101,000 bbl of oil as of January 31, 1960. This information led the discoverer into the area. The company assembled its block through lease purchase and acreage farmout. Majors held much of the acreage in the area, and most of it was nearing expiration.

The section of sediments in the Fairway field is typical of the central portion of the eastern Texas basin. Eocene and Late Cretaceous beds extend to 6,000 ft and are composed of nonmarine sands and shales of the Wilcox; the marine shales, chalks and marls of the Midway, Navarro, Taylor, Austin, and Eagle Ford groups; and the nonmarine sands, shales, and red beds of the Woodbine of East Texas field fame.

The Rodessa section is about 400 ft thick and lies immediately below the anhydrite. It produces oil from the uppermost porosity development, a porous, mealy limestone (Fig. 6–6). The James Reef is below the Rodessa and is 200 ft thick on an average at 10,000 ft. This is the field's major producing horizon. The net pay averages 70 ft; the average porosity is 11%, with many wells having 20%. The average permeability is 18 md, and many wells have some sections with 500–1,000 md. The average-size unit is 150 acres. It is thought the piercement salt domes, which lie to the east, west, and south of the Fairway field, caused the existence of

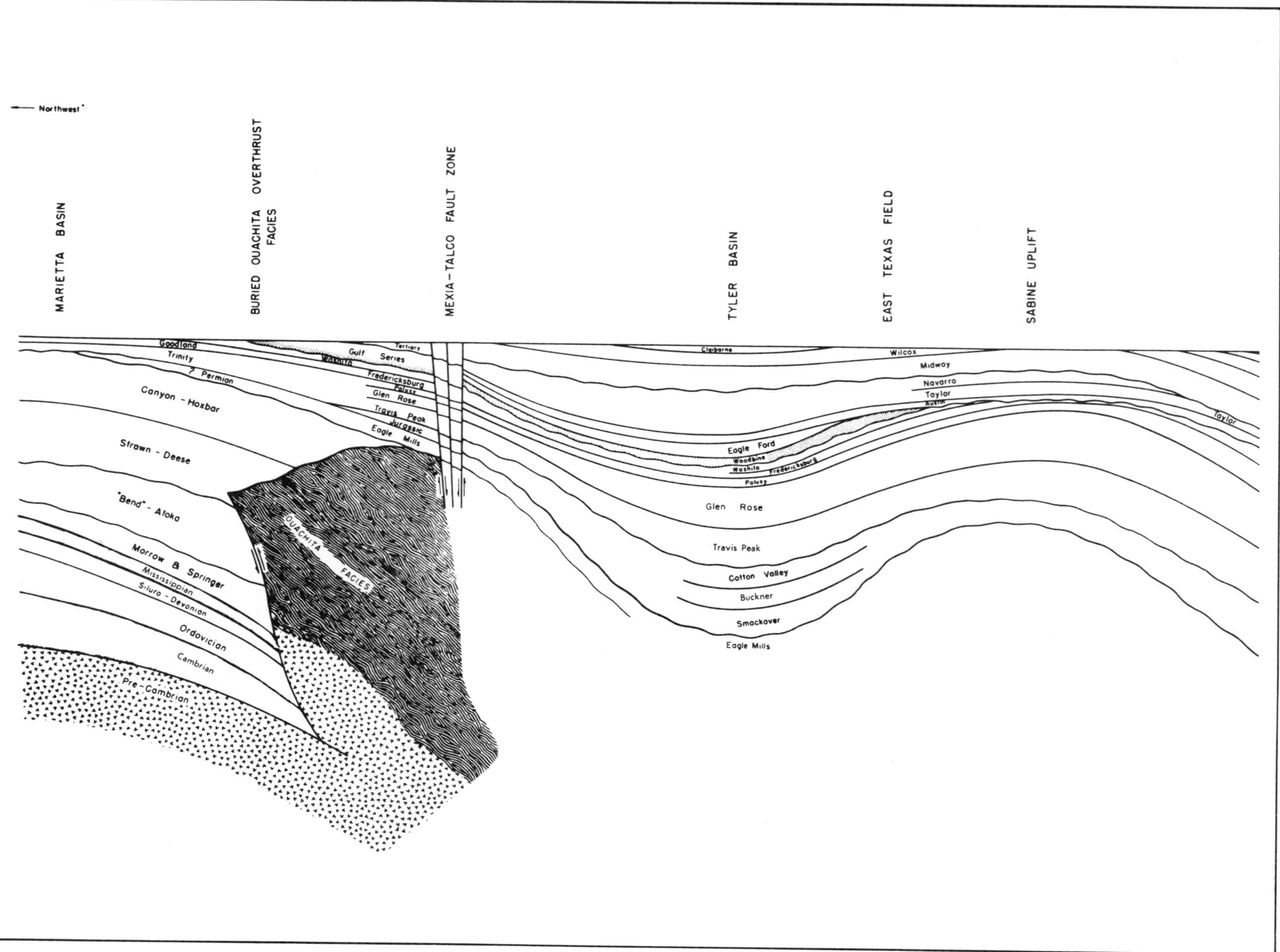

Fig. 6–3 Generalized cross section showing stratigraphic and structural relationship

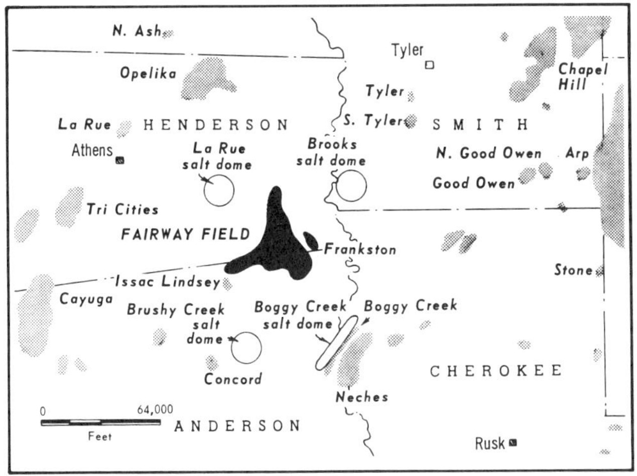

Fig. 6–4 Fairway field (courtesy *OGJ*, June 10, 1963)

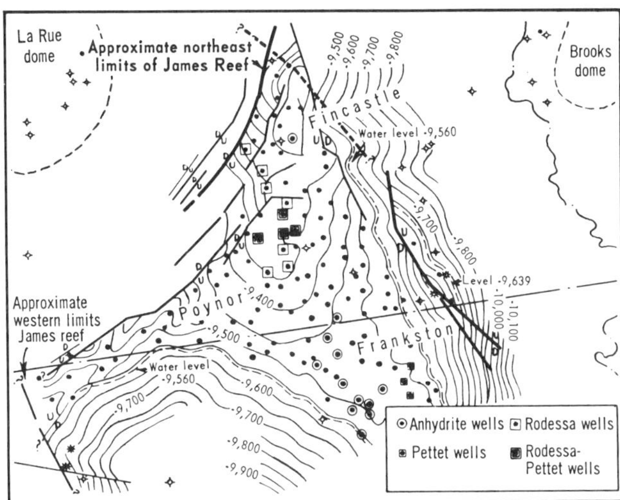

Fig. 6–6 Top of James Reef (courtesy *OGJ*, June 10, 1963)

the James Reef. There is no reason to assume that other James Reef pools will not be found in future exploration in East Texas. They may not be as large as Fairway, but look how long it took to find Fairway.

In far eastern Texas, tight gas pricing is driving much of the present Cotton Valley action. Cotton Valley and Smackover objectives also have extended the East Texas play into northern Louisiana. Pennzoil is getting higher production rates than expected from wells in an ambitious infill drilling program in the Carthage and East Bethany fields in Panola County. Production rates are in the range of 2–2.5 MMcfd/well. Pennzoil anticipated rates of 1.5–2.5 MMcfd. Fifteen of a planned 37 Pennzoil-operated wells are producing. A dozen of these wells were completed in May 1982, and more drilling is expected.

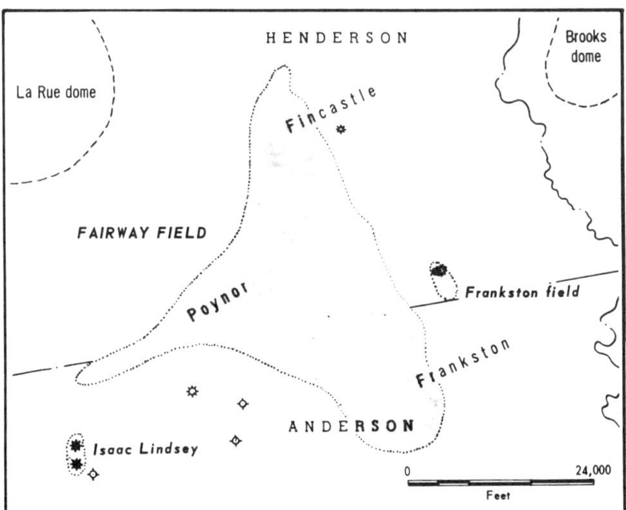

Fig. 6–5 James Reef areas prior to field discovery (courtesy *OGJ*, June 10, 1963)

Also, Getty has an infill program going in the Carthage field, and Amoco has performed massive hydraulic fracturing in the area. There were 22 active Cotton Valley lime and Smackover wells here in 1982. The Cotton Valley Bossier shale and Cotton Valley lime were designated tight gas sands in October 1980, covering 48 counties.

The Travis Peak Sym Jack West Hosston field in Cherokee County was listed as tight in 1981, and to the southwest, the Travis Peak addition (Bear Grass area), in portions of Freestone and Leon counties, also is designated tight.

ARKANSAS AND NORTH LOUISIANA

The oil industry knew all about the Smackover formation of southern Arkansas in 1946. The state's biggest oil boom had developed the final pattern of major oil fields, or so everyone thought (Fig. 6–7). This trend stretched for 50 miles east-west across the southwestern corner of the state.[4] Elongated anticlines 1 mile wide and 11 miles long capped with 100 ft of porous oolitic lime pay were typical Smackover reservoirs.

The industry was sure that the good Smackover fields had been found. Drilling money vanished, seismic work halted, and the majors and big independents moved out as production in Arkansas slowed. However, a Shreveport promoter, Harvey Broyles, dug up a geophysical mistake, sold it to an aggressive wildcatter, and drilled H.A. Chapman 1 Helms in 1968. The result was Walker Creek field.

Fig. 6–8 is a structure map on the top of the Smackover at Walker Creek. Production is draped east-west along a terrace bounded on the north by a porosity pinchout cutting across contours-10,500 to -10,400 ft and on the south

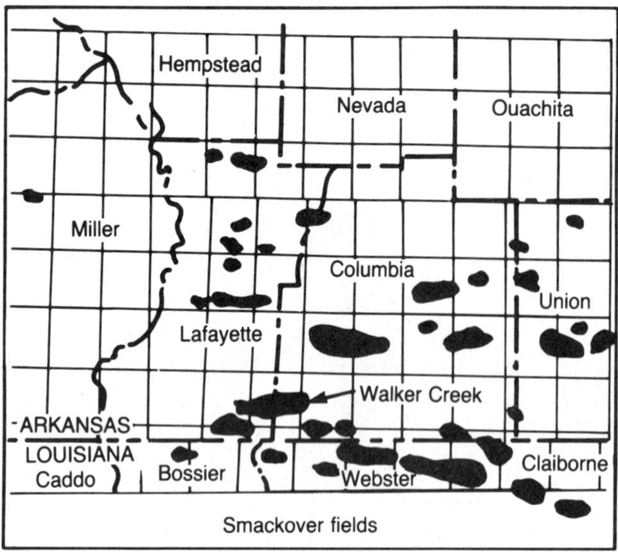

Fig. 6–7 The "final" pattern (courtesy *OGJ*, November 15, 1971)

by low porosity and an oil-water contact at -10,720 ft. Oil was selling at $3.52/bbl and gas at 19–26 cents/Mcfd in those days.

The Walker Creek discovery excited the industry, and it flocked back into southern Arkansas and northern Louisiana. Even today, the search for Jurassic Smackover and Cotton Valley pays continues at a brisk level in this region.

The year 1981 was great for Arkansas exploration and development. Drilling and exploration activity increased by 36.6% during the first half alone. Several new Jurassic fields were found. The same is true for the first part of 1982 with another banner year expected despite the lowering rig count.

Arkansas is divided into two producing regions, the Arkoma to the north-northwest and the southwestern tier

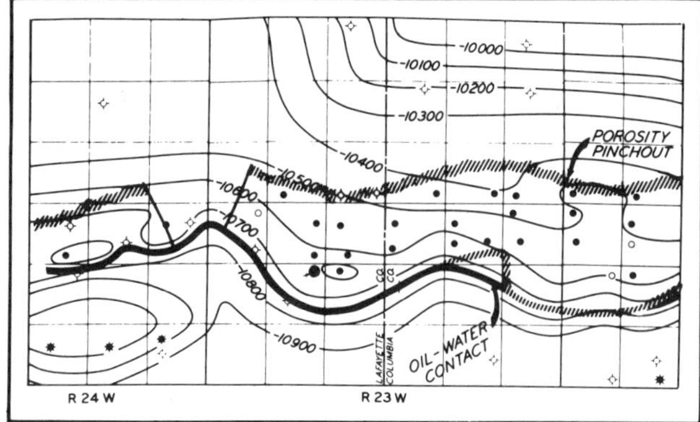

Fig. 6–8 Structural map showing the top of the Smackover (courtesy *OGJ*, November 15, 1971)

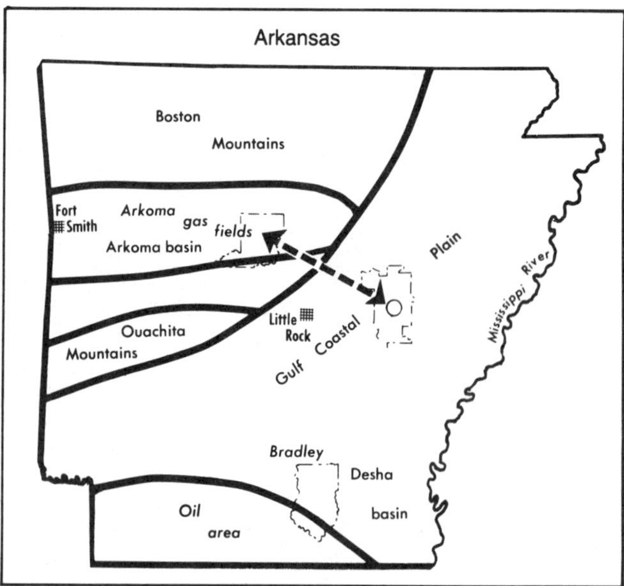

Fig. 6–9 Arkansas producing regions (courtesy *OGJ*, January 2, 1978)

of counties (Fig. 6–9). Geologically the state can be divided by a diagonal line extending northeast-southwest across Arkansas. The northwestern part is known as the Interior Highland region, or the Paleozoic outcrop area.

The southern, southeastern, and eastern sectors make up the Mississippi embayment and the Gulf Coastal plain area of the state. Here are Mesozoic and Cenozoic outcrops. The Paleozoic province in the northwestern sector is divided into the Ozark uplift, the Arkoma basin, and the Ouachita Mountains. Southeastern Arkansas contains the Cenozoic Desha basin which holds a thick Tertiary section, including the prolific Eocene Wilcox sands of Mississippi and northern Louisiana. Thick alluvial deposits from "Old Man" River, which mask the axis of the Mississippi embayment, have hindered exploration in this area. However, several deep wildcats have been drilled here.

Dow Chemical drilled one in Mississippi County in 1981. The firm also dug a deep well to 14,868 ft at another site in the area. Neither had production. But the paucity of any holes in this unproductive part of Arkansas proves that more exploratory work is needed before an appraisal can be made about the future prospects of this potential hydrocarbon province.

North Louisiana

Many old fields are still on production in northern Louisiana, but only one giant has been found in the area in many years—Black Lake. Old giants in the area include

Caddo Pine Island, opened in 1905; Delhi, found in 1944; and Haynesville, discovered in 1921. Caddo Pine Island has produced more than 329 million bbl of oil and still makes more than 3 million bbl/year of oil from 8,546 wells. Millions of barrels of oil are yet to be produced from this oldtimer. Considerable development drilling is under way in and around the old limits of the field. Pay is the Chalk.

Delhi has produced more than 204 million bbl of oil and makes 3 million/year from 114 wells. The field still has 45 million bbl of oil. Haynesville makes nearly 3 million bbl/year from 220 wells and has 8 million or more remaining. There is also activity around this giant field.

Like eastern Texas and southern Arkansas, the industry thought that all of the big ones had been caught in Ark-La-Tex. However, in 1964, the best thing to happen to the area since World War II was the discovery of the Black Lake field in Natchitoches Parish, Louisiana. Placid found this field and operated the unit. Its discovery was a long time in coming, considering how close the wildcatters came without hitting home.

Between 1947–1955, six dusters were drilled within a 7-mile radius of the field. All were drilled through the Cretaceous Pettet, the field pay. In 1967, B. Ross White and Jerry R. Sawyer of Placid told the Gulf Coast Association of Geological Societies that all the ingredients for a textbook example of a stratigraphic trap were in place at Black Lake. On the south and east, three wells had established the presence of a thick, porous Pettet reef limestone. However, updip to the north and west, three wells had found no porosity. An updip pinchout of a prospective reservoir was confirmed, along with either strong nosing, structural closure, or both.

Although some of the drillers may have developed outlines of a prospective field, they hesitated from committing themselves. The nearest Pettet (Sligo age) production was 25 miles away, and it was not a stratigraphic trap. Also the drilling reported no shows of oil in Sligo porosity, the only prospective zone.

Placid, however, saw the potential for reef production and acquired 40,000 acres in leases in 1963. Seismic surveys showed no structural closure but strong nosing. The discovery well was drilled and made 2.6 MMcfd of gas and 459 b/d of 45.8° gravity oil on an 11/64-in. choke through perforations at 7,990–8,000 ft. Subsequent drilling showed Black Lake is a southeastward-plunging structural nose with a dip averaging 120 ft/mile. The Pettet reef itself trends northeast-southwest. The discovery of Black Lake set off a flurry of drilling into the Pettet in LaSalle, Jackson, and Rapides parishes in northern Louisiana. But the fire went out, and nothing big was found.

The industry is now busy throughout North Louisiana, looking for Cretaceous and Jurassic production in and around older producing fields. This hunt is meeting with great success. Deeper drilling is also a feature of today's exploration in this province. Drilling gears are fully meshed throughout the area. There is considerable development work around Caddo Parish, Monroe field; in the Cadeville sand in Jackson Parish at the Calhoun field, South Missionary Lake; the Arkana field in Bossier Parish in the Haynesville; the Terryville field in Lincoln Parish; and in many other areas across the northern tier of parishes.

An important objective in northern Louisiana is the Mooringsport, especially in the Winn and Grant Parish areas. XB Energy and Gulf States Exploration of Houston have a play going as a result of successful completions in the Blue Lick field. The A-1 Manville Forest Products in SE NE 8-8n-3w, Grant Parish, flowed 456 b/d of oil and some gas at 9,415–9,588 ft on choke. About ½ mile away, the B-1 Manville flowed 240 b/d of oil and some gas. There were several other wildcats in the area looking for Mooringsport. Success at Blue Lick portends more action in this portion of Louisiana.

MISSISSIPPI

Many large Smackover and Cretaceous oil and gas fields have been discovered in Mississippi in the past 32 years, but no giants are among them. The state's giant fields are all of the early-day vintage but are still on production. However, large exploratory programs have been conducted sporadically during the past 10 years in the southern part. Particular interest has been directed to the Hosston, Lower Tuscaloosa, the Eocene Wilcox sands, and the Jurassic Cotton Valley and Smackover. It is from these formations that most of the state's new oil and gas has been found in recent years.

The Hosston trend stretches from west to east across southern Mississippi for over 150 miles (Fig. 6–10). This trend is more than 100 miles in width, from north to south. Generally the strike is northwest-southeast, and the dip is southwesterly.[5] The lines on the map in Fig. 6–10 represent major subdivisions of the trend. The area between lines 1 and 2 is where the first Hosston was found. Most of the oil fields that produce from various Sligo-Hosston sands are located here. The area between lines 2 and 3 is the active portion of the trend. Within the last 6 years, several major gas and gas-condensate fields have been discovered in this section.

Most of the gas found recently in the Hosston formation is between 14,000–16,700 ft. Some oil has been found between 14,000–15,000 ft. A favorite target in the Hosston is the Booth sand. This is the initial gas sand found in the Florida Gas Exloration 1 Booth, the discovery well for the Bassfield field, completed in March 1974 (well number 1 on the map).

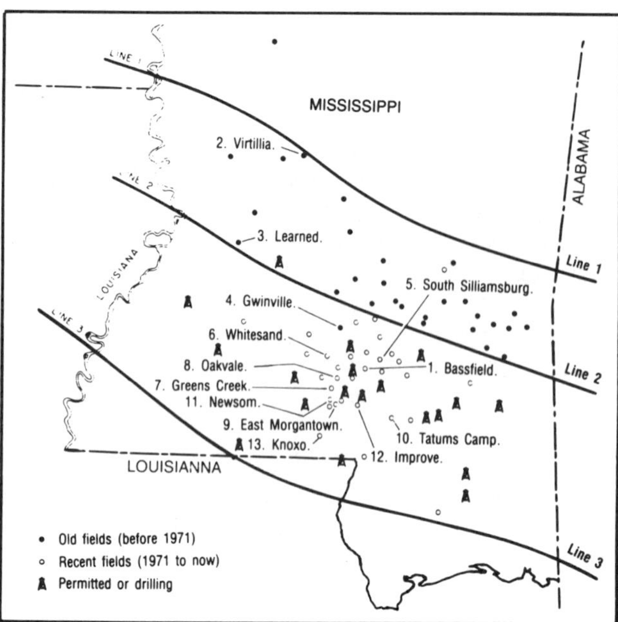

Fig. 6–10 The Hosston trend limits (courtesy *OGJ*, March 16, 1981)

The discovery of the Bassfield field in 1974 opened the eyes of explorationists to the future potential of the Hosston trend in Mississippi. More fields have been found, as shown in Fig. 6–10, and drilling continues today. Adequate gas prices will ensure a continued search for new Hosston sand reservoirs. It is not an easy task, but with careful selection of the type, age, and size of the prospect, together with the expertise of properly drilling and completing Hosston sands, the future still looks promising. There are many more fields of good caliber on the Hosston trend of Mississippi.

The Lower Tuscaloosa of southern Mississippi

This formation of Upper Cretaceous vintage produces mostly oil from more than 50 fields in the Interior Salt basin of Mississippi (Fig. 6–11). These fields lie in an area extending from eastern Louisiana into southwestern and south central Mississippi (Jefferson, Adams, Franklin, Lincoln, Amite, Pike, Walthall, Lamar, Forrest, and Pearl River counties).[6]

Stratigraphically, the lower Tuscaloosa unconformably overlies Lower Cretaceous sediments of the Dantzler formation and underlies the Upper Cretaceous shales of the middle Tuscaloosa (Fig. 6–12). The formation consists of a transgressive sequence that grades upward from alluvial plain through delta to marine deposits. The lower Tuscaloosa is divided into the Massive and the Stringer sand members in

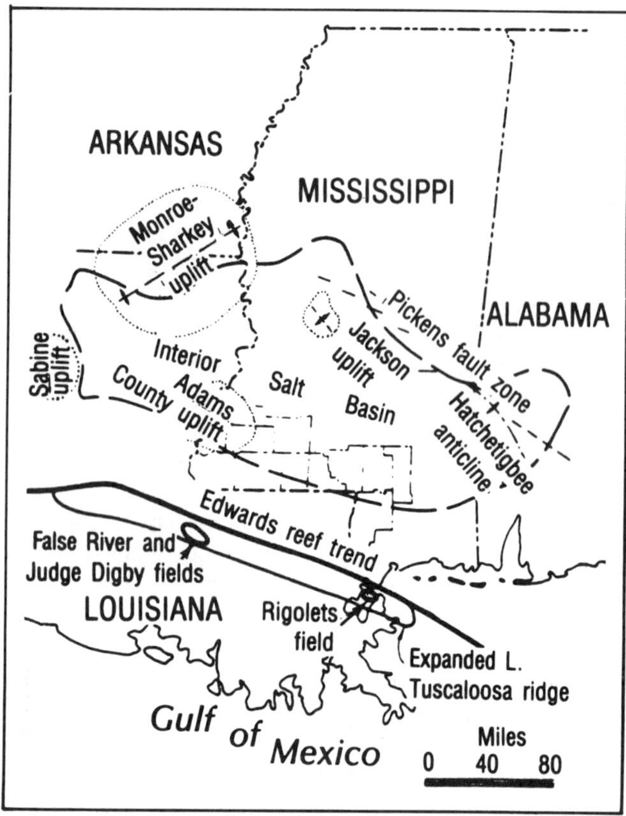

Fig. 6–11 Mississippi's Lower Tuscaloosa (courtesy *OGJ*, February 9, 1981)

Mississippi. Overlying the Massive sand is the Stringer member from where most of the state's lower Tuscaloosa production comes.

Looking for hydrocarbons in this formation can be a chore. Strong structural closure is generally lacking. The accumulation of oil and gas is controlled largely by stratigra-

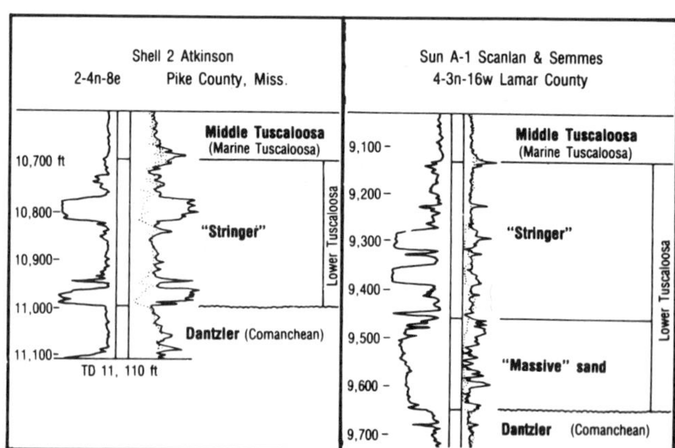

Fig. 6–12 Type logs of Lower Tuscaloosa formation (courtesy *OGJ*, February 9, 1981)

phy, with lenticular sands pinching out into shales. Electric log cross sections are inadequate for delineating the reservoirs because these channel sands are discontinuous and cannot be correlated over any sizable area. Geophysical data and lithofacies maps are the best tools for projecting sand trends. Due to the discontinuous nature of the sands, field development is often limited to four producers or less.

The industry has been looking for oil and gas in the lower Tuscaloosa since the 1940s when the Brookhaven field in Lincoln County was found. Also discovered were Baxterville in Lamar and Marion counties, Cranfield in Adams and Franklin counties, and Mallalieu in Lincoln County. Baxterville is a giant field, producing more than 208 million bbl of oil and having reserves of 26.5 million bbl.

Each of these fields has a large reserve because it is on a dome associated with a deep-seated salt structure. Most of the lower Tuscaloosa fields, however, are predominantly stratigraphic traps. Interest in exploring for production in the lower Tuscaloosa reached its peak in the 1950s and 1960s. Since then, the pace has slowed but is moderate and steady. The formation is still a viable and profitable wildcat objective.

The Jurassic province

The potential for future discoveries of hydrocarbons in southern Mississippi is enormous. However, exploration programs should be based on a knowledge of the depositional history of each stratigraphic unit. The Jurassic is a very thick and generally favorable section.

A simplified Smackover structure based on geology and geophysical data is shown in Fig. 6–13. It indicates possible areas of production. Structures such as these produce in the Ark-La-Tex province and in Alabama, Mississippi, and Florida. The Jurassic Smackover has been a big target for the wildcatter in southern Mississippi for a long time. Exploratory work has been especially active around salt domes, which probably began their growth during Smackover deposition time. There is no doubt that porous limestones and maybe even some reefs have developed around these domes. Source beds probably gathered the hydrocarbons in the interdomal areas.

The Cotton Valley formation, another prime objective in this region, is at least 2,000 ft thick in parts of the area. It is mostly a nonmarine shale and sandstone formation.

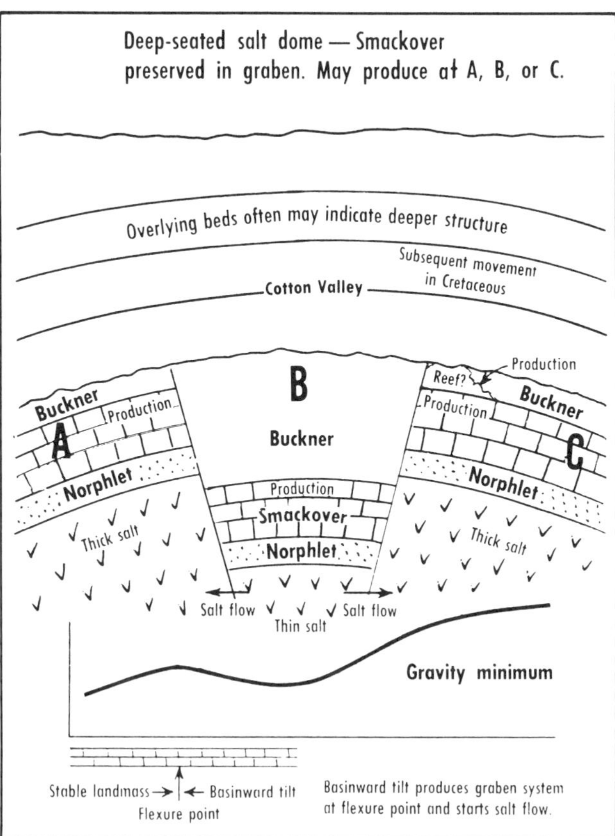

Fig. 6–13 Simplified Smackover structure (courtesy *OGJ*, February 18, 1974)

ALABAMA

Alabama's oil industry was born in 1944 at the Gilbertown field. This important strike was followed in 1952 by the opening of the Pollard field. However, the big oil find came in 1955 at Citronelle in Mobile County. The discovery of the Citronelle field proved Alabama was indeed a living member of the family of oil states. This giant still produces more than 2 million bbl/year of oil from the Cretaceous. Cumulative production is more than 133 million bbl of oil, with about 20 million bbl remaining. After Citronelle's development, the industry continued to look for Cretaceous fields in the southwestern part of the state, but no one found anything quite like the Mobile giant.

Exploration for oil and gas in the region greatly intensified after the discovery of two new oil fields in Choctaw County in 1967 (Fig. 6–14). These new strikes were of great significance because they produced the first oil in Alabama from the deep Smackover formation of Jurassic age. Exploration continued to rise, and in 1968, the Flomaton field was found in Escambia County. By 1970, new discoveries in the area brought the state's total to an even dozen fields, all in southwestern Alabama counties. This total has grown considerably, as shown in Fig. 6–14.

The latest big Jurassic Smackover is Natomas North America 11-6 Broughton in 11-5n-5e, Monroe County. The well flowed 756 b/d of 32° gravity oil and 474 Mcfd of gas

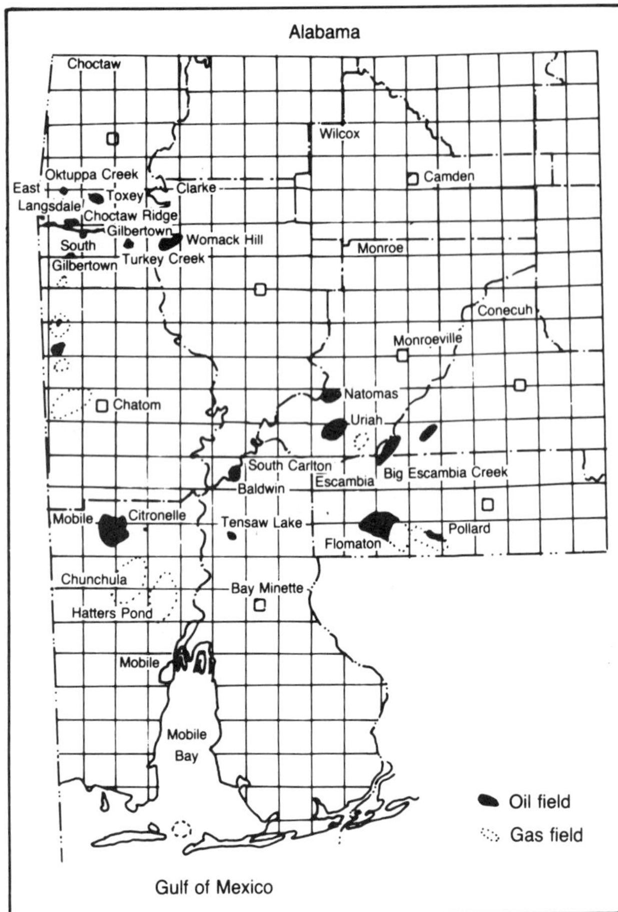

Fig. 6–14 Alabama's oil and gas fields (courtesy *OGJ*, October 16, 1972)

on an adjustable choke from Smackover pay at 13,080–13,104 ft. This well is 6 miles north of the Uriah field, also a Smackover pool.

Operators continue to show considerable interest in Smackover targets in southwestern Alabama, especially in Baldwin County. A recent Hunt Energy discovery in Hancock County heated activity considerably in that area and kicked off a good lease play.

Southwestern Alabama lies on the East Gulf Coastal Plain, near the northeastern limit of the Mississippi Interior Salt Dome basin. The Coastal Plain sediments in the subsurface range in age from pre-Jurassic to Holocene. They consist generally of alternating layers of relatively unconsolidated sand, shale, clay, and limestone that generally dip to the southwest at a rate of 40–50 ft/mile.[7] The Coastal Plain sediments range in thickness from about 5,000 ft in the northern section to more than 25,000 ft in the southern part and lie unconformably on the basement.

The deep-lying Louann salt of Jurassic age in southwestern Alabama greatly influences the oil and gas potential of the region. Structures formed as positive features by salt swells or domes and collapse-type features such as grabens and normal faults occur in the area, creating many potential oil and gas traps in overlying formations.

The major faults in southwestern Alabama were formed by movement of the Louann salt in the subsurface. This faulting is a part of a regional fault system near the updip limits of the Gulf Coast Salt basin, which extends from eastern Texas to northwestern Florida. Wildcatting along this trend during the past 40 years resulted in the discovery of many significant oil and gas fields throughout the region, even into northwestern Florida at the Jay field.

The Jurassic system

The Norphlet formation overlies the Louann salt. Completion of the Humble 1 Wessner in 14-1n-8e, Escambia County, found gas trapped in the Norphlet and opened the Flomaton field. The Smackover overlies the Norphlet. Several fields in Choctaw County produce from a porous interval in the Smackover at an average depth of about 11,500 ft below mean sea level.

The Hatter's Pond field is a classic Jurassic find, being one of the most important Smackover and Norphlet producing areas of the South. The field is in northern Mobile County and has produced more than 7 million bbl of condensate and over 30 billion cu ft of gas since its discovery in 1974. Production is from the upper Jurassic Norphlet and Smackover formations, which form a transgressive-regressive package overlying the Jurassic Louann salt. The Norphlet formation is a subarkose deposited in a coastal dune complex.[8]

The discovery of Hatter's Pond was a much needed incentive to further the cause of Jurassic exploration in the Alabama portion of the Mississippi Salt basin. In the mid-1960s, Getty Oil and others began to explore deeper into the Jurassic trend in the central Mississippi basin and shelf. As they worked eastward toward Alabama, improved seismic techniques began to reveal some interesting deeper possibilities on the Mobile platform.

Using the seismic information and gravity, Getty put together several large acreage blocks. Some of these were Hatter's Pond, Chunchula, Mauvilla, Mount Vernon, and Isabelle. Since little deep drilling had been done on the platform and since there was a question as to abnormal pressures, Getty farmed out some of its land in the Chunchula area to get a well drilled, while retaining its larger 100%-owned blocks. After Union Oil found Chunchula in 1973, Getty decided to drill its Peter Klein 3-14 No. 1 well at Hatter's Pond. It spudded in July 1974 and logged 314 ft of net pay, 100 ft into the Smackover and 214 ft into the Norphlet sandstone. The well tested that December in the

Norphlet at 3.4 MMcfd and 1,286 b/d of condensate through a 16/64-in. choke. Later, the well tested the Smackover, getting 6.2 MMcfd of gas and 2,166 b/d of condensate on a 17/64-in. choke. Immediate development followed.

In Alabama, the Cotton Valley group is the uppermost Jurassic section overlying the Haynesville and underling sediments of Early Cretaceous age. Cotton Valley produces at 8,536 ft in the Okatuppa Creek field, Choctaw County.

The Cretaceous system

Early Cretaceous sediments have yielded great amounts of oil at the Citronelle field, the state's giant in northern Mobile County. Tuscaloosa group sediments produce oil at the South Carlton field in Clarke County and in the Pollard field, Escambia County. The Eutaw produces at 3,100 ft in the Gilbertown field. Selma Chalk also produces oil at Gilbertown. Oil is from fractures in the chalk caused by faulting.

Regional structure

Prominent structural features in the salt basin area of southwestern Alabama are the Hatchetigbee anticline, the Jackson fault-Klepac dome, the Mobile Graben, and the Gilbertown, Coffeeville-West Bend, Walker Springs, Pollard, and Bethel fault zones (Fig. 6–15).

A complex north-south fault zone, the Mobile graben, extends from Jackson south to Mobile Bay. This major fault represents the west flank of the graben opposite the Jackson fault. The graben system has been a highly mobile fault zone since at least Late Mesozoic time. Many structures are favorable for oil and gas accumulation in this system.

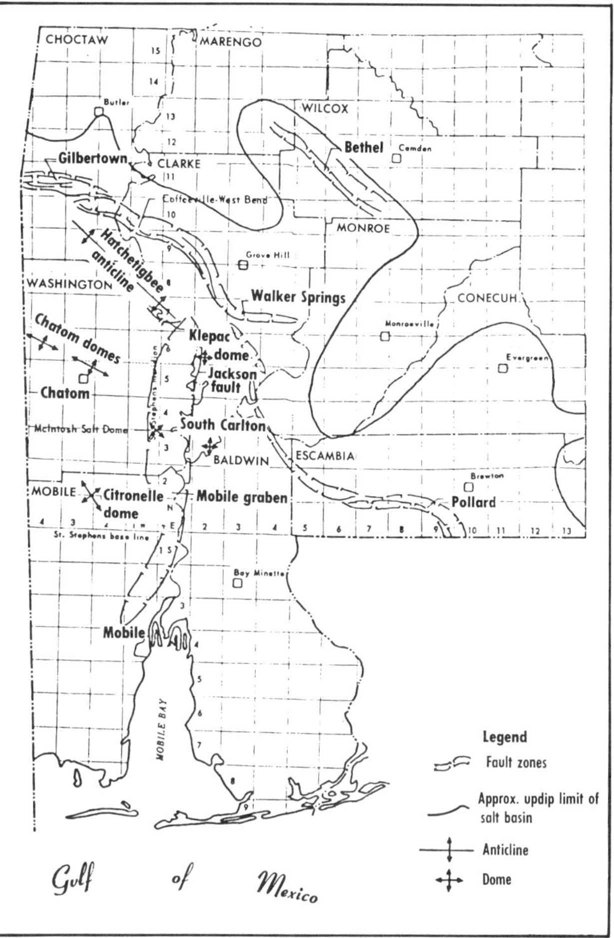

Fig. 6–15 Location of prominent structural features (courtesy *OGJ*, October 16, 1972)

Prospects

Until recently, the main exploratory target in southern Alabama has been the Late Jurassic Smackover and Norphlet formations. But the Cretaceous is a proved oil reservoir in this area (Fig. 6–16). Lower Cretaceous zones have been productive at the Citronelle field and the Tensaw Lake field, Mobile and Baldwin counties, respectively. The Upper Cretaceous is the pay zone at the South Carlton field, East Langsdale, Gilbertown, and Pollard fields (Fig. 6–17). These Cretaceous oil fields were found before 1965. Not one Cretaceous oil discovery has been made in Alabama since. With the advent of Cretaceous Tuscaloosa production in Louisiana, the deep Tuscaloosa trend has begun to spread eastward.

Salt structures in southern Mobile and Baldwin counties are excellent areas to drill for Upper Cretaceous hydrocarbon prospects. Lower Tuscaloosa pilot sand-body trends and sandstone petrography suggest that porous and permeable reservoirs should be present in this area.[9]

Geologists in Alabama say that a successful Jurassic Smackover oil prospect must include the trap, reservoir, source rock, and relationship between hydrocarbon migration and structural deformation. They think that regional geologic trends must be understood before a worthwhile Smackover probe can be undertaken. There are some excellent possibilities for new Jurassic fields on the flanks of the Wiggins-Conecuh ridge. Updip Smackover grainstones associated with salt structures are also good areas for new oil discoveries in southwestern Alabama.

The key to successful hunting in the area is the delineation of traps associated with salt movement and recognition of high to moderate energy lithofacies. They must be lithofacies that have had their primary interparticulate porosity preserved or of lithofacies that have been dolomitized and/or leached with the development of intercrystalline dolomite and/or secondary grain moldic porosity.[10]

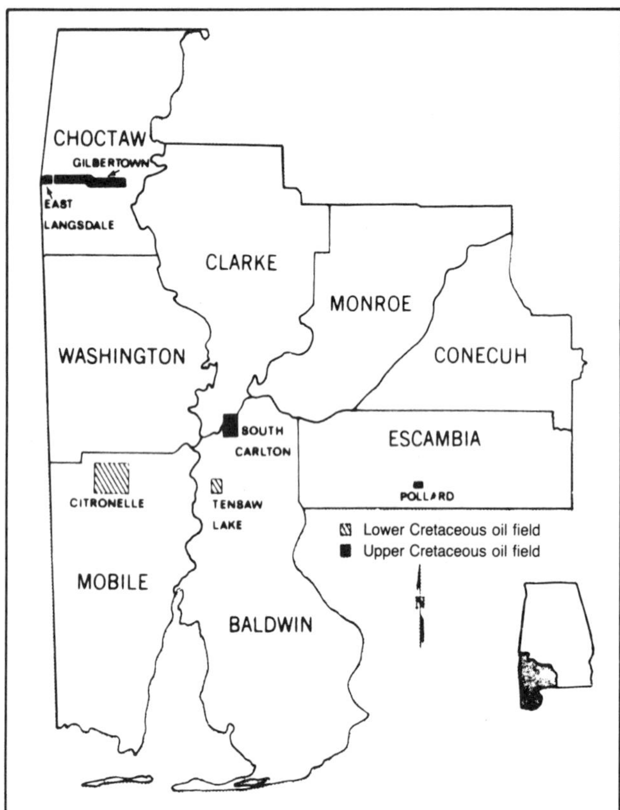

Fig. 6–16 Cretaceous oil fields in Southwest Alabama (after Mancini and Payton, courtesy *OGJ*, May 24, 1982)

		SURFACE WEST-CENTRAL AL	SUBSURFACE SOUTHWEST ALABAMA	
UPPER CRETACEOUS	TUSCALOOSA GROUP	GORDO FORMATION	UPPER TUSC.	
				"MILLER SAND"
		COKER FORMATION		"MARINE SHALE"
			LOWER TUSC.	"PILOT SAND"
				"MASSIVE SAND"

Fig. 6–17 Upper Cretaceous Tuscaloosa surface and subsurface stratigraphy in Alabama (after Mancini and Payton, courtesy *OGJ*, May 24, 1982)

The Black Warrior basin

Tucked in the northeastern corner of Mississippi and the northwestern portion of Alabama is one of the nation's leading areas for return on investment. This is due to the high percentage of producing wells in the area and because of the relatively low drilling costs at shallow depths (Fig. 6–18).

Most of the gas wells in the Black Warrior basin have become productive at depths from 2,000–4,500 ft. The Blooming Grove field in Fayette and Lamar counties and the Beaverton field in Lamar have large reserves, and these wells exhibit good deliverability. Deliverability of wells here ranges from a few hundred thousand cubic feet of gas per day up to 15 MMcfd or more, absolute open flow. The thickest pay zone found here to date is more than 40 ft.

In 1980, the Alabama portion of the basin got its first big oil find at Hughes and Hughes 1 Franks 2-14 in Lamar County, 2-14s-14w. Flow was 1,134 b/d of 26.7° gravity oil. Production is from the Mississippian Carter sand at 2,211–2,219 ft. At the time of completion, this was by far the best potential reported for any oil well in the Black Warrior basin. The basin has been drilled on and off since 1909, but for the most part discoveries have been small until 1980. Most of the basin's production is from the Mississippian Chester sandstones. Some of the best finds were at Muldon, Fairview, and Corinne.

The basin has not been fully explored, and the potential for more oil and gas is still considered excellent. Many people in the industry feel that eventually large reserves will be found in deep fault traps in the Lower Paleozoic carbonates. This section needs a good solid exploratory program. There may also be some new fields lying in stratigraphic traps along many of the basin's unconformities. These fea-

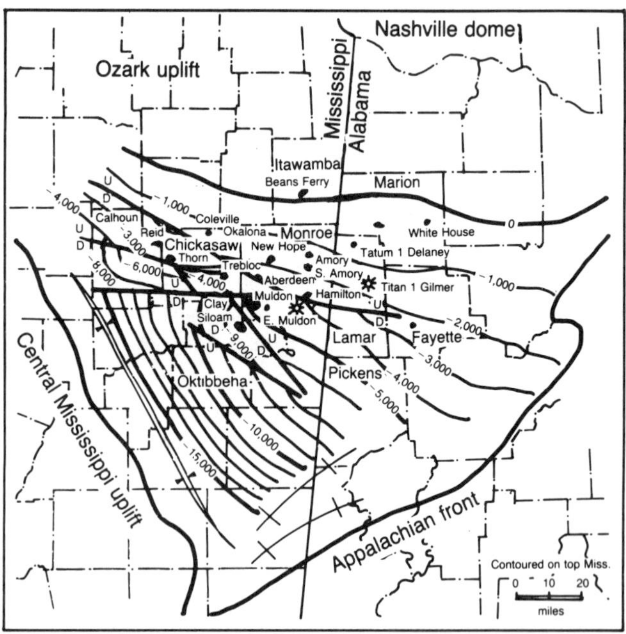

Fig. 6–18 Black Warrior basin (courtesy *OGJ*, June 18, 1973)

tures have not been drilled fully either. The Hughes and Hughes oil discovery at the Blowhorn Creek field proves there are more reasons to look for hydrocarbons in the Black Warrior basin.

SOUTH GEORGIA AND FLORIDA

Drilling in the state of Georgia has been sparse since much of the state is covered by the outcrops of the Appalachians.

Two areas of the state have the potential for hydrocarbon prospects, but chances of hitting oil are slim. However, the areas should not be overlooked. The Southwest Georgia embayment is a basin or trough between the Ocala uplift on the east and the Decatur arch on the west. It is geographically located in southwest Georgia and northern Florida. Exploratory work here is needed to uncover some clues to its possible stance as a hydrocarbon area. The Decatur arch also should be examined more closely. It is a southward-plunging fold in southwest Georgia, southeast Alabama, and northwestern Florida.

Florida

There is oil production at both extremes of Florida: in the northwestern panhandle area in Santa Rosa County at Jay and other fields and in the southern part of the state in the Sunniland trend of Cretaceous oil pools. The Jay field, tapped in 1968, was one of the most important oil finds in the U.S. since Prudhoe Bay. Current estimates indicate recoverable reserves in this Jurassic field will exceed 390 million bbl of oil. The field already has produced 300 million bbl of oil, making nearly 24 million bbl/year from 106 wells.

A year after the Jay discovery 35 miles north of Pensacola, more than 100 wildcats were drilled in adjoining Alabama and Florida counties (Fig. 6–19). Most Jurassic traps are very subtle, and the key to exploring here lies in the ability to combine favorable facies with delineation of low relief structures.

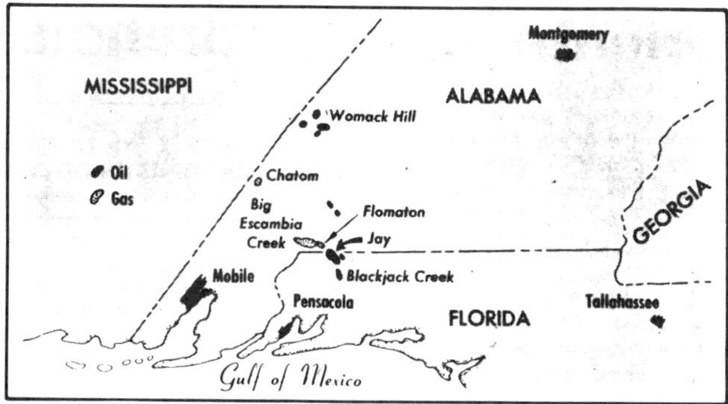

Fig. 6–19 Jay field and its environs (courtesy *OGJ*, April 14, 1975)

Interest in Smackover exploration in the Southeast was revived in 1960–68 as a result of the development of CDP seismic techniques. These techniques extended structural definition to as deep as 25,000 ft and the new ability to dig this deep. During this period, wildcatting moved to southern Mississippi where 38 new fields were found at 12,000–18,000 ft.

Exploration for the Jurassic in Alabama and Florida was a logical extension along the trend. The next period of exploration began with a Norphlet find at Flomaton, 90 miles east of the nearest existing Jurassic production. Jay came along in 1970, when dolomitized Smackover was found on a structural nose about 7 miles south of the town of Flomaton. Other fields followed.

In Mississippi, the producing Smackover is generally an oolitic facies. In southern Alabama and in the panhandle of Florida, the grainstone facies is composed of hardened pellets with oolites, contributing a very minor part of the section. Development of this lower energy facies in the Jay area can be attributed to the dissipation of wave energy over a wide shelf and to the absence of topographic features that create a focal point for oolite generation.[11]

Other Smackover prospects are available in northwestern Florida; however, more wildcatting and seismic work are needed. About 2,000 sq miles of the Apalachicola embayment are underlain by Smackover limestone. The updip limit of the Smackover has been defined partially by eight wells in Bay, Calhoun, Franklin, Gulf, and Liberty counties and by another six deep tests updip of the Smackover limit.[12]

With sparse well control in the embayment, indirect evidence indicates possible faulting, which would provide Smackover fault traps. The Smackover was deposited on igneous Paleozoic, Eagle Mills, and Norphlet formations. The presence of Jurassic salt, limestone, and marine shales warrants a continued exploration effort in this barely drilled embayment. Gravity work, and the wells drilled to date, indicate both stratigraphic and structural traps may be present. Further exploration, concentrated in those areas that have a thick Smackover section, may eventually lead to the discovery of another Jay field.

The Cretaceous fields

Continued success in exploratory drilling in recent years shows the Sunniland trend of southern Florida offers the potential for more commercially attractive oil fields (Fig. 6–20). Although the area is still not heavily drilled, recent additional well control has provided necessary data for a

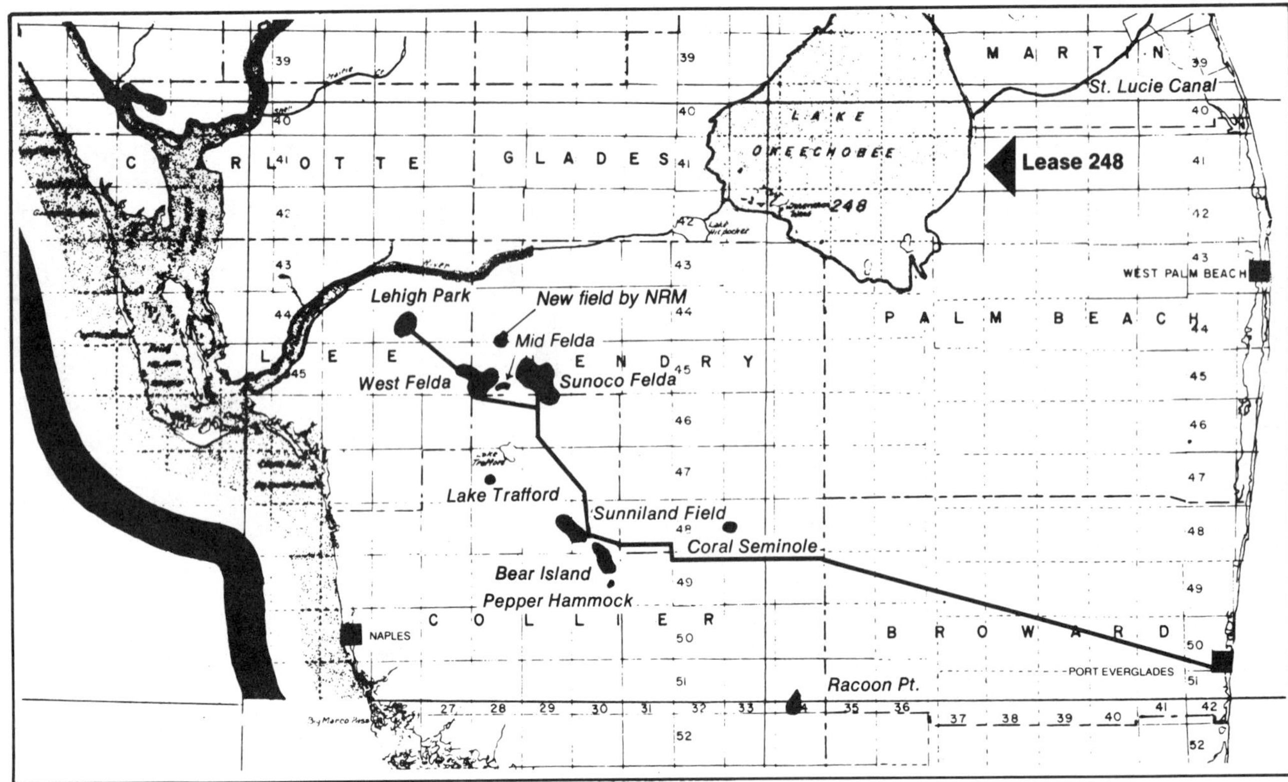

Fig. 6–20 South Florida oil fields and pipelines (courtesy Coastal Petroleum Company, April, 1981)

better understanding of the stratigraphy source, and reservoir characteristics of the Cretaceous Sunniland formation. Potentials are great; 300–500 b/d of oil are typical flow rates. Some of these wells will undoubtedly produce more than 1 million bbl of oil each. The industry still considers this area a good place to look for new oil. The latest Sunniland find was at NRM Petroleum et al. 2-3 Duda in 2-45s-28e, Hendry County. Tests had 216 b/d of oil from the Sunniland lime at 11,416 ft. Several months of testing will be necessary before it may be determined whether the well taps a new reef or is part of the Felda structure.

The Sunniland is made of a series of carbonates and evaporites of Late Trinity, Early Cretaceous age. It is the only producing zone in southern Florida and has an average depth of 11,500 ft with a net pay section of 10–40 ft. The main oil reservoir is in the upper zone, as much as 100 ft thick.

The best field found in this Sunniland trend is the West Felda, which has an estimated oil in place of 142,857,000 bbl and recoverable reserves of 50 million bbl. Productive acreage is 7,500, and the production is primarily from rudist reef deposits.[13]

Geological studies show the Sunniland formation, downdip of the reefal sediments, is too tight to enable hydrocarbon migration. Therefore, production along this trend will probably occur where the porosity units lie juxtaposed to the organic, rich, dark micrites.[14]

Exploratory work in the South Florida basin is low compared to other producing basins in the nation. New exploratory techniques and a better understanding of the regional geology should increase interest in new wildcatting in this area.

One of the most active operators in southern Florida has always been Exxon. This company, formerly Humble, has been drilling in Florida since 1943 when it found Sunniland field. The company holds leases on about 200,000 acres in an area south of Alligator Alley.

Onshore oil production has been limited to five counties in Florida: Collier, Hendry, and Lee in the south, and Escambia and Santa Rosa in the panhandle.

The South Florida basin covers an area of some 77,000 sq miles, about 30–40% of which lies onshore (Fig. 6–21). The basin contains from 15,000 to perhaps 20,000 ft of sediments composed mostly of Cenozoic and Mesozoic carbonates and evaporites with small amounts of shale. The major structural features and boundaries of the basin area shown in Fig. 6–21.

Offshore drilling has been sparse through the years. Several important wildcats were drilled off the southwest

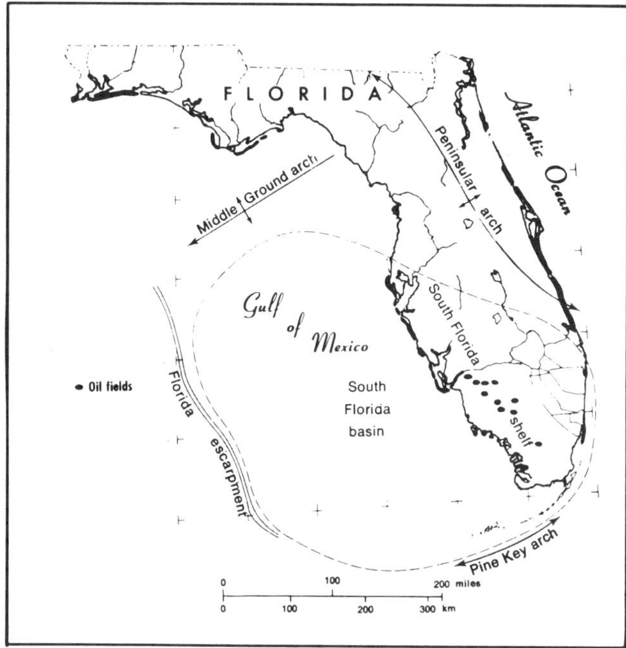

Fig. 6–21 South Florida basin (courtesy *OGJ*, December 17, 1979)

coast years ago, but found nothing to report. But the lack of well data in the offshore theater prevents us from making any appraisal of the potential of this large marine area.

REFERENCES

1. *Our Sun* (1955), p. 6.
2. Nichols, Transactions of Gulf Coast Association of Geological Societies, volume VIV (1964), 21.
3. Bruce Fox, *Oil & Gas Journal* (June 10, 1963), p. 246.
4. Dalton J. Woods, *Oil & Gas Journal* (November 15, 1971), p. 203.
5. D. Scherer, *Oil & Gas Journal* (March 16, 1981), p. 103.
6. D. Devery, *Oil & Gas Journal* (February 9, 1981), p. 152.
7. D. Moore, *Oil & Gas Journal* (October 16, 1972), p. 166.
8. Benson, Mancini, and Wilkerson, *Oil & Gas Journal* (July 20, 1981), p. 87.
9. E. Mancini and J. Payton, *Oil & Gas Journal* (May 24, 1982), p. 110.
10. E. Mancini and E. Benson, Transactions of Gulf Coast Association of Geological Societies, volume XXX (1980), 163.
11. R. Ottman, P. Keyes, and M. Ziegler, *Oil & Gas Journal* (April 14, 1975), pp. 114–115.
12. A. Applegate, F. Pontigo, and J. Rooke, *Oil & Gas Journal* (January 23, 1978), p. 80.
13. *Oil & Gas Journal* (July 30, 1979), p. 226.
14. Ibid., p. 232.

7
Gulf Coast

Wildcatting along the Gulf Coast spread furiously after the Spindletop discovery in 1901.

Texas production soared to 50,000 b/d of oil in 1902, second only to Ohio. The coastal salt domes of Texas and south Louisiana produced 174,354,000 bbl of oil from 1901–1910. The Conroe gas field was discovered in south Texas prior to World War II and was the harbinger of a major oil field on the same site. The Old Ocean field was found in 1934, the forerunner of a wave of geophysical successes along the Gulf Coast such as the Hastings and Anahuac finds. Creole, discovered in 1938, was the first offshore field, pioneering revolutionary methods for exploration and development of underwater reserves.

However, the major finds in southern Louisiana fields have occurred since 1940. The Black Bay West field opened in 1953. This field has produced 127 million bbl of oil and has 28 million bbl remaining. The South Pass Block 24 field was discovered in 1950. It is onshore and offshore with 425 million bbl of oil produced and 53 million bbl to go. South Pass Block 27 followed in 1954 with 298 million bbl of oil produced and 89 million left, Eugene Island block 330 in 1971 with 200 million produced and 124 million bbl left, Grand Isle Block 43 in 1956 with 231 million bbl produced and 126 million left. Main Pass Block 41 came next in 1957 with 187 million bbl of oil produced and 92 million bbl remaining, Main Pass Block 306 in 1969 with 60 million bbl produced and 89 million bbl left, Mississippi Canyon Block 194 in 1980 with 12 million bbl produced and 95 million bbl plus left. Ship Shoal Block 204 followed in 1968 with 54 million bbl of oil produced and 50 million bbl left, Ship Shoal Block 207 and 208 in 1967 and 1962 with 76 million bbl produced and 50 million bbl left, and 132 million bbl produced and 92 million bbl left, respectively.

Big Texas Gulf Coast and South Texas fields discovered recently include Lake Pasture in District 2 with 71 million bbl of oil produced and 29 million bbl remaining and Giddings field in 1971 with 125 million bbl of oil produced and 300 million bbl left. Most of the big fields in Texas were discovered prior to and during World War II.

The Gulf Coastal province is indeed one of the most important petroleum provinces in the U.S. and, for that matter, in the world (Fig. 7–1). Back in 1957, Grover E. Murray (the famous geologist and professor at Louisiana State University) noted that, although small oil fields are numerous in the Gulf Coast province, the 77 major fields (at that time)

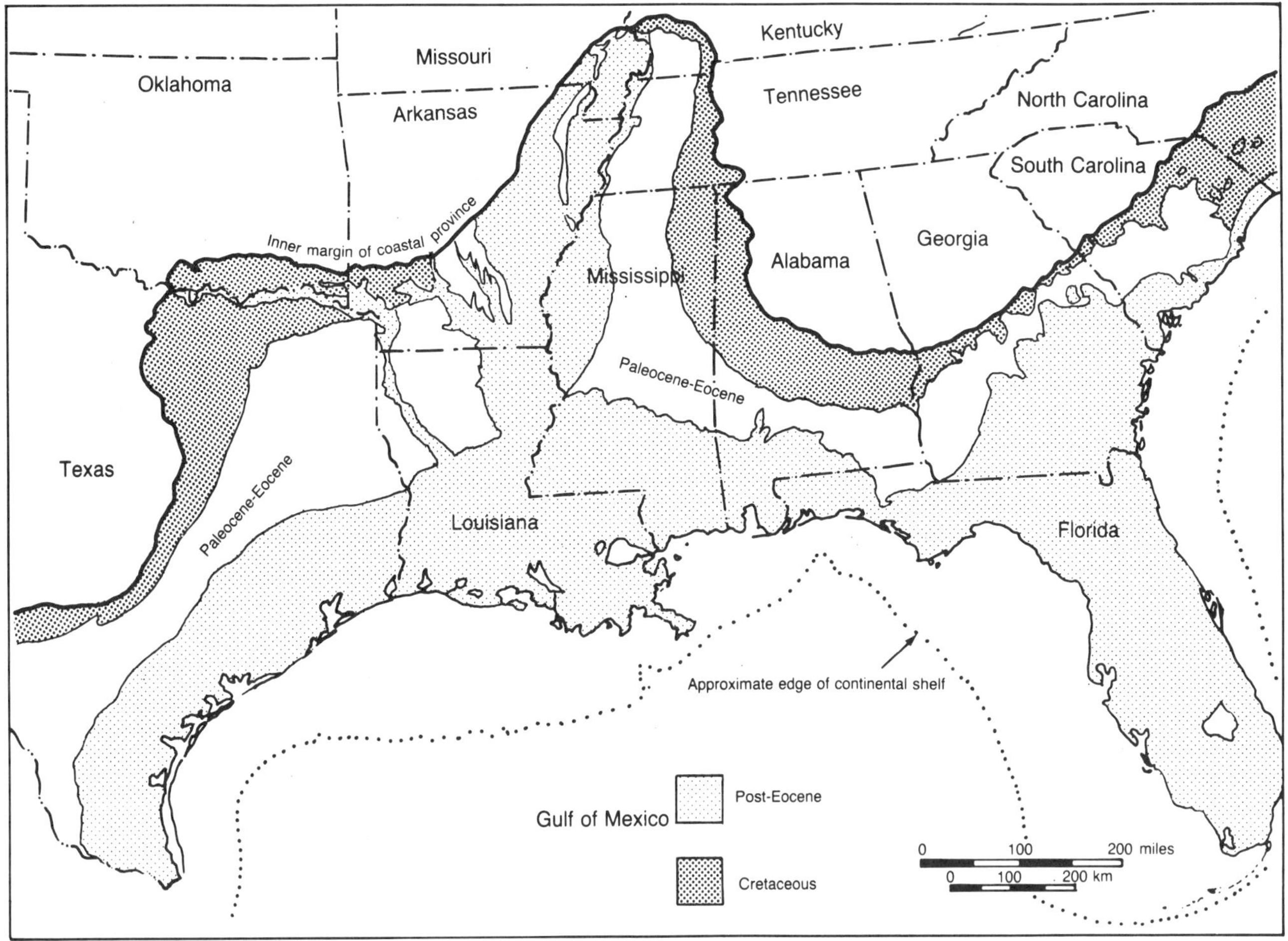

Fig. 7–1 Areal geology of Gulf Coast (courtesy *OGJ*, November 4, 1957)

would eventually yield about two-thirds of the total ultimate production of the province. Production then was estimated at 23 billion bbl of oil.

Occurrence

Production occurs in arenaceous and carbonate rocks, but the clastic rocks are by far the most important oil-bearing zones, especially the sandstones of Miocene age. Much of the oil production is directly related to structures resulting from the movement of salt, probably the Louann. However, the traps are usually the result of two or more geological factors, involving regional and local structure, stratigraphy, porosity, and permeability.[1]

Oil traps run the gamut from simple anticlines to complex faults on the flanks of large salt domes. Regional fault trends play a major role in trapping oil locally. This is most apparent along the Balcones, Mexia-Talco, Luling, Arkansas, and Pickens-Quitman-Gilbertown trends.

The Gulf Coast geosyncline forms the north coastal flank of the Gulf of Mexico and has a tremendously thick geologic column ranging to as much as 40,000 ft of Mesozoic-Cenozoic rocks. Most of the important oil fields occur in areas of 30–50% sand and in areas of structural terracing updip of monoclinal flexing. Changes in porosity and permeability control oil accumulation locally and may account for production.[2]

The overall regional form and shape of the Gulf Coastal province are controlled by Paleozoic and Precambrian structures and alignments. Mesozoic and Cenozoic sediments reach a thickness of 40,000 ft, forming a vast coastal geosyncline. Igneous rocks of Mesozoic age are present locally, especially in southern Arkansas, northern Louisiana, and western Mississippi.

Structural features

Structurally the province is one of vertical movements, and most of the local features are related to salt movement.

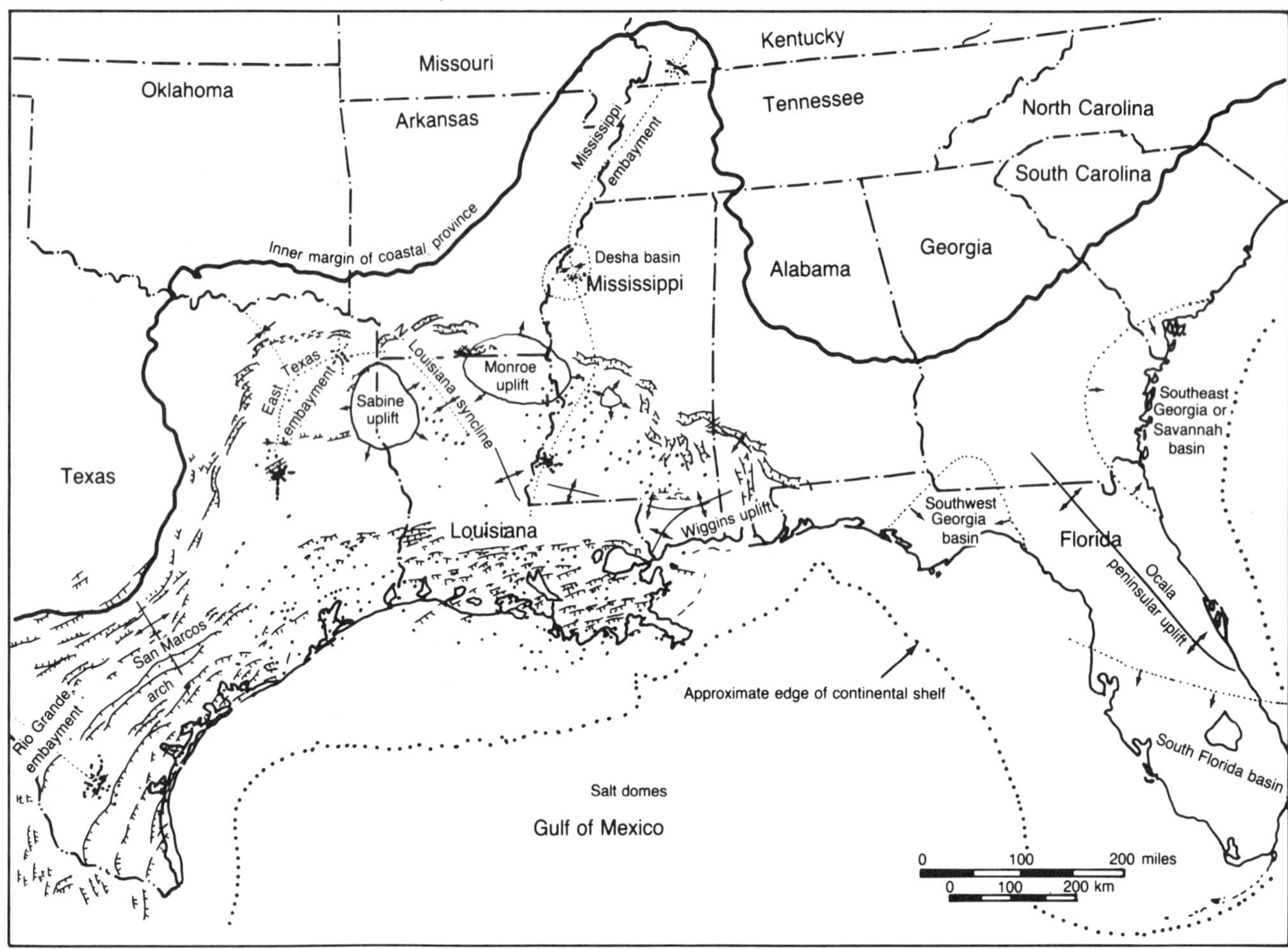

Fig. 7–2 Major structural features of Gulf Coast area (courtesy *OGJ*, November 4, 1957)

Deposits of the northern Gulf Coastal element possess an overall homoclinal dip toward the Gulf of Mexico and constitute a great sedimentary and structural arc from Florida to Mexico (Fig. 7–2).[3] Appreciable alterations of the regional dip result from large positive and negative anomalies.

Several major fault trends, probably the result of salt movement, are regional in extent. So-called salt basins exist in a belt across western Alabama, central Mississippi, and east central Louisiana; in the north Louisiana syncline; in the northeast Texas syncline; in the Rio Grande embayment; and in a belt across lower coastal Louisiana and southeast Texas.

The larger anomalies trend northwest-southeast and northeast-southwest as a rule, although secondary features are present that trend north-south or east-west. Major zones of faulting, or of pronounced downflexing, exist in the inner portions of the province.[4] These zones roughly parallel the trends of the Ouachita fold belt. These fault zones are as follows:

1. The Balcones system, consisting mostly of down-to-the-coast normal faults with complementary faults that combined form a graben system.
2. The Luling system of Texas, containing mainly up-to-the-coast faults.
3. The Mexia-Talco of Texas, the Arkansas, and the Pickens-Quitman-Gilbertown of Mississippi, Alabama, and northern Florida, consisting of a graben system.

Piercement salt domes are present in the various salt basins. Clusters of these domes appear to be associated with thick salt and thicker-than-normal sedimentary accumulations. Scattered wells have encountered Paleozoic and Precambrian rocks beneath the Mesozoic and Cenozoic sediments. Well data reveal the surface of the precoastal rocks slopes generally toward the Gulf of Mexico, like the overlying Cretaceous and Mesozoic beds but at a greater rate.

The Ouachita fold belt can be traced from its outcrop in Oklahoma southwestward in the subsurface to the

Marathon uplift of Texas. It also can be traced southeastward across Mississippi into Alabama.[5] These subsurface lines of folding should not be ignored because rocks of the same age (although of different facies) produce hydrocarbons in Oklahoma and Texas.

Fig. 7–3 is a generalized geologic column of the northern Gulf Coastal province. Geologists recognize the Morehouse-Eagle Mills clastics, the Werner formation, and the Louann salt in ascending order. None of these to date has produced oil or gas. The significance of the widespread Louann salt cannot be overemphasized because so many, if not all, of the structural features are related to salt movement, either directly or indirectly.[6] Most of these features are evident as gravity anomalies, confirming the salt origin.

Jurassic rocks are known only in the subsurface in the Gulf Coastal province. Updip the beds are mostly clastics, but they grade downdip into a dominantly marine facies. Evaporite and reef facies are present in northeast Texas, southern Arkansas, north Louisiana, Mississippi, Alabama, and northwestern Florida. There are numerous fields producing from Jurassic rocks, mostly from the Smackover limestone and sandstones of the Cotton Valley section throughout the region from East Texas to Florida.

System	Series	Stage		Texas	Arkansas	Louisiana	Mississippi	Alabama	Florida
				Oil and gas producing areas					
Quaternary	Recent Pleistocene					●			
Tertiary	Pliocene			●		●			
	Miocene			●		●			
	Oligocene	Vicksburg		●		●			
	Eocene	Jackson		●		●			
		Claiborne		●		●	●		
		Sabine	Wilcox Group	●		●	●		
	Paleocene	Midway		?		?	?		
Cretaceous	Gulfian	Navarro		●	●	●	●	●	
		Taylor		●	●	●	●		
		Austin		●	●	●	●		
		Eagleford		●	●	●	●	●	
		Woodbine (Tuscaloosa)		●	●	●	●	●	
	Comanchean	Washita		●	?	●	●		
		Fredericksburg		●	?	●	●		
		Trinity		●	●	●	●	●	
	Coahuilan	Nuevo Leon and Durango Groups of Mexico (Hosston-Sligo)		●	●	●	●		●
Jurassic (Upper)		Cotton Valley		●	●	●			
		Haynesville	Louark Group	?	●	●			
		Smackover (Including Norphlet formation)		●	●	●	●		
Jurassic (Lower and Middle)		Louann Salt Werner formation							
Pre-Jurassic		Morehouse-Eagle Mills formations							

Data on oil and gas producing areas principally from AIME, *Statistics of Oil and Development and Production;* and National Oil Scouts & Landman's Association, *Yearbooks of Oil and Gas Development in U.S.*

Fig. 7–3 Generalized geologic column of northern Gulf Coast province (courtesy *OGJ*, November 4, 1957)

Early Cretaceous rocks are mainly red clastic facies in the updip areas of the region, from northeastern Texas to the Florida panhandle. Downdip, the dominant facies is a dark argillaceous-calcareous type of deposit. Production from these rocks is minimal in the region.

Older Gulfian Late Cretaceous deposits (Tuscaloosa-Woodbine-Eagleford) are mainly fine to coarse clastics, except in Florida and southern Texas. Best known of these pays is the Woodbine of East Texas. Gulfian rocks account for at least one-third of total oil found in the province.

Cenozoic rocks are primarily deltaic in origin. These beds reach a thickness of more than 25,000 ft in coastal Louisiana and Texas. Miocene and post-Miocene sediments of the Gulf province comprise one of the great deltaic accumulations in the world.[7] These beds are more than 20,000 ft thick. Neogene rocks, predominantly Miocene age sands, contain a fourth of the total oil found to date in the Gulf Coastal basin.

Jurassic production is found in north Louisiana, south Arkansas, east Texas, Mississippi, Alabama, and northwestern Florida (Fig. 7–4). Smackover production is associated with structures, but stratigraphy is very important. Often porous zones are developed on the flanks of the structures that are absent over the crest.[8]

Cretaceous production (Coahuilan-Comanchean-Gulfian) occupies an arcuate belt, extending from the Rio Grande into Alabama (Fig. 7–5). Many of the oil fields are multipay fault-line traps, such as Talco and Mexia. The best-known Cretaceous fields are in east Texas and Alabama. Stratigraphy is very important as a controlling factor for oil accumulation. Future explorationists must continue to look for the strat traps in these oil- and gas-bearing rocks.

Tertiary oil occurs in a narrow belt along the Texas and Louisiana coastal plains. Inland, the older rocks of Paleogene age produce. Downdip or toward the coast, the Neogene rocks, mostly sands of the Miocene age, are the big reservoirs (Fig. 7–6). Almost all Tertiary production is related to salt movement, and the high reserves (especially in the Miocene sandstones) have made the Gulf Coast what it is today—a highly active area for oil exploration and development through the years with a long future ahead of it.

Giant oil fields account for a huge percentage of the total reserves of the Gulf Coastal province. Most of these major fields produce from Neogene rocks and are concentrated in a narrow arc belt along the Gulf Coast. There are also large reserves of oil and gas in offshore fields in both Texas and Louisiana, with much more to be found.

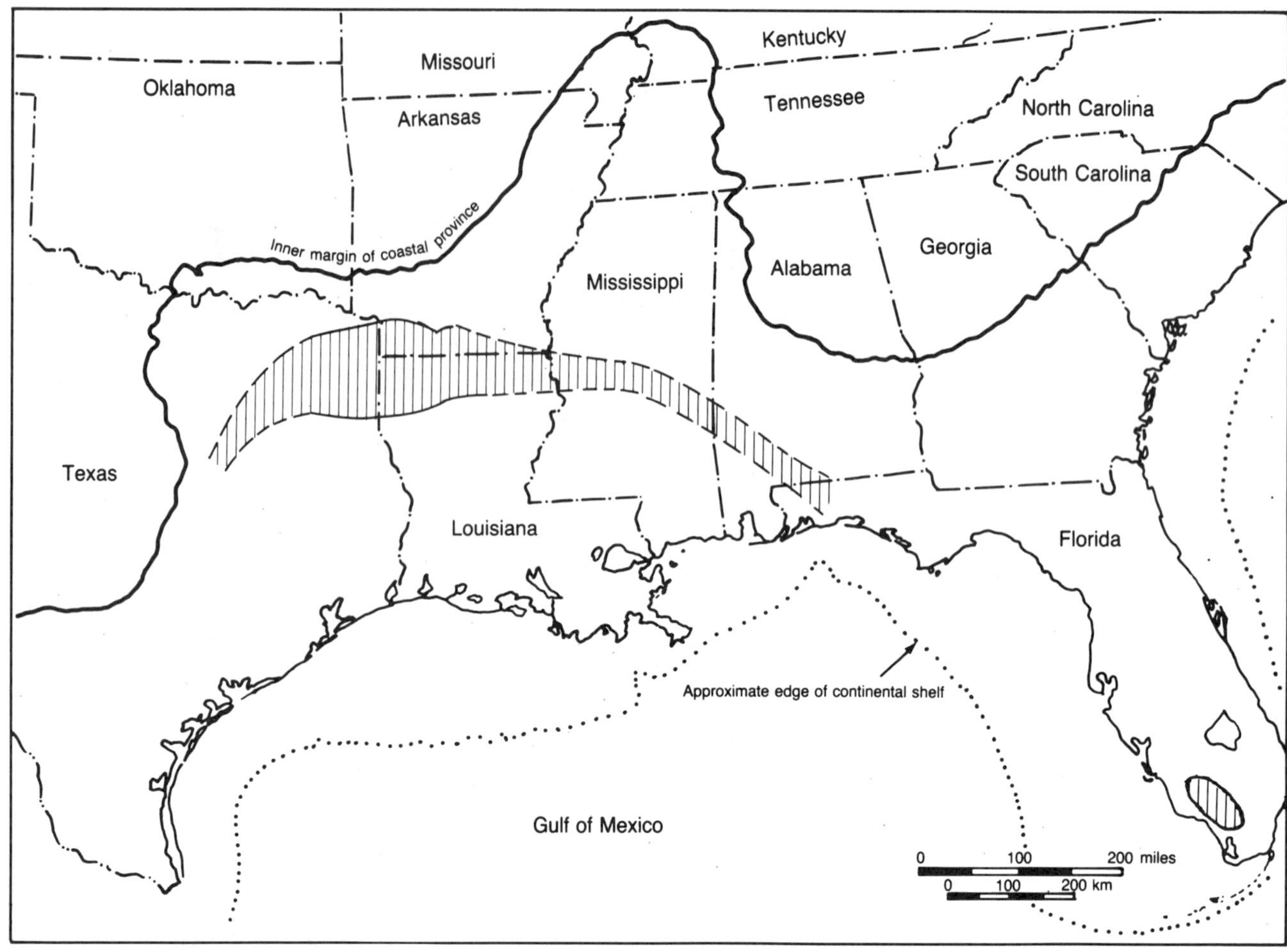

Fig. 7–4 Approximate limits of discovered Upper Jurassic production—Gulf Coast area (courtesy *OGJ*, November 4, 1957)

Texas portion

Production of crude oil within the Texas Gulf Coastal province alone amounts to nearly 400 million bbl/year, some 40% of the total state output. About 60% of the state's and 25% of the U.S. production of natural gas come from this province. Producing formations include all ages of rock, from near-surface asphaltic sands in East Texas to pays at 20,000 ft and deeper in a variety of traps and reservoirs.

Table 7–1 is a general grouping of various types and areas of oil and gas production in the Texas Gulf Coast province.[9] Most of the discovered reserves of oil and gas in the region occur in a few large fields. Of the more than 6,000 fields and pools, at least 15 discovered reserves of more than 200 million bbl; about 35 found reserves of more than 100 million bbl. However, East Texas is still the largest field in this province.

Of the major fields in the province, 70% are in late Cenozoic rocks (mostly Oligocene Frio), 18% occur in late Cretaceous rocks (the Woodbine for the most part), and about 12% produce from the early Cenozoic rocks of Wilcox-Carrizo, Yegua, and Jackson vintage. About 60% of the major oil and gas fields in the province are associated with salt structures. Most of the others, except the East Texas field, are associated with regional strike fault systems. East Texas is mainly a classic stratigraphic trap.

South Texas today

There are several important happenings in southern Texas that portend a good future for hydrocarbon exploration in this part of the Gulf Coastal province. Aminoil is testing at least one fault block in its East Chapa field in Live Oak County, Texas. This is the site of the company's first south Texas discovery. Other operators have drilled in the field area but with no success. This new field, between Corpus Christi and San Antonio, produces natural gas and con-

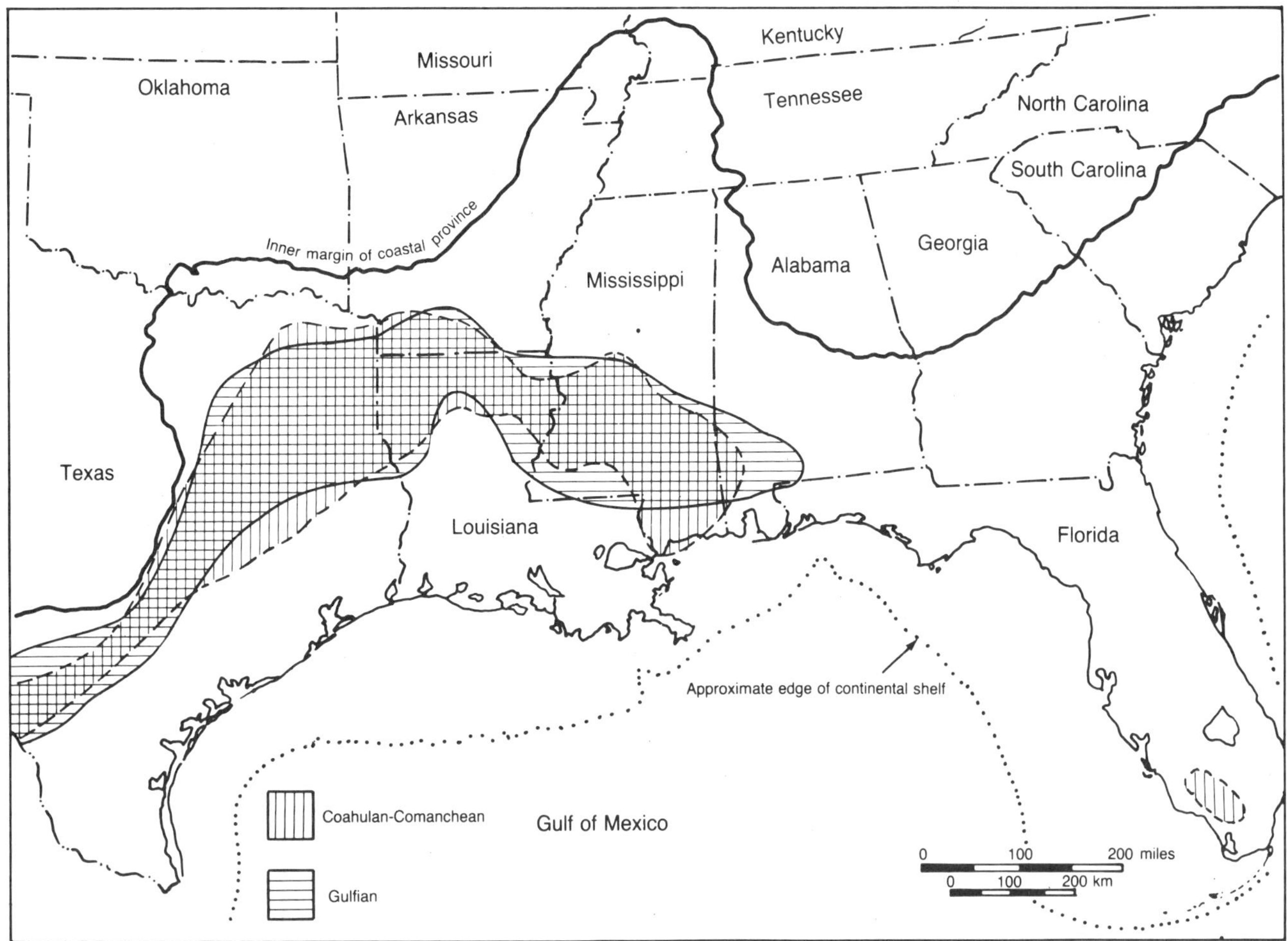

Fig. 7–5 Approximate limits of discovered Coahuilan-Comanchean-Gulfian production (courtesy *OGJ*, November 4, 1957)

densate from Early Eocene Wilcox sands bounded by dry wildcats. There is a fairly small anomaly between dry holes here. This kind of exploration is typical of South Texas today.

The field is about 1½ miles long and ¾ mile wide. As of May 1982, Aminoil had produced about 12 billion cu ft of gas and 60,000 bbl of condensate from the field. It estimates total ultimate gas recovery of 90 billion cu ft. The field discovery was drilled to 10,500 ft, completed at 9,834 ft in the Luling sand of the upper Wilcox. Flow was 7.1 MMcfd of gas and 104 b/d of condensate.

Dry holes were drilled in the surrounding area in 1961, 1972, and 1976. Atlantic Richfield drilled a 1972 test to 14,838 ft, ½ mile west of the Aminoil discovery. That well gave Aminoil a clue to the site of its subsequent discovery. The ARCO well was on the upthrown side of a fault. Aminoil's play was on an anticline on the downthrown side. It was believed that the Wilcox sand would be more permeable and better developed there. There should be more of these types of fields in southern Texas.

Lobo trend

Deep in the South Texas brush country, 140 miles west of Corpus Christi, is the Lobo trend of the South Laredo area. Since its discovery in 1973, hundreds of wells have been drilled with an estimated 75% success ratio. The producing area still has not been defined, continuing to spread east and south across an indicated productive area of 500,000–600,000 acres in Webb and Zapata counties.

The Lobo trend is thought to be one of the most complex geologic provinces in the entire Gulf Coast region (Fig. 7–7). The major reserves occur beneath the angular unconformity. One or more periods of intense faulting and structural activity occurred prior to the angular unconformity, and another period of faulting followed, subsequent to the erosion and later deposition.[10]

The geology has been further complicated by more recent regional gulfward tilting and faulting, which affected the Lobo section as well as the overlying middle and upper

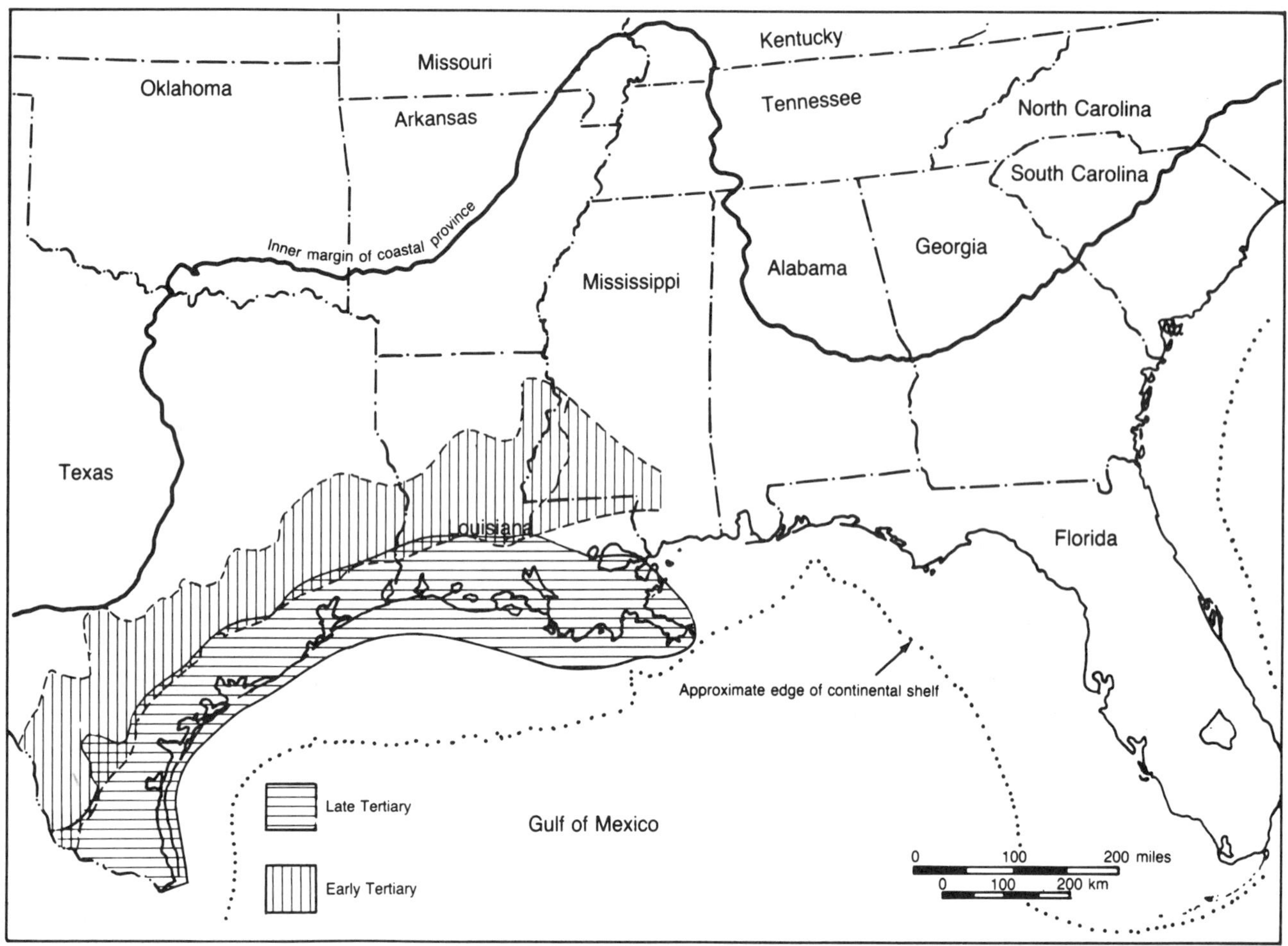

Fig. 7–6 Approximate limits of discovered Early and Late Tertiary production (courtesy *OGJ*, November 4, 1957)

Wilcox sediments of Eocene age. Production on the Lobo trend is from a series of geopressured, low permeability, lower Wilcox sandstones at 4,400 ft, updip in Mexico to nearly 12,000 ft and downdip toward the east in Zapata County, Texas. Individual sandstones in the series are as much as 300 ft thick and contain few water contacts.

Trapping conditions are believed to be provided mostly by counterregional, westward tilted fault blocks bounded by northwest-southeast striking, down-to-the-coast normal faults. Lateral seals commonly appear to be due to shaleout, permeability barrier, or local closure. Displacements across the faults are commonly as much as 700–1,000 ft. Since the stratigraphic sequence containing the major producing zones is commonly no more than 1,000 ft thick, the exact location of a large fault becomes critical in certain areas. This type of problem is most accurately solved when adequate seismic and well-control data are available.[11]

Development and exploration of this huge Lobo gas trend continues today. Geologists believe there will be more such trends in south and southwest Texas and across the river in Mexico.

The Austin Chalk

In southern Texas, the frustrating but exciting Austin Chalk play, with action centered in the Giddings field, continues to woo operators (Fig. 7–8). The trend's activity has declined at a faster rate than activity for the rest of Texas in 1982, but the play is still one of the most interesting and challenging. There seems to be no end to the trend. How long the chalk activity level will last depends on tha price of oil and the investment climate.

Giddings is the center of activity with 1,105 well completions in 1981 alone. As of February 1, 1982, there were 2,061 producing wells in this fantastic field. The average rate of production is 48 b/d of oil and 190 Mcfd of gas. Texas officials think the Giddings field boom could last

Table 7–1

TEXAS GULF COAST OIL AND GAS PRODUCTION AREAS

INNER COASTAL PROVINCE

East Texas embayment and Sabine uplift

1. Strike fault systems fields (Mexia-Talco). Production from Cretaceous and Jurassic rocks; e.g., Mexia, Powell, Corsicana, and Talco fields.
2. Stratigraphic-structural fields on flanks and crest of Sabine uplift. Production from Cretaceous rocks; e.g., East Texas and Carthage fields.
3. Structural fields associated with interior salt domes along axis and southern margin of East Texas embayment. Production chiefly from Cretaceous rocks; e.g., Hawkins and Van fields.
4. Stratigraphic-structural fields on flanks of East Texas embayment. Jurassic production in dolomitic facies of Smackover; e.g., New Hope, West Yantis, and Bryans Mill fields. Cretaceous production; e.g., Rodessa, Trawick, and Fairway fields.

Central Texas and Upper Rio Grande embayment

1. Strike fault systems (Luling, Charlotte). Production from Cretaceous and Lower Eocene rocks; e.g., Luling, Salt Flat, Somerset, Pearsall, Big Foot, Charlotte, and Pleasanton fields.
2. Serpentine fields. Production from Upper Cretaceous rocks; e.g., Lytton Springs field.
3. Stuart City Reef Trend. Production from Lower Cretaceous biohermal limestones; e.g., Stuart City field.
4. Stratigraphic (strand line) fields of Rio Grande embayment. Production from Yegua and Jackson sands (Upper Eocene); e.g., Government Wells field.

OUTER COASTAL PROVINCE

1. Flexure and fault systems fields (Mirando-Provident City, Sam Fordyce-Vanderbilt, McAllen, Willamar, Bancroft). Production from thick sand accumulations on slope of large flexures and downdropped blocks of strike fault systems. Accumulation controlled by various types of structural and stratigraphic traps. Production from Eocene and Upper Tertiary sands; e.g., Seeligson, Agua Dulce-Stratton, and Tom O'Connor fields.
2. Coastal salt struture fields. Production from Eocene and Upper Tertiary stands (Wilcox, Yegua, Frio). Salt dome basins within and independent of regional flexure systems; e.g., Hull, West Ranch, Webster, Old Ocean, Hastings, Conroe, Anahuac, and Katy fields.

OFFSHORE PROVINCE

1. Flexure and fault systems with salt structure. Production chiefly from Miocene sands.

another 5 years with another 2,000 wells being drilled—if prices and the economy improve.

The play covers five counties, but Burleson, Lee, and Fayette still command most of the attention, with lesser action in Bastrop and Washington counties. Good production also is being found above the chalk from the Edwards, Georgetown, and the Buda lime, and some action occurs in Brazos and Gonzales counties.

The Austin Chalk trend extends in an arcuate trend from Maverick County on the Rio Grande all the way to Louisiana. It is about 50 miles wide. The trend crosses a line of counties just west of Houston and east of San Antonio. The chalk is a band of grey and black limestone about 800 ft thick. It is found at 8,000 ft in south Texas.

Pennzoil drilled 1 City of Giddings in 1961, but it was a borderline producer. The well was later sold to

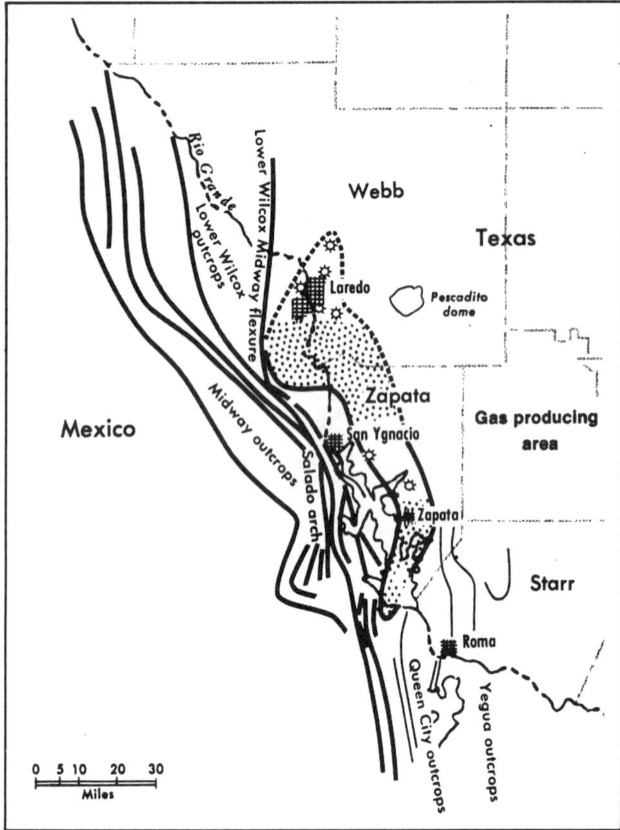

Fig. 7–7 Lobo trend (courtesy *OGJ*, August 6, 1979)

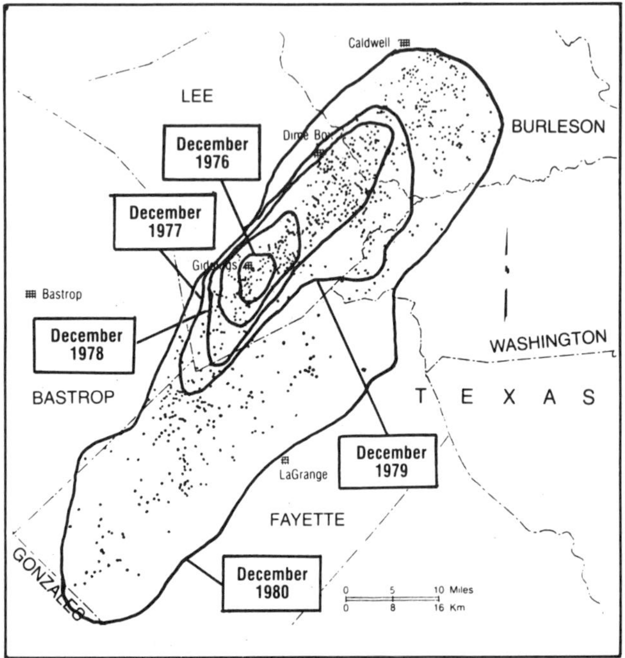

Fig. 7–8 Giddings area development (courtesy *OGJ*, March 23, 1981)

C.W. Alcorn Jr., a Victoria geologist. A frac job and recompletion of the discovery well reopened the play in 1973. But drilling was tough in this chalk play until new techniques were developed in 1976. The proven oil production to date is so great that this area can be tagged one of the world's large oil fields. It will sustain costly operational requirements until many millions of barrels of oil are produced.

Other plays

More South Texas plays may be found south and east of the Giddings action. Operators have been drilling the old Sour Lake salt dome in Hardin County, pursuing objectives from 10,000–13,000 ft, as well as deeper formations in Jefferson County. They also have been finding some Edwards gas production and to the east some Woodbine and Georgetown success. Oligocene and Miocene action can be found in Matagorda and Brazoria counties on the South Texas coast. Also receiving increasing interest is deep Wilcox gas, and the outlook is bright for stepped up activity as gas prices are allowed to rise to parity with other fuels.

Fields such as Charline in Live Oak County, Seven Sisters and Rosita in Duval County, and the Fandango field in Zapata County—where ARCO 2 Gorman Gas Unit is a new important find—could hold large volumes of deep gas. A 20,000-ft wildcat was drilled by Texas International Petroleum on its West San Diego prospect, Duval County, to test a rollover anticline. There will be more such wildcats throughout the region as time goes on.

The Frio formation of Oligocene-Miocene age is one of the main progradational wedges of the Texas Gulf Coast Tertiary basin. Several hundred feet of vertical aggradation of the ancient coastal plain accompanied progradation of a continental platform wedge more than 15,000 ft thick and up to 50 miles in width. This wedge (which consists of interfingering marine and nonmarine sands and shales) has produced more than 16 billion bbl of oil, making it one of the most important petroleum reservoirs of the Gulf Coastal Plain.[12]

More than 50 years of hard wildcatting and extensive exploration into deeper Frio zones make the area a mature objective that may well have given up most of its reserves. However, even a small remaining fraction of undiscovered hydrocarbons would make a sizable, major target for more exploration and development.[13]

The Tuscaloosa trend

The Tuscaloosa, for all of its sour gas conditions and high pressures, is still the hottest play in Louisiana. Chevron is one of the key operators in this program. In April 1982 there were 10 wildcats and 36 development wells being drilled in the deep Tuscaloosa trend of Cretaceous age compared with 24 wildcats and 35 field wells at the same date in 1981.

Amoco has an acreage position totaling about 110,000 acres and has entered the development and production stage of its Tuscaloosa trend operations. As of early May 1982, 37 wells had been drilled, including producers, shutins, and those drilling at that time. Amoco had 19 producers, 10 shutins, and eight drilling or in the process of completion. Fields with activity include Lockhart Crossing, Port Hudson, Moore-Sams, Morganza, Schwab, Comite, and others.

More than 20 gas fields have been discovered in this gas play since 1975. Wildcat success in the total play area is 26%. Production is found on faulted anticlines with up to 1,300 ft of structural closure and is from Upper Cretaceous Tuscaloosa, deltaic and pro-delta sands at 16,000–22,000 ft. Good porosity is present at these depths. Formation evaluation with electric logs alone is difficult due to changes in rock types and formation water salinities.

Normal to abnormal pressure regimes in this area result in formation pressures from 8,000–17,000 lb. Sand thickness, pressure, and condensate content are similar within regional fault blocks. Drilling problems are due to depth and pressure changes. Potential reserves found to date exceed 5 trillion cu ft of gas. Seven fields each have more than 100 billion cu ft of gas in reserve. The average daily gas production from the deep trend in 1980 was 198 MMcf. Cumulative production from all deep Tuscaloosa fields through March 1980 was 82.6 billion cu ft of gas and 1.7 million bbl of condensate. The ultimate reserves could double or triple if the activity continues. Good geophysical data will be needed to find the remaining, more subtle structures. Other stratigraphic zones and geographic areas will be tested as a result of this play.

The trend covers a band about 30 miles wide and 200 miles in length from the Texas state line on the west and extending past Lake Ponchartrain on the east (Fig. 7–9). The objective Tuscaloosa sands are of Upper Cretaceous age and overlain by Eagleford shale and the Austin-Taylor-Navarro chalk section. Underlying the Tuscaloosa sequence are Lower Cretaceous carbonates and shales of Lower Cenomanian-Albian age.[14]

Regional exploration studies by Chevron that eventually led to the False River field discovery were begun in the 1960s. These data confirmed the presence of an undrilled stratigraphic unit (called the wedge), existing between the Upper Cretaceous chalk formation and the more steeply dipping forebank Lower Cretaceous section (Fig. 7–10). Leases were taken along the wedge trend where an opportunity for stratigraphic accumulation existed, possibly equal in age to the East Texas field.

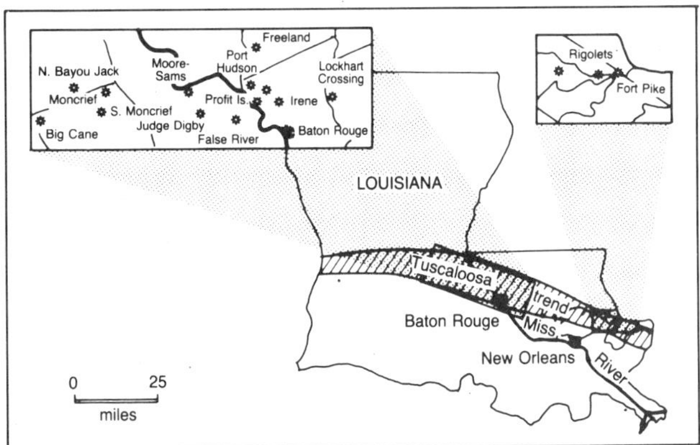

Fig. 7–9 The Tuscaloosa trend (courtesy *OGJ*, September 8, 1980)

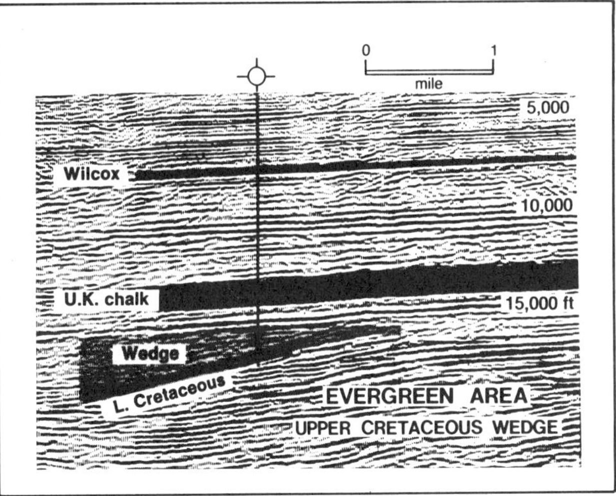

Fig. 7–10 The Evergreen area cross section (courtesy *OGJ*, September 8, 1980)

The Chevron 1 Alma Plantation was drilled on the False River prospect as a structural test of a growth fault closure, developed within the wedge interval. This was the discovery well for the deep Tuscaloosa trend. It was completed in May 1975 for 20 MMcfd at 19,800 ft. Later, the Rigolets field was opened at the other extreme end of the trend, the second discovery for the deep trend. Fig. 7–11 is a block diagram, showing the interpreted environment of deposition of the Lower Tuscaloosa sediment in the False River-Port Hudson area.

In contrast to the Tuscaloosa reservoirs of Mississippi, the lucrative deep Tuscaloosa gas trend of South Louisiana has a slightly different depositional history. At the end of Lower Cretaceous time, northern Louisiana, Arkansas, and Mississippi were uplifted, while southern Louisiana remained submerged. During this time, the Tuscaloosa sediments of

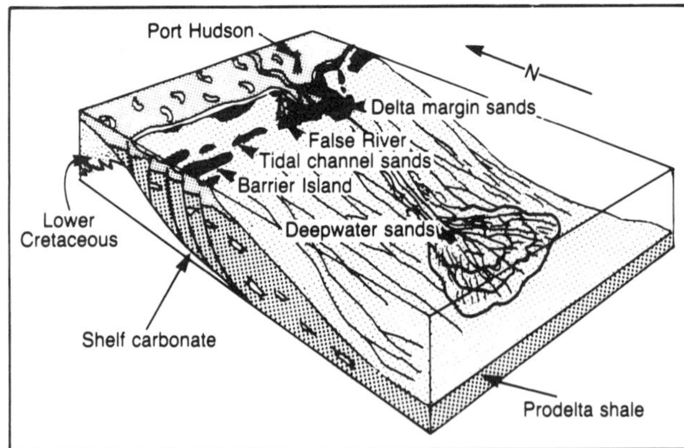

Fig. 7–11 Tuscaloosa depositional environments (courtesy *OGJ*, September 8, 1980)

southern Louisiana were carried by several fluvial systems over the Lower Cretaceous Edwards reef complex, which marked the edge of the continental shelf. The sediments were then deposited in deltaic to near-shore marine environments. When the river breached the reef, Tuscaloosa sediments were carried onto the steep incline of the continental shelf and deposited in a series of deltas. On the forebank side of the reef, stratigraphic traps for hydrocarbon accumulation were formed in the deltaic areas where Tuscaloosa sediments pinch out against impermeable Lower Cretaceous sediments. Structural traps formed by growth faulting.

The following are some of the more important Tuscaloosa wells completed in the southern Louisiana trend in the past few years:

1. Amoco 1 Hess et al., 95-4s-f8e, Pointe Coupee Parish, 1980. IPF 14,500 Mcfd of gas and 126 b/d of condensate from lower Tuscaloosa at 18,377–18,432 ft.
2. Amoco 1 Barnett Heirs, 20-6s-3e, Livingston Parish, 1980. IPF 8,406 Mcfd of gas and 792 b/d of condensate from lower Tuscaloosa at 16,895–16,923 ft.
3. Louisiana Land & Exploration 1 Crown Zellerbach A, 48-4s-2w, West Feliciana Parish, 1980. IPF 40 b/d of oil and some gas, lower Tuscaloosa 15,500–15,749 ft.
4. LL&E 1 Crown Zellerbach, NE NW 36-2s-12e, Washington Parish, 1979. IPF 7 MMcfd of gas from lower Tuscaloosa at 19,735–20,577 ft.
5. Chevron 3 Lowrey Heirs, 10-3s-7e, Elba field, St. Landry Parish. IPF 9,600 Mcfd of gas and 60 b/d of condensate, lower Tuscaloosa 18,356–18,389 ft in 1981.
6. Hunt Energy 1 Georgia Pacific Corp., 22-3s-4w, West Feliciana Parish, 1981. IPF 5,643 Mcfd of gas, opens St. Francisville field. Lower Tuscaloosa at 17,094–17,140 ft.
7. Exxon 1 Robert H. Strain, 38-6s-2e, East Baton Rouge Parish, 1981. IPF 6,800 Mcfd of gas and 144 b/d of condensate, lower Tuscaloosa 18,570–18,587 ft.
8. Chevron 1 Hurst-Ricard et al., 49-5s-11e, Pointe Coupee Parish, 1980. IPF 9,921 Mcfd of gas and 411 b/d of condensate, lower Tuscaloosa 18,640–18,834 ft.
9. Stone Petroleum 1 Rita R. Stephens, 57-2s-5e, St. Helena Parish, Greensburg field, third oil producer. IPF 590 b/d of oil and 136 Mcfd of gas, lower Tuscaloosa 12,925–12,936 ft, 1982. Stone opened the field in 1981 at 1 Robert Riggs in 25-2s-5e flowing 616 b/d of oil and 225 Mcfd of gas from the lower Tuscaloosa at 12,912–12,936 ft. This was the first commercial oil discovery in the parish, and it was the last of the 64 parishes in the state to establish production.

The Miocene is still one of the hottest Gulf offshore targets, and federal and state waters off Texas and Louisiana hold the domestic offshore spotlight. Pleistocene and Miocene wildcats are under way off Cameron Parish in the East and West Cameron areas, and past state sale interest for Louisiana tracts has centered on acreage in the Eugene Island and West Cameron areas, as well as Main Pass and South Pass. There were 29 new field discoveries made in the Gulf of Mexico in 1981, and most of these were off south Louisiana—17 gas and 4 oil.

Freeport-McMoran is building a 14-slot drilling/production platform on Matagorda Island Block 555-L. It will be used for exploration on adjacent blocks. Initial production, expected in 1984, could reach 58 MMcfd.

Another high dollar Sale 62 item was High Island Block 108, 15 miles off Texas. ARCO is drilling here and got the lease for $31.1 million. Texasgulf, in East Cameron Block 65 off Louisiana, is drilling a second Miocene wildcat and plans a third near the West Cameron Block 192 field.

Pennzoil is working on the Pliocene-Pleistocene on High Island Block A-551, and Exxon is drilling on High Island Block 19-L and Block A-34. Exxon drilled a deepwater test on the East Breaks in 1981 then moved to the Mississippi Canyon Block 397. After drilling there, Exxon is back on East Breaks Block 247.

Shell Offshore is drilling wildcats on four Gulf of Mexico blocks, including West Cameron 291 and Eugene Island Block 32. Sohio Petroleum plans to confirm on Block 20 of Mississippi Canyon and will study whether a platform is warranted after a recent strike on a high-dollar tract it acquired in 1981 during OCS Sale 66. Sohio owns adjacent blocks to the south and west and is continuing evaluation of those after drilling two wildcats on South Pass and East Addition Block 73.

South of the Destin dome, Sohio was drilling to 21,500 ft and the Jurassic in Block 563 in 275 ft of water. This well is about 130 miles east-southeast of Mobil's Jurassic discovery in Mobile Bay off Alabama and south of where other operators have drilled nothing but dusters in previous activity.

The outlook remains bright for the future federal Gulf leasing. Federal acreage under study for the three area-wide sales in 1983 totals about 130 million acres. A draft environmental statement was completed in August 1982.

However, observers expect the Gulf will experience an activity decline, partially due to an oversupply of gas and to price uncertainty. Many of the fields are marginal and are highly sensitive to gas prices.

The Gulf Coastal province will continue to be one of the most lucrative regions in the world to look for new oil and gas. This vast region has everything going for it—demand, geology, traps, seals, reservoirs. The industry is fully aware of the vast potential still awaiting the explorationists throughout the province, from the Rio Grande to the tip of Florida.

REFERENCES

1. Grover Murray, *Oil & Gas Journal* (November 4, 1957), p. 109.
2. Ibid., p. 109.
3. Ibid., p. 110.
4. Ibid., p. 111.
5. Ibid., p. 111.
6. Ibid., p. 112.
7. Ibid., p. 114.
8. Ibid., p. 115.
9. W. Fisher, Bureau of Economic Geology, Texas University, speech given to Interstate Oil Compact Commission in 1965.
10. B. O'Brien and R. Freeman, *Oil & Gas Journal* (August 6, 1979), p. 156.
11. Ibid., p. 156.
12. W. E. Galloway, D. Hobday, and K. Magara, *AAPG Bulletin* (July 1982).
13. Ibid.
14. L. Funkhouser, F. Bland, and C. Humphris, *Oil & Gas Journal* (September 8, 1980), p. 96.

8

Texas and the Southwest

West Texas and the states of New Mexico and Arizona make up one of the most prolific oil-and-gas-producing areas of the world.

West Texas

Scores of giant fields exist in the vast West Texas region, most discovered from 1920–1940. The largest field is Yates in District 8. Cumulative production here is 874 million bbl of oil. The field, discovered in 1926, still has more than 1 billion bbl of oil to be produced from more than 570 wells. The 1981 production at Yates was 54 million bbl of oil.

Kelly-Snyder, discovered in 1948, is another large field. It is in District 8-A and has produced 1.1 billion bbl with 245 million bbl of oil remaining. There are 797 wells in this vast complex. The author worked on a seismic crew preceding this giant oil discovery. The find turned a quiet West Texas flatland into a modern boomtown scene in a matter of hours. As a young trainee, little did I know I was setting jugs in an area soon to be one of the largest oil fields in the world. Kelly-Snyder produced 28.8 million bbl of oil in 1981.

The Wasson field, discovered in 1936, has produced 1.4 billion bbl of oil and has 179 million bbl left to flow from more than 2,000 wells. Many other fields in the area already have produced more than a half billion barrels of oil. The pays are prolific, and the zones are many. In 1981, the total daily output from all of West Texas was 1.441 MM b/d of oil. This figure is higher than that of Louisiana with 1.233 MM b/d and more than California (1.052 MM b/d), but it is less than Alaska with 1.611 MM b/d of oil.

The history of West Texas oil actually began in 1921 at the Westbrook field in Mitchell County where 3,000-ft oil was discovered. Then came the Big Lake field in Reagan County in 1923, another prolific and shallow field. But the big news came when Cambro-Ordovician Ellenburger oil, the big daddy producer in West Texas, was found at Big Lake at 8,500–8,600 ft.

More discoveries followed, leading to the 1948 Scurry-Snyder reef strike which had an impact on oil finding throughout the world. Geophysics was notably successful in finding similar producing trends in the vast Permian basin empire as well.

The setting

Most of West Texas lies in what has been called the Permian basin empire. This huge region, also including southeast New Mexico, is made up of several basins, shelves, atolls, and platforms. The northern extreme of the area is bounded by the Matador arch (Fig. 8–1). The central point for this empire is the Delaware basin where most of the ultradeep drilling took place before the birth of the Anadarko basin trough play in recent years.

Prominent shelves in the region, all of which harbor giant oil fields, are the Northwestern and the Eastern. Production exists on the Ozona platform, the Central Basin platform, all along the Horseshoe atoll, in the Val Verde basin, and along the Matador arch. Not all of this area has been explored, despite the multitude of producing areas. Brisk exploration is under way in all of southeast New Mexico, particularly into the Morrow sands of Pennsylvanian age in Eddy and Lea counties. Gas is the objective, and the zones are deeper than in previous drilling plays.

The Val Verde basin has always intrigued geologists. Drilling through the years has been anything but prolific, and successes have been few. Here, too, deep drilling is needed. Few wells have been drilled in the Marfa basin and on the Diablo uplift.

Current activity

Current action is centered in the Permian and Delaware basins, where operators are getting the most they can from infill drilling, secondary recovery, and enhanced recovery.

Deep Delaware basin activity is concentrated in Loving, Winkler, Ward, Reeves, and Pecos counties in Texas, and in Lea and Eddy counties, New Mexico. The deep and prolific Ellenburger is still actively pursued. Gulf completed 1 Gulf Northwest Hamon unit in the basin in Reeves County and plans further work with offsets. The 21,900-ft discovery well, 19 miles southeast of Pecos and 2 miles west of Hamon field, flowed 5.2 MMcfd of gas on a 13/64-in. choke with 5,413-psi flowing tubing pressure. The unit, totaling 3,200 acres, was formed by Gulf, Texaco, and Phillips. Gulf has about 500,000 net acres under lease in the Delaware basin.

Superdeep drilling has returned to the basin in Pecos County. Exxon was drilling three wildcats, two of which were tagged for 26,000 ft and a third for 24,000 ft. The firm also is drilling a field well at Mary McComb Gas Unit B-1 to 25,000 ft.

An important new discovery in the Val Verde basin marks this area for further exploration. McCormick Oil and Gas completed 1 E.P. Royalty in Val Verde County on its Alma prospect, flowing 1.3 MMcfd of dry gas from the Ellenburger at 13,956–14,006 ft after acidizing. This zone has been designated tight, as has the Wilcox Eocene Lobo (geopressured) in Zapata and Webb counties in southwest Texas.

Thousands of wells are drilled in West Texas each year. *Petroleum Information* reports that this area had 6,225 well completions in 1981 alone, mostly from the Permian basin. This was a 19.5% increase over the 5,206 completions recorded in 1980, according to the Denver publication.

Crockett County was the most heavily drilled area in the District 7-C region with 435 completions in 1981. Andrews County led District 8 with 366 completions, and Gaines County in District 8-A with 276 well completions.

Deep wildcatting in 1981 came up with 16 discoveries in the Delaware basin, all below 15,000 ft. Ten of these were recorded in Loving County. Pays are the Pennsylvanian Atoka, the Permian Wolfcamp, and the Cambro-Ordovician Ellenburger and Pennsylvanian Morrow. Deep action also was under way in Ward, Pecos, and Reeves counties, inside the trough of the Delaware basin. In the summer of 1981, Hunt Energy of Dallas set a Texas drilling depth record at 1-9 Cerf Ranch, 31 miles southeast of Fort Stockton in Pecos County. Penrod Drilling's Rig 121 drilled to total depth of 29,670 ft on June 25.* Gulf set the previous Texas depth record at 2 Emma Lou about 7 miles north of the new well. This well was plugged back to 23,875 ft and completed in the Pennsylvanian Bend producing gas from perforations at 21,546–23,381 ft.

Many large gas fields have been discovered in West Texas in the years since 1950. While oil production soared in older giants and other large fields in the region, the industry was busy drilling to deep zones, particularly the Ellenburger, to find needed gas reserves (Fig. 8–2). This search continues today but with less exploratory furor than during the past 2–3 decades.

The future

West Texas and its environs, containing some of the largest oil fields in the world, definitely have a future in the exploratory game. The basins, shelves, and platforms of this area are well endowed with all that it takes to be a hydrocarbon province—source rocks, reservoirs, traps, reserves, and markets.

More Abo reefs and large gas fields will be found in the deeper zones of the Delaware basin. The Val Verde basin

* Only two holes have been drilled deeper in the Western Hemisphere. The Lone Star Producing 1 Rogers in Washita County, Oklahoma, went to 31,441 ft, while Lone Star 1 Baden, Beckham County, Oklahoma, drilled to 30,050 ft.

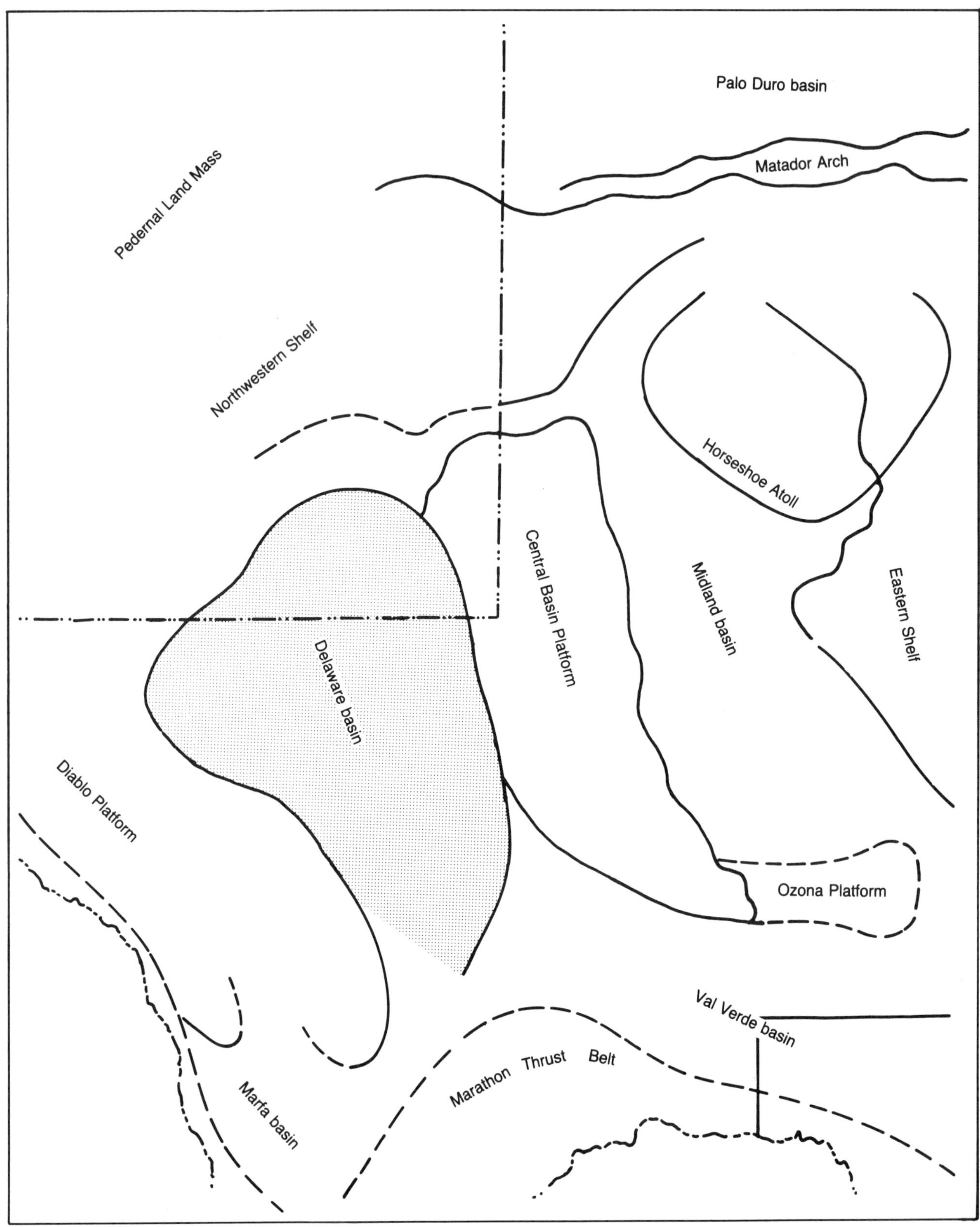

Fig. 8–1 The Delaware basin of Texas and New Mexico

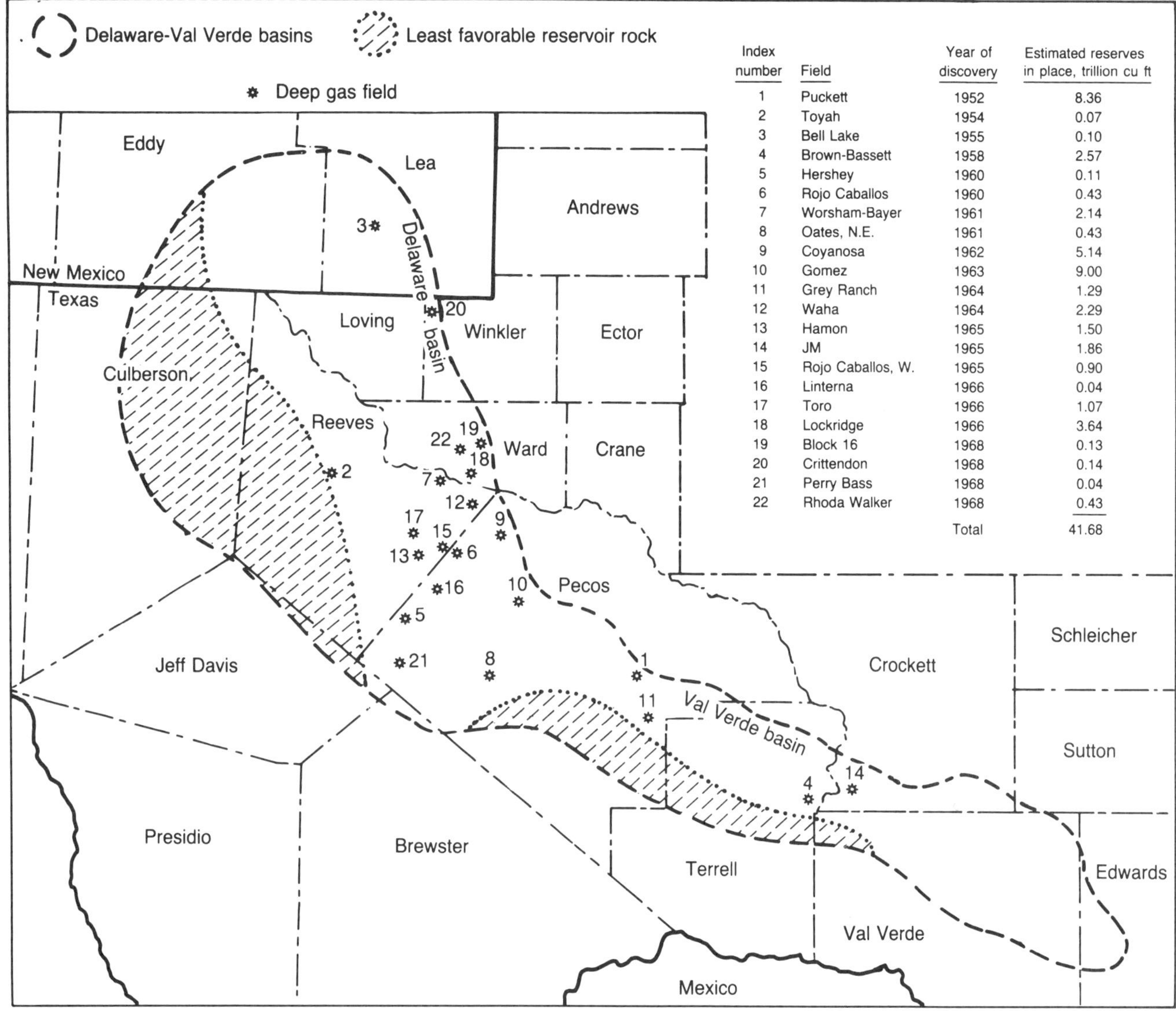

Index number	Field	Year of discovery	Estimated reserves in place, trillion cu ft
1	Puckett	1952	8.36
2	Toyah	1954	0.07
3	Bell Lake	1955	0.10
4	Brown-Bassett	1958	2.57
5	Hershey	1960	0.11
6	Rojo Caballos	1960	0.43
7	Worsham-Bayer	1961	2.14
8	Oates, N.E.	1961	0.43
9	Coyanosa	1962	5.14
10	Gomez	1963	9.00
11	Grey Ranch	1964	1.29
12	Waha	1964	2.29
13	Hamon	1965	1.50
14	JM	1965	1.86
15	Rojo Caballos, W.	1965	0.90
16	Linterna	1966	0.04
17	Toro	1966	1.07
18	Lockridge	1966	3.64
19	Block 16	1968	0.13
20	Crittendon	1968	0.14
21	Perry Bass	1968	0.04
22	Rhoda Walker	1968	0.43
		Total	41.68

Fig. 8–2 Deep gas fields in Delaware and Val Verde basins

remains for all purposes untouched. There is always the chance exploratory work will uncover something in the Marfa basin at the southeastern end of the Diablo platform.

The Fort Worth basin

Long overdue for more exploration is the Fort Worth basin in North Texas (Fig. 8–3). The basin is in the foreland of the Ouachita Mountain folded belt. It is limited abruptly on the southeast by the Ouachita Front and on the east and north by the highstanding elements of the Muenster and Red River arch systems. The basin shoals gradually toward the axis of the Bend flexure.

The Cambro-Ordovician sequence in the basin is relatively thin, less than half the estimated maximum thickness of equivalent section in the Marietta and Ardmore basins of southern Oklahoma to the north. Mississippian rocks truncate the Viola-Simpson and overlap onto the Ellenburger group from the center of the basin westward.

The Mississippian section in the basin reaches a maximum thickness of more than 1,300 ft adjacent to the Muenster arch. It thins westward, primarily due to truncation by overlying Pennsylvanian, until the basal formation (the Barnett shale) is in contact with the base of Pennsylvanian in Central Jack County.[1] Thickening resumes westward into Young County, suggesting the presence of a broad, low-relief, north-south-trending ridge of truncated Missis-

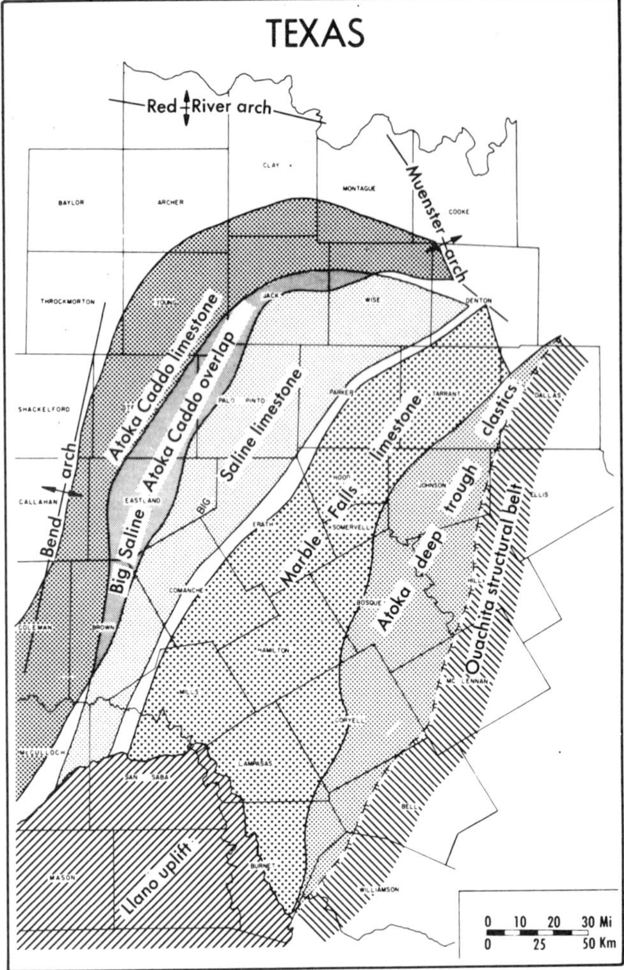

Fig. 8–3 The Fort Worth basin (courtesy *OGJ*)

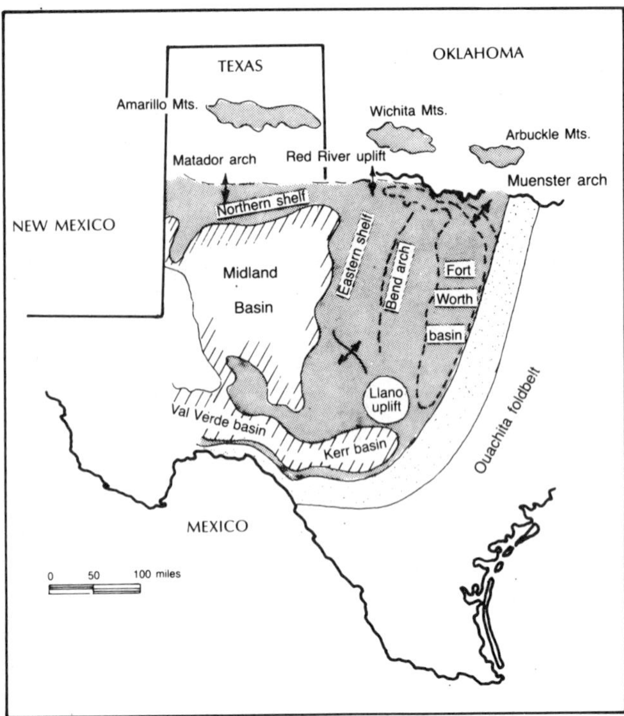

Fig. 8–4 The Tectonic area (courtesy *OGJ*)

sippian within the basin. The lower part of the Atokan includes the Davis sandstone, known in wells drilled in the Fort Worth, Ardmore, and Marietta basins, and correlative of the Hartshorne sands of the Arkoma basin.

The Pennsylvanian sediments of the Fort Worth basin are the main objectives of oil and gas exploration since the turn of the century. In recent years, increased emphasis has been placed on exploiting the Atokan Big Saline (Bend) conglomerates.[2] The basin is a structural and depositional feature developed by orogenic movements occurring from early Pennsylvanian through Permian times (Fig. 8–4). Fort Worth basin is about 200 miles long, extending south from Oklahoma to the Llano uplift of central Texas. Several good fields have been found through the years. Increasing gas prices and demand are the main factors in keeping action here alive.

Last frontiers

Texas still has a few last outposts for exploratory work. One of these is the Kerr basin, south of the Llano uplift and north of the Balcones fault zone. To the west lies the Edwards Plateau where Pennsylvanian gas exploration has gone on through the years in stages. Recent discoveries in Sutton and other regional counties in this part of Texas point to the need for further exploration throughout the province. Focal point of the basin is Kerrville.

Stratigraphic traps may exist in northwestern Bandera County updip from multiple, oil-stained sandstone units penetrated at depths above 8,000 ft by several wells. Some productive structural or stratigraphic/structural traps may lurk in this little-drilled basin. Geologists believe lower Pennsylvanian sandstones are potential as well as high-risk objectives for oil and gas reservoirs on the north flank of the basin.

NEW MEXICO

The Land of Enchantment is also a land of many sedimentary basins (Fig. 8–5). Two of these basins have some of the largest oil and gas fields in the world. But most of the other basins in New Mexico are still on the dry list or

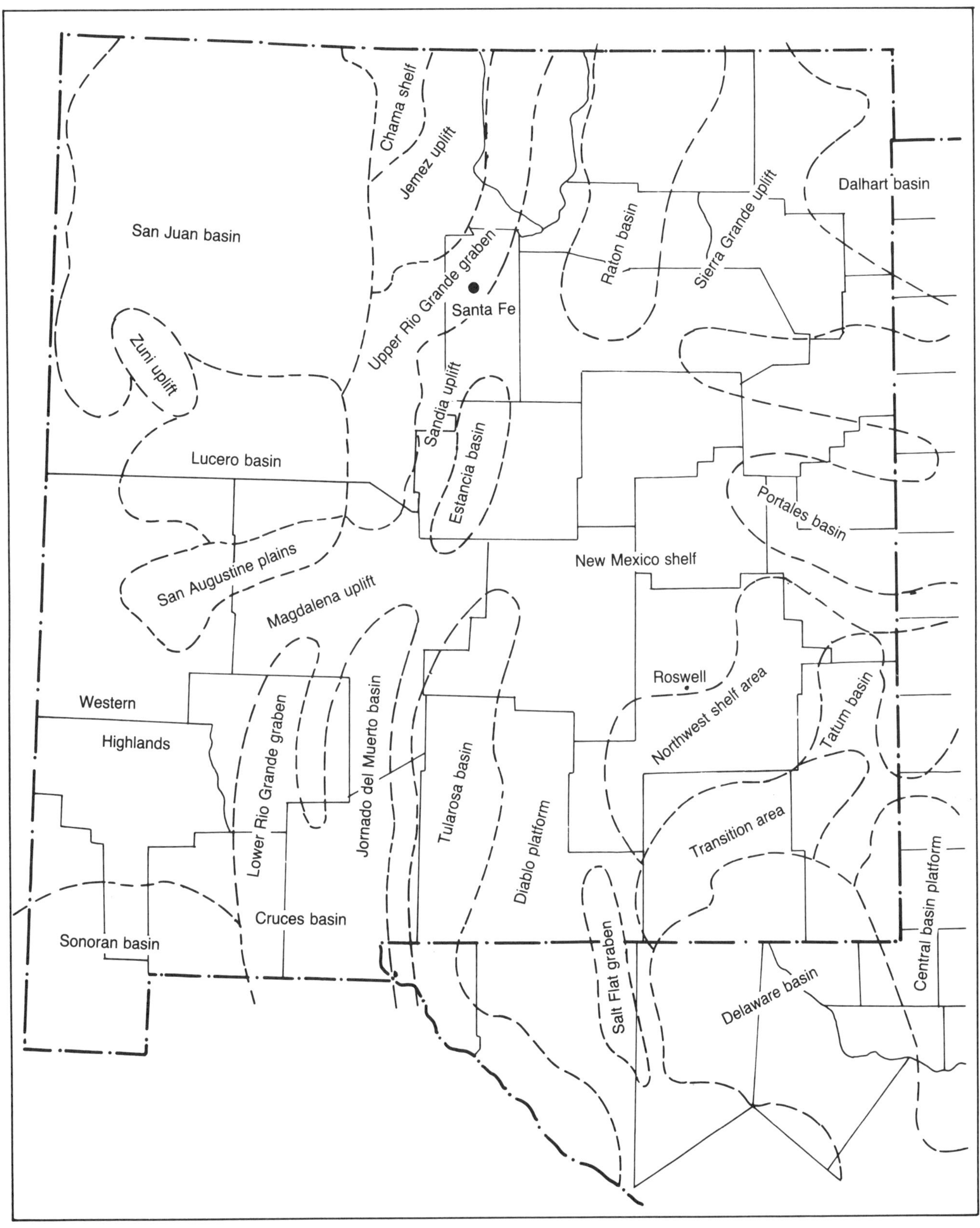

Fig. 8–5 New Mexico's sedimentary basins

have had only a smattering of drilling success through the years.

New Mexico has a long petroleum history. Oil was discovered near Artesia in 1902, but the supply was not commercially attractive nor exciting. Natural gas was first produced in salable quantities near Aztec in 1922. The first real commercial oil discovery came in the same year in the San Juan basin. The state now ranks seventh in the production of crude oil. Crude oil and natural gas accounted for 74% of the value of New Mexico's total mineral production in 1980, and the daily output of oil is 195,000 b/d. Cumulative production of oil to late 1982 was 3.6 billion bbl.

New Mexico is the fifth largest state in area with more than 77.7 million acres. The U.S. Geological Survey has classified about 84% of the state (65 million acres) as prospective oil and gas lands—lands where geologic conditions offer some possibility for the occurrence of oil and gas.

Fifty-five percent, or 42.6 million acres, of the land in New Mexico is owned by federal or state governments or by Indian nations. About 48% (20.5 million acres) of this government land is leased for oil and gas exploration and production. The total number of wells drilled in New Mexico to January 1, 1981, was 49,935. Of these, 50.7% were for oil, 27.3% were gas, and the remaining 22% were dusters.

Four of the 100 largest oil fields in the U.S. are in New Mexico. They were discovered before 1930. The state hosts six giant oil fields: Denton, Empire Abo, Eunice-Monument, Hobbs, Maljamar, and Vacuum. The largest oil field is Vacuum with 358 million bbl of oil produced and 47 million remaining. The Hobbs field has produced 251 million bbl of oil and has 36 million bbl left. Empire Abo, the Abo pay found in 1957, yielded 203 million bbl of oil through the years and has 66 million remaining. Most notable recent discoveries have been gas, both in the San Juan basin and in the Delaware basin.

Proved reserves of oil in New Mexico as of January 1, 1980, were set at 461.7 million bbl, down 23.9 million from the previous year. The all time high was 1.1 billion bbl in 1962. Proved gas reserves reached 13.5 trillion cu ft, up 200 billion from 1979.

Petroleum Information says drilling for oil and gas in southeast New Mexico in 1981 increased by 25%. Total well completions were 1,247, up from 997 the previous year. Field development accounted for almost 80% of all drilling in the area, however.

Eight of New Mexico's 33 counties have a history of oil and gas production. Those counties are Chaves, Eddy, Lea, and Roosevelt in the southeast, and McKinley, Rio Arriba, San Juan, and Sandoval in the northwest. Mora County has some gas, and Harding has carbon dioxide production.

One of the best of 52 discoveries in Eddy County, the leader in the region, was Yates Petroleum 1 Eastern Shore LV Federal, 22 miles southwest of Loco Hills in 17-19s-27e. This discovery well flowed 6.8 MMcfd of gas from the Pennsylvanian Morrow at 9,908–9,924 ft. Many discoveries with similar flows were reported, and the play continues today throughout Lea and Eddy counties.

An array of basins

Historically, most exploratory and development work in New Mexico has been centered in two large areas, the San Juan basin of the northwest and the Permian basin of the southeast. Through the years, however, wildcatting has been in myriad smaller basins throughout the state. Wildcatters have looked at the prospects of the Pedregosa basin, the Hidalgo basin, the Estancia and Albuquerque basins, the Raton and Palo Duro basin, and in the Tularosa basin. Operators have met with little success in most of these areas through the long years of frustrating exploratory work, but interest keeps rekindling.

The Permian and Delaware basins of southeastern New Mexico have long been among the major oil and gas producing areas of the nation and the world. More than 90% of the state's production of oil comes from the southeastern corner of the area. Prolific oil and gas fields are associated with Paleozoic production. Most of the oil and gas in the San Juan basin is produced from reservoirs of Pennsylvanian and Cretaceous age, much of it in Cretaceous rocks.

New Mexico's first oil discovery was made at Hogback field in 1922, San Juan County. Gallup sands of Cretaceous vintage have been the major producers in the San Juan basin. This production comes from sandbar-type stratigraphic traps and from fractured zones of the Mancos shale.

The first major discovery in southeast New Mexico was made in 1924 at Artesia in Eddy County. Most early day production came from rocks of Permian age where reservoirs were shallow and very prolific. Later, oil was discovered in deep structures in the Pennsylvanian, Mississippian, Devonian, Ordovician, and Silurian strata. Devonian rocks have been particularly rewarding to oil producers.

Most gas in the San Juan basin comes from two large stratigraphic reservoirs, the Blanco Mesaverde and the Basin Dakota gas fields. Major reserves of dry gas have been found in the southeastern part of the state in the Pennsylvanian formations. The Morrow of Lower Pennsylvanian age has been the big target in recent years, with Eddy County reaping the best crop of fields. Success also has been high in Lea County. With such good drilling prospects being tested in the three basins, the industry has become bolder, moving out into the smaller and nonproductive areas.

With a Paleozoic stratigraphic section that is 11,000 ft thick and similar in facies to that of the productive Permian basin, the Pedregosa basin of southeastern Arizona, south-

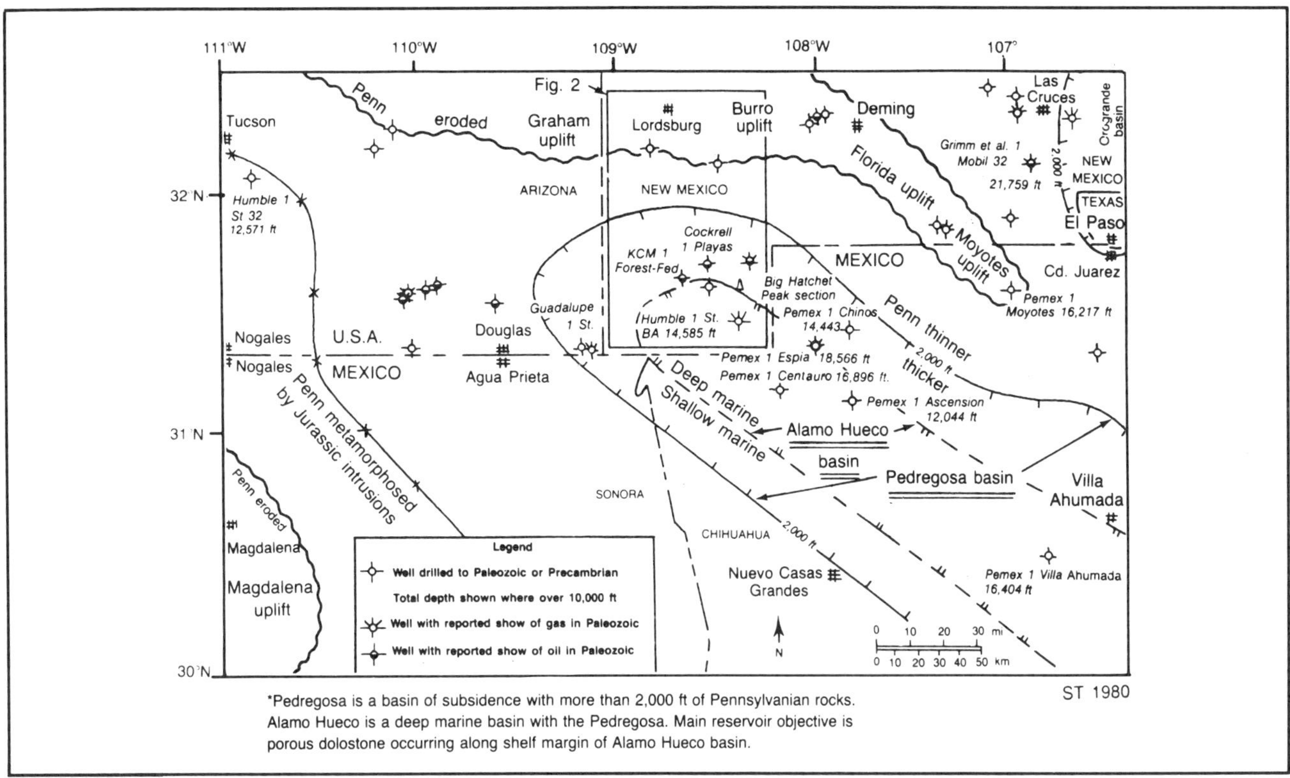

Fig. 8–6 Pedregosa basin area (courtesy *OGJ*, October 20, 1980)

western New Mexico, and northwestern Chihuahua and northeastern Sonora has a high potential for petroleum prospecting (Fig. 8–6). The prospect is a basin of subsidence where more than 2,000 ft of marine sedimentary rocks (mostly shallow marine limestones) were deposited in Pennsylvania time. It is 70 miles wide, and its northwest-southeast trending axis is at least 200 miles long. Only a little more than 100 wells have been drilled in the basin, and only a few reached the Paleozoic or Precambrian section.

The Las Vegas subbasin is another potentially productive area in the state. The first commercially producible dry gas discovered in northeast New Mexico is in this basin. Pay is the Cretaceous Dakota sandstone. Following the wildcat discovery well in September 1973 by Brooks Exploration, the Wagon Mound Dakota-Morrison gas field was set up. It lies near the town of Wagon Mound in 14-21n-21e, Mora County (Fig. 8–7). The pool is unusually interesting because it has unique low-formation pressures, averaging around 5 psi.

This area of the Las Vegas subbasin was designated as favorable for oil and gas exploration on the basis of sedimentary rock thickness in excess of 9,000 ft and the presence of source rocks and good reservoir rocks. This subbasin is a Tertiary geologic feature. The existence of an earlier basin, the Paleozoic Rowe-Mora, is recorded by a thick Pennsylvanian section. The Rowe-Mora basin was an intervening depositional area between the San Luis, Wet Mountain, Apishapa, and Sierra Grande uplifts during Paleozoic time. After the uplifts were reduced by erosion,

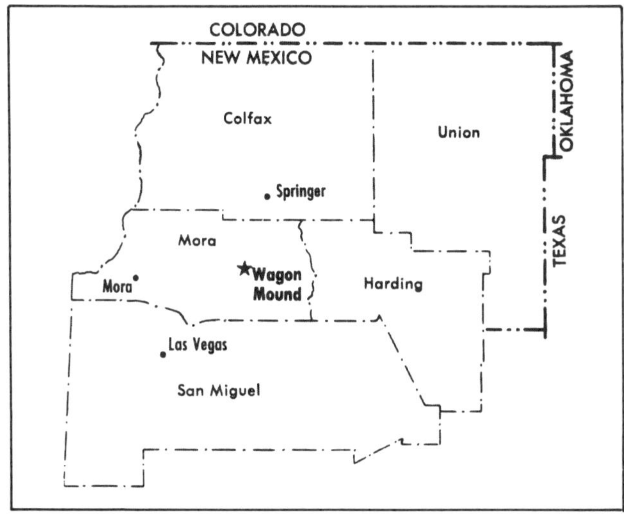

Fig. 8–7 Wagon Mound area (courtesy *OGJ*, March 27, 1978)

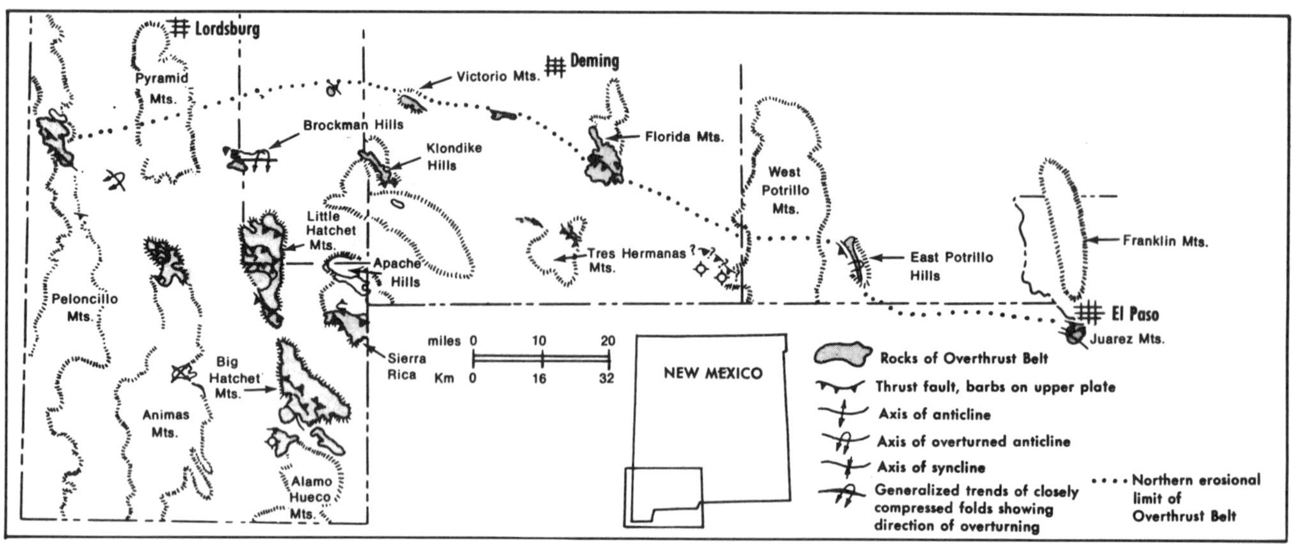

Fig. 8–8 Generalized tectonic map, showing Overthrust Belt in Southwest New Mexico (after Corbitt and Woodward, from AAPG, courtesy *OGJ*, April 21, 1980)

the area was covered by sediments ranging in age from Permian through Jurassic.

These uplifts were rejuvenated during Early Cretaceous time and are believed to be the source area for the braided alluvial interval of the Dakota sandstone. Explorers continue looking for production in this basin with the primary target being the Dakota and Morrison sandstones, although production chances are good in other zones such as the Entrada (Exeter) sandstone.

Overthrust Belt

The Overthrust Belt forms a major tectonic zone stretching from Montana to New Mexico (Fig. 8–8). From the southern New Mexico border, the Overthrust Belt trends west-northwesterly through southwestern New Mexico. Its northeastern edge extends from the central Peloncillo Mountains on the New Mexico-Arizona border to the Juarez Mountains of Mexico on the east.[3]

The best reservoir rocks in this region are probably of Paleozoic age and include dolomites of the upper Horquilla, the Fusselman, El Paso, Conch, and Epitaph. The Overthrust Belt of southwestern New Mexico presents special problems for the explorationist. The techniques used successfully for oil and gas exploration in the Wyoming and Utah portions of the Overthrust Belt are more difficult to apply in New Mexico because the structure is more complex. However, the region does have merit for further exploration, and only wildcats will tell the final chapter in this corner of the state.

The south flank of the San Juan basin is another important area. Here, shallow, multiple-stratigraphic potential in the Cretaceous is underlain by major petroleum potential of the Pennsylvanian where conditions form a classic mode of entrapment. This large area of the state still remains relatively well-free.

The primary interest is that portion of the San Juan basin south of the Bisti-South Blanco fields and north of Ambrosia Lake-Cabezon Peak, west of the Nacimiento uplift and east of the Navajo Indian Reservation (Fig. 8–9). This

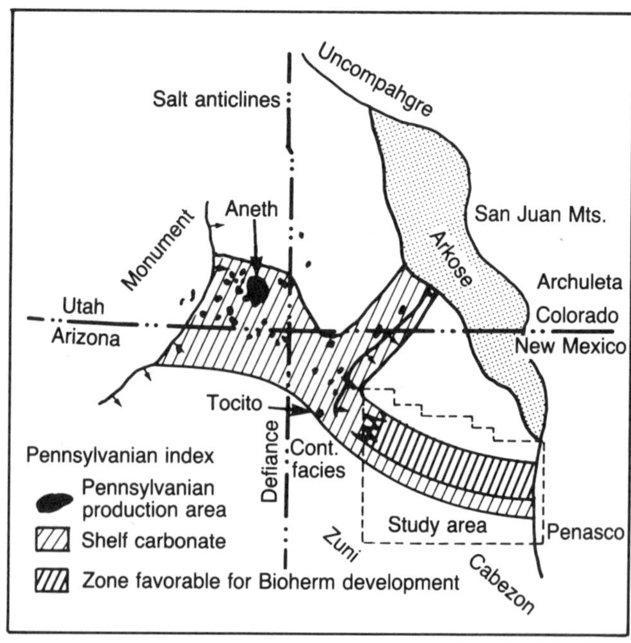

Fig. 8–9 Neglected south flank area (courtesy *OGJ*, March 6, 1967)

area has been one of the most neglected and yet one of the most promising regions in the basins of the Rocky Mountains because Cretaceous and major Pennsylvanian reserves have been found at reasonable depths.[4]

Recent discoveries in the Texas portion of the Palo Duro basin and continued interest in the prospects of the western lobe of the basin in New Mexico promise further exploration work in this area.

The exploratory outlook for the entire state of New Mexico looks good. Continued drilling programs in the Delaware basin Morrow play will continue, perhaps a bit less active than in the past. In-field drilling and field-edge development work continues in the San Juan basin. Some interesting new wildcat programs are working in this old area, since the industry discovered it had not found everything in New Mexico.

ARIZONA

Wildcatters always have been excited about the hydrocarbon prospects of the state of Arizona. There have been many wildcats drilled over the state through the years. Success has dimmed for the most part, except in the northeastern corner of the state in the Black Mesa basin and the southern end of the Paradox basin. Yet each time a new wildcat location is announced in Arizona, there is a wave of enthusiasm that wafts through to those interested in the prospects of an oil discovery.

Current exploratory interest in the Overthrust Belt of America's West spilled over into Arizona in 1980. Two deep wildcats were drilled, and operators spent large sums of money in the state. The results were never really elaborated upon, and the industry still hopes to find a big field along the Hingeline that sweeps down from Utah into Arizona and into New Mexico and Mexico (Fig. 8–10).

In 1982, Phillips did some strat testing along the southern edge of Arizona with depths from 6,000–9,000 ft. Locations were in Cochise, Graham, Pima, and Yavapai counties.

Although often used interchangeably, the terms Overthrust Belt and Hingeline actually refer to very different geologic phenomena. The Hingeline represents a former stable shelf or transition zone on the eastern edge of the Cordilleran geosyncline, similar to the present continental shelf. It contains lower Paleozoic rocks thickened to the west off the edge of the continent into the geosyncline. The Hingeline is depositional in nature. Geologists feel that because of the abundance of organically rich marine sediments that are typically deposited in a hingeline environment, it is an ideal place to look for hydrocarbons.

The Overthrust Belt, on the other hand, resulted from violent stresses in the earth associated with compressional

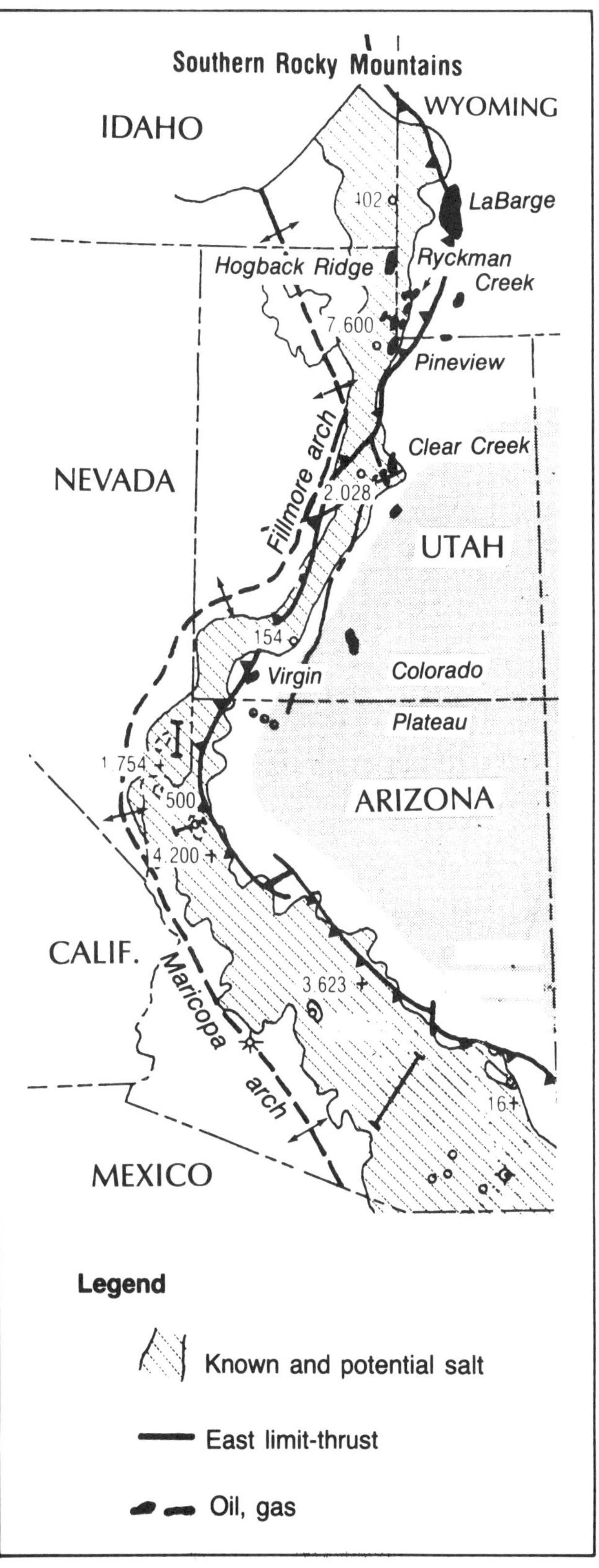

Fig. 8–10 Arizona action (courtesy *OGJ*, April 19, 1982)

deformation during the Sevier and Laramide orogenies beginning in late Jurassic and late Cretaceous time, respectively. The anomaly drilled in 2-7s-10e, Pinal County, is 12 miles long and 6 miles wide, with more than 5,000 ft of closure. Similar anomalies exist along this trend that could be of great importance to the industry.

Despite years of drilling throughout the state, Arizona production is confined to the northeastern corner. One field, Dineh-bi-Keyah (where 20 wells were completed in a Tertiary sill intruded into Pennsylvanian marine strata) is on the flank of the Defiance uplift.

The setting

Arizona sits astride two major geologic provinces, the Colorado Plateau to the north and the Basin and Range to the south and west (Fig. 8–11). The provinces are characterized by quite different stratigraphic framework and structural patterns and are separated by a narrow area of transition. This area is commonly considered a third structural province called the Transition zone or Central Mountain region, an area of rugged mountains and erosional remnants of Paleozoic age.[5]

The Plateau Province includes the northern third of the state bounded on the south by the Mogollon Rim and on the west by the Grand Wash cliffs in the western Grand Canyon of the Colorado. It extends into southern Utah, southwestern Colorado, and northwestern New Mexico. The Basin and Range province includes the southwestern half of the state. It also extends into southern California and into the states of Sonora and Chihuahua in Mexico.

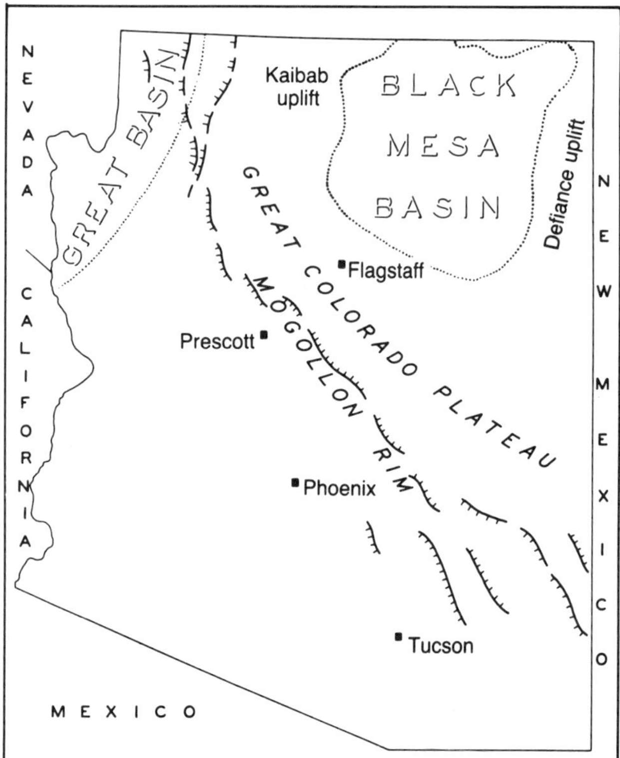

Fig. 8–11 Basin and Range provinces of Arizona

The northeast

Based on today's knowledge, the best oil and gas potential in Arizona is in the Plateau Province in the northern portion of the state. All oil and gas production and known reserves are in Apache County. About 40 wells have produced oil in this area of 10 fields. Production is from the Pennsylvanian in most of the wells; however, four have Mississippian production, and one has Devonian pay. The foreland facies zone of the Paradox basin in the Four Corners area will continue to offer the most attractive potential for unfound oil and gas.

Too few wells within the remainder of the southwestern shelf portion of the basin have been drilled to evaluate the potential of the Paleozoic rocks. Those drilled in the Mogollon slope area (mostly on surface structural anomalies in the general Holbrook area of Navajo County) have been dry holes.

Based on exploratory and development drilling to date, the productive carbonate zones of the Pennsylvania Hermosa group are the best targets in the northeastern part of Arizona. Past exploration has been for traps on anticlines.

Recent additions to the producing list in Arizona include the Kenai Oil and Gas well at 34 Navajo-7 in Apache County which pumped 161 b/d of oil from the Mississippian at 5,562–5,567 ft. This field is 6 miles southeast of the East Boundary Butte field and 10 miles southwest of Teec Nos Pos field. Both fields produce from the Pennsylvanian. The Black Rock field, location of the Kenai well, was discovered in 1975 and has produced oil and gas from the Pennsylvanian and Mississippian.

Oil flowed at 290 b/d at a deeper pay discovery recently in Teec Nos Pos field at Mountain States Resources 12 Navajo O in 22-41n-30e, Apache County. The pay is the Pennsylvanian Desert Creek at 5,108–5,118 ft. This field has produced more than 400,000 bbl of oil since 1972.

Other areas of hope

One area in southern Arizona that deserves a closer look is the Sonoita basin, about 50 miles south of Tucson. If this basin proves productive, it will open up possibilities in much larger basins in southern Arizona.[6] The basin is especially attractive because of the Cretaceous section in the area. The rocks correlate in a general way with the Early Cretaceous rocks of Texas.

In addition to the Cretaceous section, rocks of Permian, Pennsylvanian, and Mississippian age are well developed and hold promise for oil and gas production. The Permian section is over 3,000 ft thick and contains limestone, dolomite, and anhydrite, with interbedded sandstone beds. The Permian section can be roughly correlated with units in the Permian basin of West Texas.

Northwestern Arizona has not been explored enough to prove its merit for commercial oil or gas production. Some of the major folds have been tested but none adequately. There are about 16 well-defined surface structures and many smaller ones in the region. Arizona geologists feel that these all represent a variety of distinct potential stratigraphic traps. These structures probably formed long after hydrocarbons had developed in the Paleozoic source rocks. Migration and subsequent local entrapment of the primary hydrocarbons may have been first dominated by stratigraphic entrapment during the tectonically quiet Paleozoic and most of the Mesozoic eras.[7] Also, it is conceivable that the development of anticlines during the Laramide orogeny may have locally caused secondary migration into the higher structural closures. But definitive conclusions derived from testing several of these structures have been clouded by low porosity or subnormal pressures in zones exhibiting good oil shows, inadequate followup testing in some structures, and general failure to drill into the deeper parts of the section. Future exploration should direct greater emphasis to delineation of stratigraphic traps in light of the regional shows. These shows suggest regional hydrocarbon migration, combined with a conducive updip wedgeout of the stratigraphic section.[8]

Although shelf sediments deposited near the Paleozoic hingeline have classically offered excellent potential for stratigraphic traps, their location at this time can be inferred only from a broad appraisal of the regional stratigraphic relationships. Identification of specific traps will require careful review of existing petrologic/stratigraphic reports in an area of interest, supplemented by more subsurface geophysical and downhole data.[9]

The Holbrook basin is centered around the town of Holbrook in east-central Arizona, extending northwest almost to Flagstaff and southeast into west central New Mexico (Fig. 8–12). The basin is 200 miles long and 100 miles wide. It is a true sedimentary basin, expressed in subsurface rocks of Permian age. Generally speaking, the overall Paleozoic section thins from 3,500 ft to the west to less

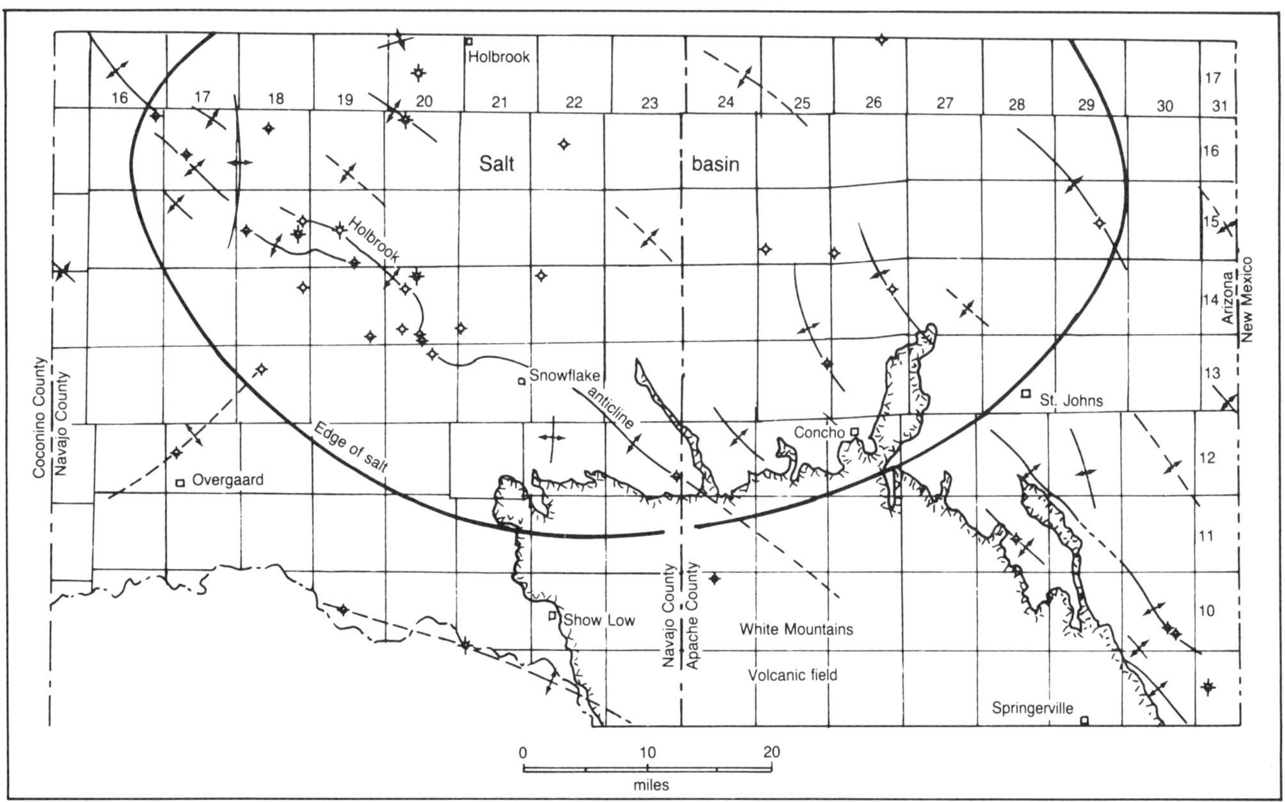

Fig. 8–12 Principal anticlines and deeper oil and gas tests in the Holbrook basin. The salt basin is developed in the Permian Supai formation. (Courtesy *OGJ*, January 5, 1981)

than 1,500 ft in the eastern part of the area. With such thinning, a variety of stratigraphic traps for oil and gas can be expected to occur in the subsurface.

The system that probably holds the most promise for commercial oil and gas production in the basin is the Permian. During the Permian age, a closed marine-evaporite basin formed in the area, resulting in the deposition of considerable gypsum, anhdyrite, and salt. In ascending order, the Permian includes the Supai formation, the Coconino sandstone, and the Kaibab limestone. All have had oil and gas shows in the basin.

The Supai is as much as 3,000 ft thick in the heart of the basin, thinning outward in all directions. It grades eastward into the Abo and Yeso formations of New Mexico. The Coconino sandstone underlies the Supai and ranges up to 1,000 ft in thickness toward the west. It generally has good porosity and permeability. It grades eastward, transgressively, into the Glorieta sandstone of New Mexico. The Kaibab, where adequately buried, may be an important objective zone for oil and gas exploration in the basin.

Since the Holbrook basin is a sedimentary rather than a structural basin, there is little evidence of the basin-like nature of the region on the surface. The most characteristic feature is a series of northwest-striking anticlinal features. The first well for oil and gas in the Holbrook basin was drilled in 1921. The deepest well in the basin was drilled in 1927, just south of Holbrook. It went 4,675 ft into the Precambrian, looking for helium. Since then, over 200 tests have been drilled for oil, natural gas, and helium in the basin. But most of these wells were clustered around Holbrook, Pinta Dome, and Navajo Springs helium fields. There were also some wells drilled on the Holbrook anticline where good oil and gas shows were reported.

This all means that large areas of the basin have not been tested thoroughly, if at all. Only 30 wells tested the entire section to the basement. Some 40 wells had oil and natural gas shows. The deepest oil show was in the Devonian at 3,812 ft. The shallowest oil show was in the San Andres Permian at 180 ft.

All evidence indicates that the Holbrook basin offers the wildcatter some excellent opportunities for finding hydrocarbons at relatively shallow depths.

REFERENCES

1. David Gee, *Oil & Gas Journal* (April 26, 1976), p. 199.

2. G. Lovick, C. Mazzini, and D. Kotila, *Oil & Gas Journal* (February 1, 1982), p. 181.

3. L. Woodward, *Oil & Gas Journal* (April 21, 1980), p. 114.

4. R. Scott, *Oil & Gas Journal* (March 6, 1967), p. 140.

5. Dale Nations, Northern Arizona University, in an Interstate Oil Compact Commission paper.

6. E. Heylmun, *Oil & Gas Journal* (May 5, 1979), p. 185.

7. S. Giardina, *Oil & Gas Journal* (March 17, 1980), p. 205.

8. Ibid.

9. Ibid.

9

The Great Basin and the West

The western U.S. holds some of the world's largest undrilled provinces as well as many of the largest oil fields. The industry has known oil exists in the West since the last century. Early-day fields in California are still on production. Exploratory work has been under way for many years, from Nevada and the Great Basin west-northwest through most of California, Idaho, Oregon, and Washington. However, except for production in Nevada's Railroad Valley, a gas field at Mist, Oregon, and the huge oil and gas fields of California, an extensive area remains in which to explore for hydrocarbons.

THE GREAT BASIN

The geographical Great Basin is bounded by the Colorado and Columbian plateaus, the Sonoran Desert section of the Basin and Range province, the Sierra Nevada of California on the west, and the Wasatch Range on the east (Fig. 9–1).[1] The explorer Frémont named this region the Great Basin in 1844, after recognizing that from this major area

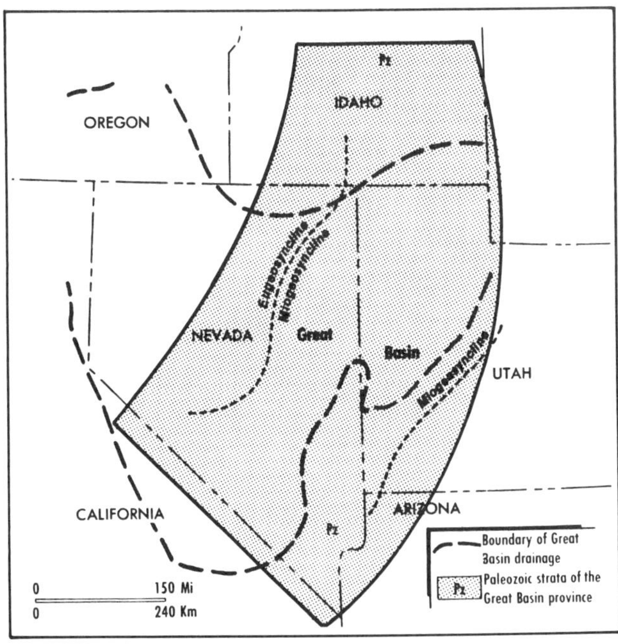

Fig. 9–1 America's Great Basin (courtesy *OGJ*, January 2, 1978)

no water drained to the ocean. The Great Basin is a major part of the Basin and Range province, the latter extending across the border into Mexico. More than 150 mountain ranges traverse the basin, nearly all along north-south trends. Most are tilted horsts with basins of undefined grabens in between.

Much of the faulting is geologically young, often as late as Oligocene and even as late as the Quaternary. The upper layer of valley-fill is also young. Lower layers may include pre-Oligocene Tertiary and Mesozoic sediments. Drilling has indicated the presence of Paleozoic strata below the valley-fill, demonstrating that the Paleozoic is an uninterrupted depositional mass.[2] Sedimentary volume is 200,000 miles.[3]

Post-Paleozoic tectogenesis of Basin and Range type has caused wildcatters to turn away from Paleozoic prospects in the Great Basin province. Perhaps this skepticism is unwarranted. Paleozoic strata are almost totally marine and probably generate and accumulate hydrocarbons.

The discovery of four Tertiary oil fields in Nevada caused a return of exploratory interest in the basin. A Late Paleozoic (possibly Pennsylvanian) limestone, below the Tertiary reservoir at Eagle Springs oil field in Nye County, Nevada, actually produced small amounts of oil. It is believed, however, that while the reservoir is Paleozoic the oil may have migrated into it from the overlying Tertiary. Other evidence in the basin indicates hydrocarbons in the Paleozoic. Wildcats in both Utah and Nevada have reported oil staining and gas shows in the Silurian, Devonian, and Permian rocks.

Despite much tilting, tight folding, local overfolding, thrusting, and gliding, it has been possible to locate structures with Paleozoic strata in Nevada and Utah unaffected enough to drill. More structures of this type can probably be found by careful study. Structural situations also exist where the major anticlinal axis of an area has been converted into a thrust by compressive stress, leaving only the associated syncline as the preserved element. The flanks of such syn-

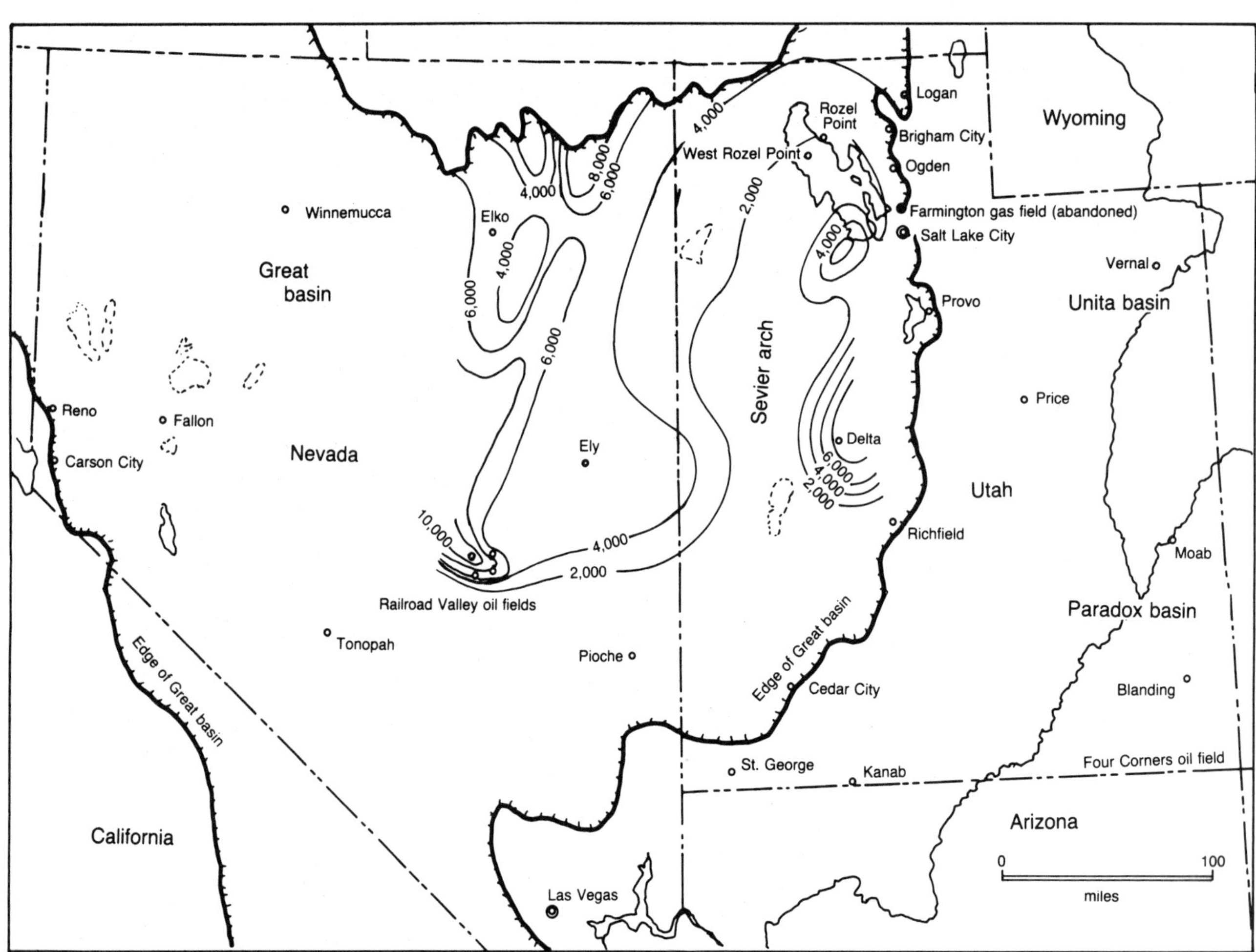

Fig. 9–2 Great Basin fields and environs (courtesy *OGJ*, December 8, 1958)

clines could provide good gathering ground and migration paths for hydrocarbons.[4]

Strike faults on the flanks may certainly be prospective stores of oil in the Great Basin province. It seems likely the petroleum possibilities of Paleozoic rocks in the Great Basin, especially of those in its miogeocline, can be more attractive than the structural style of the rocks might imply.[5]

The 1976 discovery of Trap Spring oil field in Railroad Valley, the Currant and Bacon Flat finds, and the 1979 discovery of West Rozel oil field in Great Salt Lake have focused new attention on the Great Basin as a probable petroliferous province of note. To date, six oil fields have been found in the basin, and there is a chance for two more. These include Eagle Springs, Trap Spring, Currant, Bacon Flat, Rozel Point, and West Rozel Point (Fig. 9–2). All of these fields produce from either Tertiary lake sediments or fractured volcanic rocks. Accumulations occur in truncation-fault traps or in drape-over faulted structure.

Looking for these accumulations requires much work and careful study. Exploration consists of mapping basin source rocks with proper depth for maturity, presence of good reservoir rocks, and delineation of traps by photogeologic-geomorphic methods, gravity surveys, and seismic shoots.[6]

It took half a century of wildcatting and 77 dry holes in Nevada before the Eagle Springs field discovery. Trap Spring was not found until 1976, and another 100 wells were drilled between the two discoveries' completion dates. The exploratory task ahead in the huge Great Basin is a big one. Continued exploration for both Tertiary and Paleozoic prospects in the Great Basin should result in significant oil and gas discoveries.

A long history

The first drilling on record in the Great Basin was in 1892 when several shallow wells were drilled 3 miles southwest of Farmington, Utah, near the foot of the Wasatch Ranch. Inflammable gas was reported in at least nine different sands between 400–700 ft in Quaternary lake sediments. About 20 shallow wells were drilled between 1892–1896 in a 1-sq-mile area, and in 1895 a pipeline was laid to Salt Lake City, 12 miles south. This field supplied the town with gas for a time but soon dwindled and died out.

The drillers went to Rozel Point in 1904 on the north shore of Great Salt Lake. The largest and best known asphalt seeps in the Great Basin are under the waters of the lake at this site. Amoco moved back into this area during the last few years and found West Rozel Point field.

The Great Basin hosted very little drilling between 1931–1948. Oil companies and independents moved into the basin after 1948, and several significant wildcats were drilled. Wells drilled near Las Vegas and Elko reportedly had oil shows at shallow depths. Several shallow wells near the Bruffy oil seep, 50 miles southwest of Elko, had shows in Ordovician rocks.

Shell drilled Nevada's first commercial oil well in 1954 in Railroad Valley, 55 miles southwest of Ely in Nye County. Initial production was 373 b/d of 27.8° gravity oil from a welded rhyolitic tuff of Oligocene age between 6,430–6,730 ft. Other wells followed, opening the Eagle Springs oil field.

CALIFORNIA

California is noted for more than 40 oil giants. Most were found years ago, two of them in the 1880s. Brea Olinda field in the Los Angeles basin was discovered in 1884. Total production in the field in 1981 was 2.9 million bbl of oil. Cumulative production in this old-timer is 366.5 million bbl, and the field still has 19 million or more to go. The other oil field is McKittrick in the historic San Joaquin Valley, which was found in 1887. Output in 1981 was 4.7 million bbl of oil, and the cumulative output is 240 million bbl of oil with 30 million or more to go.

The search for oil in California began as early as 1854. There were findings all the way from Los Angeles to 200 miles north of San Francisco in Humboldt County. But the commercial history of California oil fields did not get under way until 1875. More than 30 oil fields were discovered between then and 1905, all of these near oil seeps.

Two large oil fields were found in the 1940s at San Ardo and South Cuyama. Other giants followed in the state's various basins—Los Angeles, Santa Maria, Ventura, the Sacramento and San Joaquin valleys. Then, in the late 1960s, the offshore California theater opened with important findings in areas off Los Angeles and Santa Barbara.

Supergiant oil fields in California include Midway-Sunset with 1.5 billion bbl of oil cumulative and 469 million yet to produce. The 1981 output was 44 million bbl of oil. Huntington Beach is another California giant. It is in the Los Angeles basin with 1 billion bbl of cumulative oil production and 10 million produced in 1981. Huntington Beach has 68 million bbl yet to be produced. Wilmington is the big one, with 2 billion bbl already produced and 367 million yet to come. Production in 1981 was 39 million bbl of oil.

The total oil production in California amounts to 1.052 million b/d of oil. To date, more than 19 billion bbl have been produced in the state. Of the 58 California counties, 32 are producers.

Large basins

The prolific oil fields in California are in the Los Angeles, the Santa Maria, the Cuyama-Salinas, and the San Joaquin Valley basins. Large gas fields are found primarily in the Sacramento Valley in northern California.

Oil hunting in the Los Angeles basin goes back to 1892. But the basin's fame did not arrive until the 1920s. These were the days of Long Beach, Santa Fe Springs, and Huntington Beach fields. Then came Wilmington in 1932.

This basin, for its size, is the most prolific oil-producing basin in the world. Overall initial reserves of proven fields in this basin average 125,000 bbl/acre, and Signal Hill will produce a total of 500,000 bbl/acre in primary production, a world record. On the basis of known reserves per cubic mile of prospective sediments, the basin is the world's most prolific with about 9 billion bbl of ultimate primary reserves overall, or 5.5 million bbl/cu mile of sediments.[7] Los Angeles basin exploration is divided into three eras (Fig. 9–3).

The first was pre-Signal Hill days (1900–1920), when 1.5 billion bbl were found on the basis of oil seeps, topography, basic geological considerations, and random drilling. Included in this group of field discoveries were Beverly Hills (Pliocene) with 10 million bbl, Salt Lake with 50 million, L.A. City with 25 million bbl, Whittier with 60 million bbl, Brea Olinda with 350 million, East Coyote with 100 million, West Coyote with 300 million, Montebello with 200 million bbl, and Richfield with 150 million bbl of oil. During these days, little or no restrictions on oil drilling activity were imposed on exploration and producing companies.

Discoveries during the second era are those found from 1920–46 as a result of improved exploratory methods, including more sophisticated geology, fault analysis, and seismology (after 1930). During this time, about 7 billion bbl of oil reserves were found. These fields included West Newport with 50 million bbl, Huntington Beach with 1 billion, Belmont with 20 million, Seal Beach with 200 million, and Santa Fe Springs with 600 million. Also included are Sansenina with 50 million, Yorba Linda with 20 million, East L.A. with 15 million, Long Beach with 900 million, Dominguez with 300 million, Rosecrans with 100 million, Potrero with 15 million, Inglewood with 400 million, Torrance with 200 million, and Wilmington with 2.6 billion. Many of these fields are on topographic highs.

Wildcatting slowed in the basin from 1946–1953 because of the fast growth of the city of Los Angeles and its suburbs. New regulations resulted and a new era of drilling began.

In 1953, Universal Consolidated Oil opened the third era with a deeper oil find on the lot of 20th Century-Fox Studios in West Los Angeles. Miocene oil was discovered below the old Beverly Hills field. This discovery forced more regulations and drilling operations. With the new set of rules, the industry set out to find more oil. Operators made new discoveries with reserves totaling about 300 million bbl of oil, most of this new oil in city areas. New large deposits were discovered at West Beverly Hills with 20 million bbl, Cheviot Hills with 30 million, Las Cienegas with 70 million, Downtown L.A. with 15 million, East Beverly Hills with 100

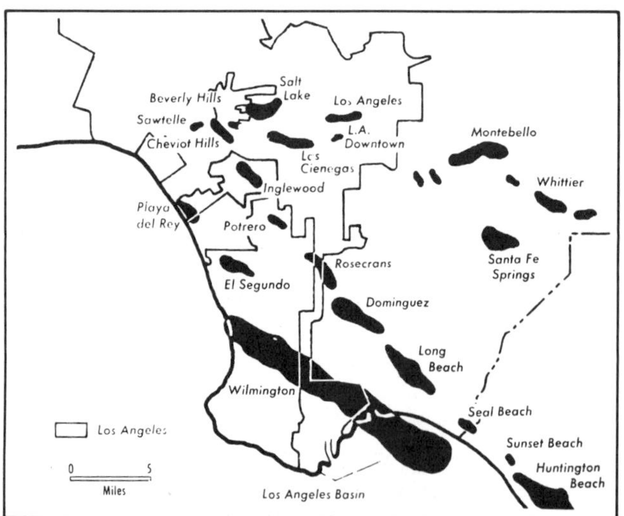

Fig. 9–3 Home of the giants (courtesy *OGJ*, August 16, 1971)

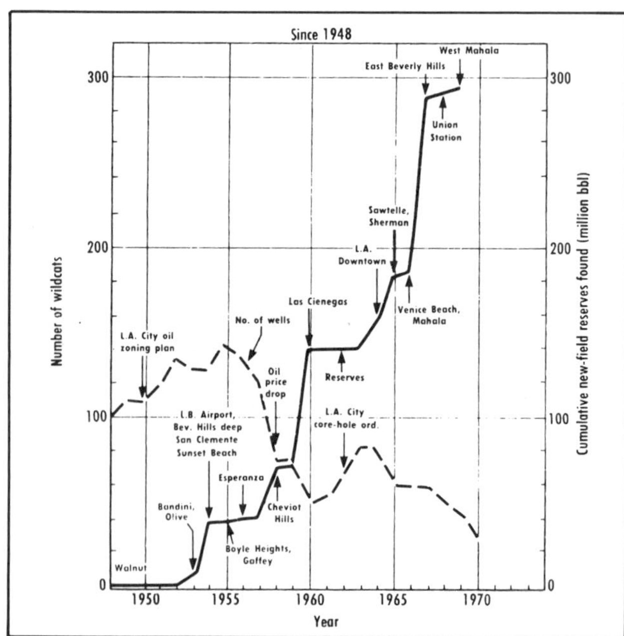

Fig. 9–4 Los Angeles basin wildcatting and success (courtesy *OGJ*, August 16, 1971)

million, Sawtelle with 15 million, Salt Lake-Sherman with 11 million, Venice Beach with 3 million, Bandini with 6 million, Olive with 2 million, Long Beach Airport with 11 million, and Sunset Beach with 7 million. Most of this oil was developed in heavily congested city neighborhoods. Exploration during the third era was limited, although the success ratio was fairly good (Fig. 9–4).

New areas

There should still be some more new oil to be found in the basin. It seems likely that oil can be found along the northern margin of the basin in connection with the Santa Monica-Raymond fault system. More accumulations along the Whittier fault system are probable. Deep holes should be drilled in the actual basin with its center near the confluence of the Rio Hondo and Los Angeles rivers. This area is 20 miles long by 5 miles wide with more than 18,000 ft of Puente, Repetto, Pico, and younger sediments. The area has seen very little exploration and very few deep holes.

Little exploration has been made of so-called prebasinal sediments, particularly at the east end of the basin where Lower Miocene, Eocene, and Cretaceous rocks offer potential and relatively unexplored targets. Exploration has been sparse in the surrounding San Fernando and San Gabriel Valley basins. But extensive marine sediments are present, and oil and gas shows have been reported from several wells. More exploration also is needed in the Imperial Valley.

Santa Maria basin

This basin occupies an area of the Coast Ranges just north of the Transverse Ranges. Production is confined to parts of Santa Barbara County and the southwestern corner of San Luis Obispo County.

Three giant oil fields are in the area—Santa Maria Valley, Orcutt, and Cat Canyon West. Production at Santa Maria Valley was 3 million bbl in 1981. Cumulative production here is 181 million bbl of oil with 43 million bbl yet to be produced. Orcutt has an annual output of 1.1 million bbl of oil. The field already has produced 158 million bbl of oil and has 18 million left. The Cat Canyon complex produced 4.8 million bbl of oil in 1981. Cumulative production here is 215 million bbl of oil with 40 million or more yet to be produced. The Santa Maria basin is a megasyncline cut by east-trending elongate folds. Most of the trapping is anticlinal.

Cuyama-Salinas basin

The Cuyama-Salinas basin's largest field was discovered in 1949 at South Cuyama. Annual output is about 590,000 bbl of oil. Cumulative oil production stands at 214 million bbl with about 3 million bbl of oil left to be produced. The basin's San Ardo field produces 10 million bbl/year, has cumulatively produced 358 million bbl of oil, and has 157 million bbl remaining.

The San Joaquin Valley basin

There are 19 giant oil fields in this huge basin with famous names like Elk Hills, Coalinga, Midway-Sunset, and Kern River on the producing agenda. Midway-Sunset is the largest with 1.5 billion bbl of oil already produced. Elk Hills is a big one, too, with 573 million bbl of oil produced and 843 million bbl yet to be produced (Fig. 9–5).

The San Joaquin Valley is a part of the Great Valley, a structural trough between the Sierra Nevada and the Coast Ranges of California. Thick Tertiary deposits in the southern end of the Great Valley produce vast amounts of oil. Reservoirs range in age from Pleistocene through Tertiary and Upper Cretaceous to crystalline, basement schists. Many large folds are in the basin, some several miles long with thousands of feet of closure.

Elk Hills is one of the most interesting of a host of similar fields in the San Joaquin Valley. This field constitutes the largest part of Naval Petroleum Reserve 1. It is about 20 miles southwest of Bakersfield. The field extends over part of a large anticlinal trend whose surface expression is a line of hills about 17 miles long and 7 miles wide. Elk Hills field is classed among the giant oil fields of the world. Ultimate recoverable reserves are estimated to be more than 1.3 billion bbl of oil of which 574 million bbl have been produced.[8]

Oil seeps and asphalt deposits in the southwestern San Joaquin Valley led to oil drilling as early as 1864. Elk Hills was discovered in 1911, but the Standard Oil of California 1 Hay well in 1919 was the official discoverer of this big field.

Regional tectonic elements in central California are included in three major structural provinces: the Sierra Nevada, the Central Valley, and the Coast Ranges. These are elongated belts more than 400 miles long, subparallel to the Pacific coastline. The Sierra Nevada is an immense block of granitic rock that has been faulted upward along its east edge and tilted gently to the west.

The Coast Ranges, which make up the westernmost province, are an anticlinorium in which Mesozoic and Cenozoic sediments were complexly folded and faulted. The southern part of the province is cut obliquely by the San Andreas fault. The Great Valley lies between the Sierra Nevada and the Coast Ranges.

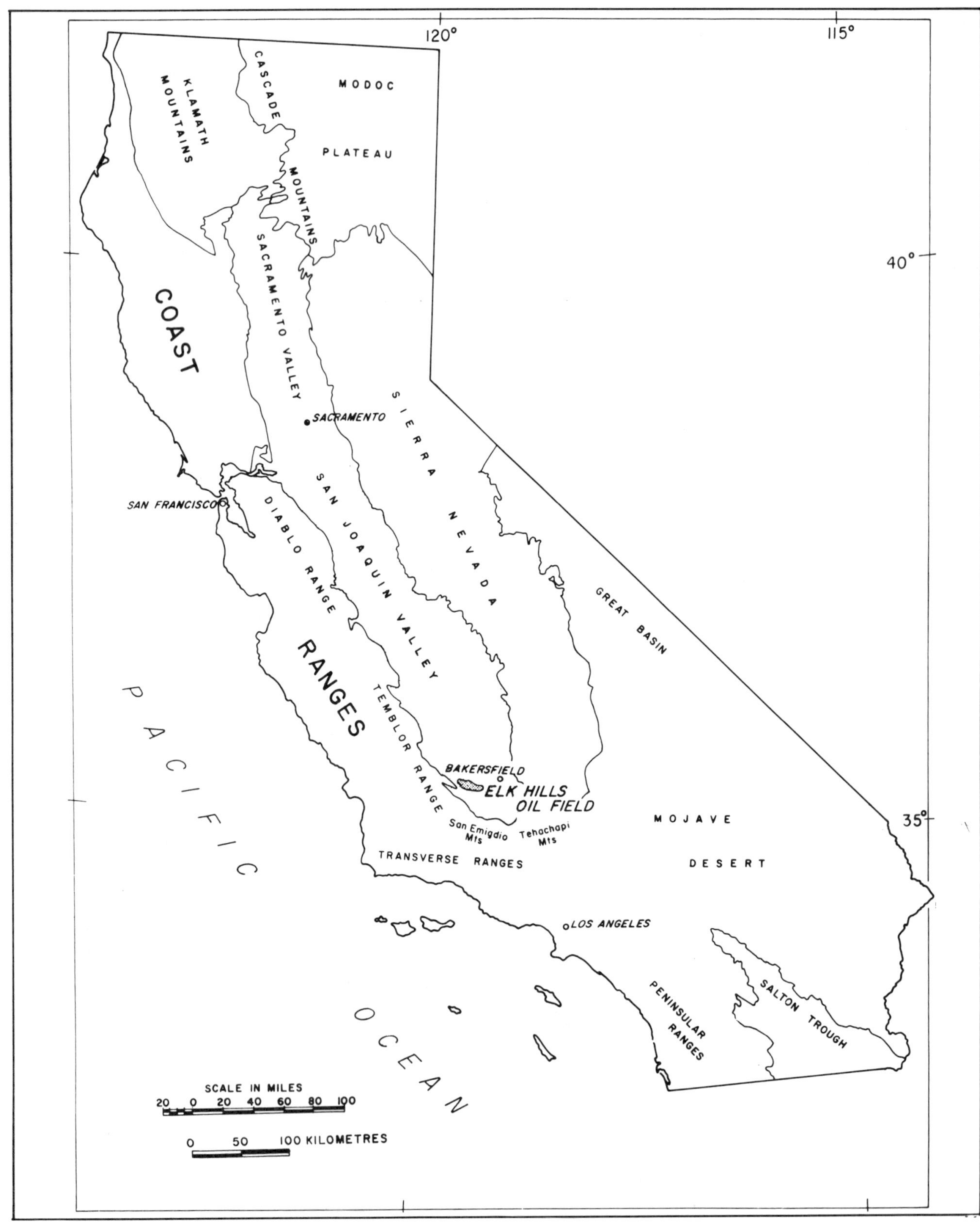

Fig. 9–5 Elk Hills oil field (courtesy Naval Petroleum Reserve No. 1)

Tertiary and Quaternary sediments as much as 17,000 ft thick underlie Elk Hills and nearby oil fields. The Temblor of Oligocene and Miocene age is the oldest formation drilled at Elk Hills. Some wells nearby did reach the Eocene, and one well in North Coles Levee field stopped in metamorphic basement. Naval Petroleum Reserve 1 oil and gas reservoirs are shown in Table 9–1, as well as other pertinent information.

The Sacramento Valley basin extends 200 miles north of the Stockton arch, which divides the Great Valley into the two areas. The Rio Vista gas field, found in 1936, is the largest of many gas fields in the basin. There are other large gas deposits at Willows-Beehive Bend. Exploratory work continues in this part of California, and other large gas accumulations probably will be found in the future.

The Ventura basin

Estimates of 20–30 billion bbl of remaining oil in place do not appear unreasonable for this California basin. The Ventura basin is divided into two general geographic portions, the onshore area and the Santa Barbara Channel offshore area, which includes the continental shelf extending some 70 miles west from San Miguel Island and Point Conception.

The basin is a complexly folded and faulted synclinorium formed by north-south compressional forces.[9] Major structures are west-trending, and the 67,000-ft sedimentary section ranges in age from Cretaceous through Holocene. Including the onshore, the Santa Barbara Channel, and the continental shelf, the basin averages about 215 miles long and 30 miles wide. Total volume of sediments is estimated at ±40,000 cu miles. Many favorable geologic conditions are conducive to the accumulation of hydrocarbons within the extensive thick sedimentary section in the Ventura basin.

The relationships of the various stratigraphic units in the basin as compared with the subprovinces are shown in Fig. 9–6. The most promising areas for future exploration will be in the Santa Clara trough, the Oxnard Shelf, and the eastern Ventura basin. Fig. 9–7, B-B′, shows deep prospect possibilities in the Santa Clara trough and the undeveloped Pliocene tar reserves of the Oxnard Shelf area. Fig. 9–8, C-C′, in the eastern part of the westerly plunging Santa Clara trough province, shows where more chances are provided for updip stratigraphic entrapment with deeper drilling and suggests that more potential exists in and along major zones of faulting. Possibilities also exist in the eastern part of the Ventura basin in the older unexplored sediments in and along zones of major faulting and by stratigraphic entrapment.

Exploration for the ultimate potential of the onshore part of the Ventura basin is far from complete. Most of the section has been drilled in only a few places on the margins of the basin. Large volumes of marine Lower Tertiary and Upper Cretaceous rocks have been undrilled. Virtually nothing below 15,000 ft has been tested. The presently estimated 1,750 million bbl of recoverable reserves will come from rocks of Pliocene and Upper Miocene age.[10]

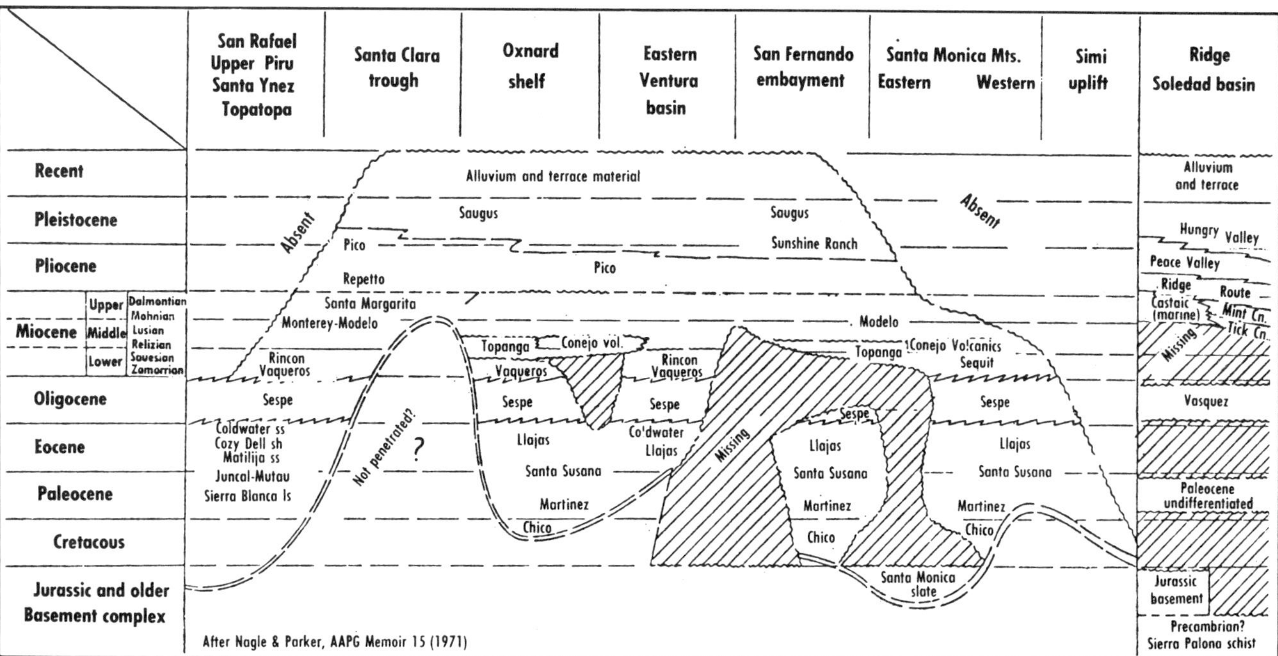

Fig. 9–6 Regional correlation (courtesy *OGJ*, September 2, 1974)

Table 9–1

OIL AND GAS RESERVOIRS, PRODUCING ZONES, PRODUCTION, AND RESERVES OF NAVAL PETROLEUM RESERVE NO. 1

Oil and gas zones	Oil and gas production zones of Elk Hills Engineering Committee	Cumulative production to December 31, 1972	Estimated remaining recoverable reserves	Estimated original recoverable reserves	Stratigraphic position		
					Member	Formation	Series
Mya sand zone	Dry gas zone, 5 producible wells	99,766,848 Mcf gas	220,192,762 Mcf gas	319,959,610 Mcf gas		San Joaquin formation	Pliocene
Scalez sand zone, Mulinia sand zone, Wilhelm sand zone, Gusher sand zone, Calitroleum sand zone, Olig sand zone	Shallow oil zone, 722 producible wells[1]	280,169,912 bbl oil (adjusted to unit production records)	318,489,715 bbl oil	598,659,627 bbl oil	Carman sandstone member Tupman shale member	Etchegoin formation	
						Reef Ridge shale	Miocene
N zone (24Z sand), A zone (26R sand), B zone (B sands), C zone, D zone	Stevens oil zone, 253 producible wells[1]	14,092,070 bbl oil	683,988,638 bbl oil[2]	698,080,708 bbl oil[2]	Elk Hills shale member	Monterey shale	
					Gould and devil-water shale members		
					Media shale member	Temblor formation	
Carneros oil zone	Carneros oil zone, 7 producible wells[1,2]	368,572 bbl oil	4,901,428 bbl oil[2]	5,270,000 bbl oil[2]	Carneros sandstone member		
Santos oil zone	Santos zone (Railroad Gap field only), 1 standing well	21,494 bbl oil	No estimate	No estimate	Santos shale member		

[Source: Annual report calendar year 1972, Naval Petroleum Reserves 1 and 2, Naval Petroleum Reserve, Elk Hills, California]

[1] Including production from Asphalto field.
[2] Does not include Railroad Gap field; reserve studies have not been made.

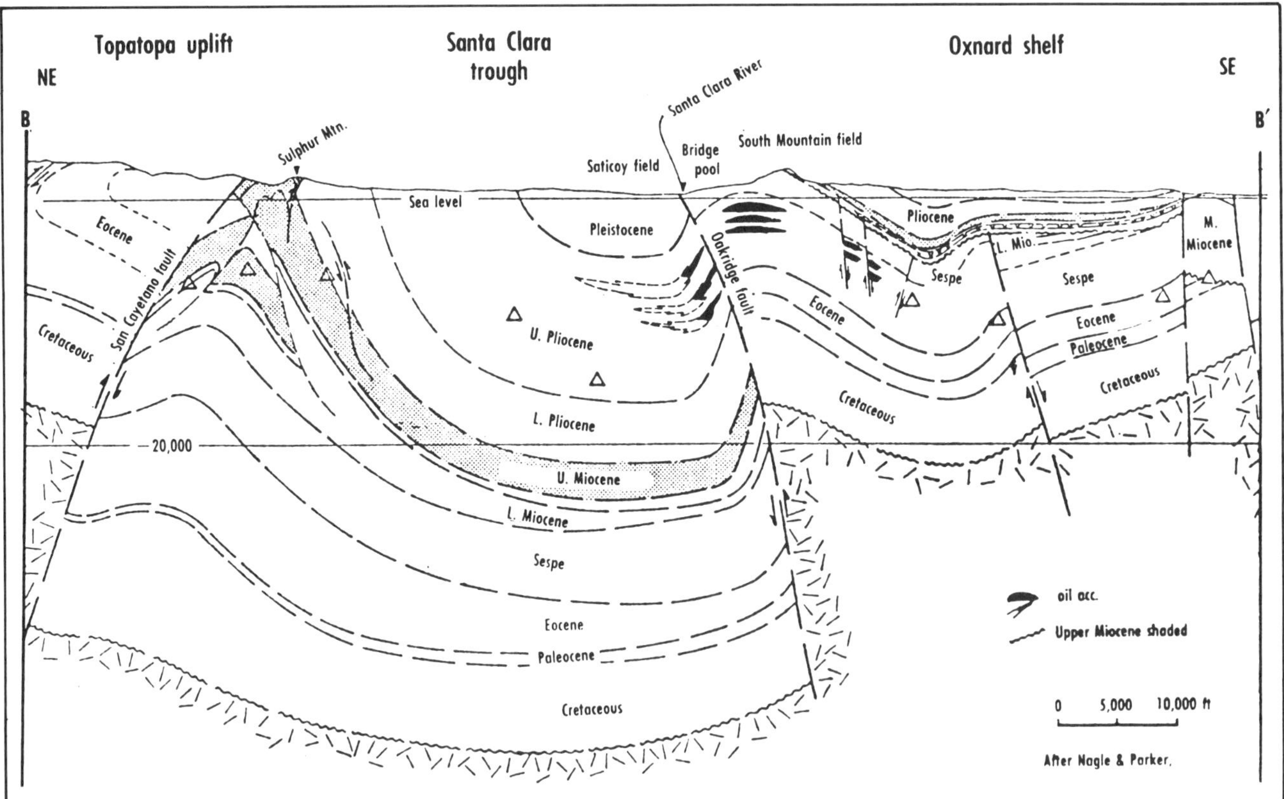

Fig. 9–7 Cross section B-B', Ventura basin (courtesy *OGJ*, September 2, 1974)

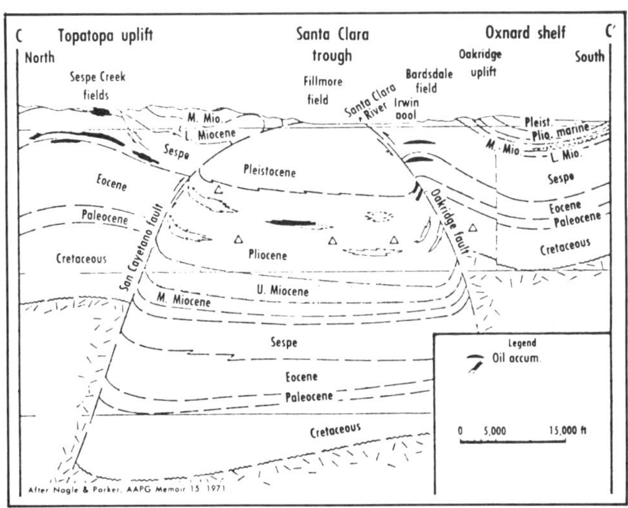

Fig. 9–8 Section C-C', Ventura basin (courtesy *OGJ*, September 2, 1974)

The Sespe formation, the reservoir for 240 million bbl of oil to date, has been tested thoroughly only in a small part of this area. Major structures are known to exist where many thousands of feet of Eocene, Paleocene, and Cretaceous marine rocks lie undrilled.

The Santa Clara trough has an ultimate recoverable developed reserve of 1.1 billion bbl. The Oxnard Shelf has 330 million bbl of developed oil reserves. This area will provide 40% of the future reserves in the area. The eastern Ventura basin has 296 million bbl of developed oil reserves. Other future reserves will come from relatively unexplored areas of the basin.

Total sedimentary section in the Santa Barbara Channel varies in thickness from 19,000–67,000 ft. Maximum thickness of potential reservoir rocks ranges from 5,000–25,000 ft.[11] All Tertiary rocks are potential objectives for future exploration. Both structural and stratigraphic opportunities for oil and gas fields lie in the western portion of the basin. There are several thousand feet of Cretaceous sand with good reservoir characteristics on and near San Miguel Island. The Santa Ynez Unit near Point Conception should produce at least 2 billion bbl of oil from Eocene and Miocene rocks.

Comparison of the Santa Barbara Channel with other oil basins in coastal California resulted in estimates of 10–15 billion bbl of original oil in place. However, in view of the estimates of 2 billion bbl of recoverable oil in the Santa Ynez Unit, perhaps earlier estimates of 25–35 billion bbl are more nearly correct.[12]

Development due

A group led by Texaco plans to develop a Santa Barbara Channel discovery off Point Conception, leading to possible production of as much as 100,000 b/d of oil by 1986. Preliminary studies are under way to develop the confirmed strike at Texaco 1 OCS P-0315, announced in June 1982. Texaco, as operator for itself, Pennzoil, Sun, and Koch, probably will drill 20–50 wells to develop the structure in Outer Continental Shelf Tract P-0315.

Estimates of reserves range from 100–500 million bbl of oil. The discovery well is 3 miles west of Tract 316 on which Chevron and partners opened Port Arguello field. It is possible Texaco has tapped the same structure that produces in Point Arguello field. Both fields have pays in the Miocene Monterey at comparable depths.

Texaco gauged rates of more than 3,000 b/d of clean oil from multiple zones during tests at the confirmation well and 4,200 b/d of oil from the discovery well. Pays in each well occurred within an interval more than 1,000 ft thick. Water depths range from 500–1,500 ft on the tract, which covers 4,269 acres.

Another big Miocene Monterery oil strike in the channel could lead an Atlantic Richfield unit to more than double production of 7,500 b/d from South Elwood field. Monterey sands have been proving prolific oil pays off California recently—Point Arguello, Texaco's new strike, and Exxon's development of Big Hondo and other fields in the Santa Ynez Unit.

ARCO plans to develop its Coal Oil Point field discovery in state waters, 2 miles offshore, extending South Elwood to the east. The 8-309 discovery flowed 4,138 b/d of oil from three zones. ARCO plans to test another structure in the area with a second extension. ARCO and Aminoil hold the same interests in adjacent State Tract 308, where ARCO earlier announced plans to reenter three South Elwood field wells looking for Miocene Monterey pay. In addition, ARCO and Aminoil applied to the state for permits to drill wildcats in 3242, 3140, and 208. Drilling should get under way in the 1980s.

The future

Most of the industry's major exploration and production efforts will focus offshore for the rest of the 1980s. Offshore California should prove to be one of the hottest exploration/development arenas in the U.S. during the 1980s. Operators are confident they will find several giant oil fields on the Outer Continental Shelf off southern California.

The most significant California oil news in 1981 was the record $333 million Chevron and Phillips paid for a Santa Maria basin tract off southern California in OCS Sale 53. Chevron later opened Point Arguello field.

Development work is heavy off California at three OCS fields—Exxon's Hondo, Shell's Beta, and Chevron's Santa Clara—all on stream since early 1981, accounting for about two-thirds of California's oil production increase during the year. Combined total is about 50,000 b/d of oil. Union's Platforms Gina and Gilda off Ventura County produce 20,000 b/d of oil and 7.6 MMcfd of gas.

Exxon also is developing Sacate and Pescado fields, which share Santa Ynez Unit with Hondo, and probably will announce another platform for Hondo soon. Chevron hopes to produce 8,000 b/d of oil from its Beta pool. Texaco will start production at Platform Habitat off Pitas Point with a goal of 60 MMcfd of gas.

Imperial Valley

Another area in California that will be getting future exploratory interest is the Imperial Valley. In 1981, operators scrambled for acreage position in the valley following a significant gas discovery in deep zones in the Gulf of California by Pemex of Mexico. The discovery well was 1 Extremeno drilled in 130 ft of water. Potential deep Miocene prospects in the Imperial Valley and adjacent areas may be on trend with the Pemex strike.

There have been a number of gas shows on the flanks of the valley, but nothing commercial. Objectives in future exploration should be at 13,000–20,000 ft on the flanks of the basin. Geothermal production in the center of the valley precludes oil success there. Hot waters probably have been flushing the hydrocarbons updip along the flanks of the Imperial Valley.

IDAHO

Almost two-thirds of Idaho is automatically verboten for oil and gas exploration because a vast expanse of igneous and highly metamorphosed Precambrian rocks covers the central and northern portion of the state (Fig. 9–9). It would seem that most exploration people would shy away from Idaho drilling prospects, but this is not so. Despite the lack of drilling sites and the negative completion reports through the years, the industry continues to explore in those areas of the state, looking for a break in the dry habit.

Even with the score still at zero, Idaho wildcatters have been looking for hydrocarbons since 1903, when the state's first drilling took place in Teton County. Based on the presence of a possible sandstone reservoir, Fremont Oil and Gas drilled 660 ft, coming up with a distinct dry hole.

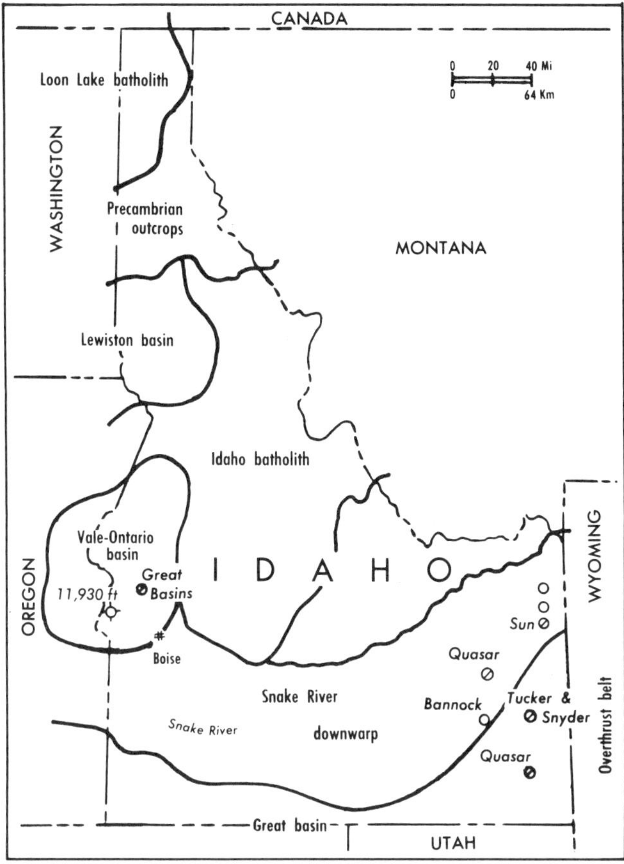

Fig. 9–9 Idaho production areas (courtesy *OGJ*)

Following this first failure, very few wells were drilled during the next 20 years. Some early-day gas shows in Ada, Canyon, Gem, Owyhee, Payette, and Washington counties did fire up a small gas-hunting program. A few wells were tapped with enough reserves to light ranch lamps and farm buildings. However, there just was not enough gas to go around, and the interest died.

The hydrocarbons search came to a virtual halt in Idaho from the middle 1940s to the middle 1950s. Another gas play ignited in the late 1950s in the Vale-Ontario basin, northwest of the capital at Boise. But this gas hunt blew out too. After the brief Vale-Ontario search, drillers shifted their interest to the southeastern counties along the Wyoming border. This is still where most of the wildcatting is today. The magic words, Overthrust Belt, have caught the interest of the industry from Canada to Mexico. Eastern Idaho lies in this belt, which has proven to be so prolific in northeastern Utah and southwestern Wyoming.

Several important wildcats have been drilled up the Idaho-Wyoming line in the past 3 years. Deep tests have been made into the thrusts, but to date the score is still zero for the industry. The industry, however, has not thrown in the towel, and new sites are being announced regularly in this part of Idaho.

OREGON AND WASHINGTON

Oregon and Washington first attempted oil and gas production in the early 1900s, but significant exploratory programs with adequate financing and technical know-how did not fire up until the 1940s.

Since then, almost all of the major companies and a number of high-grade independents have had their names on dry-hole efforts throughout the Pacific Northwest. Only one field has been found in this vast region of beautiful mountains, lakes, rivers, deserts, and forests. Mist gas field in Columbia County is a shining example that the industry has not given up (Fig. 9–10).

Reichold Energy is the dominant firm in Oregon's young gas play. The company drilled four wells in Columbia County's Mist field in 1981. The company has two other wildcats on tap and contingency for another eight wells, including the St. Louis prospect in the Willamette Valley and another in the Telemuck area along Oregon's coast.

Oregon's gas outlook is improving with the Clark and Wilson pay turning up in the Willamette Valley and the Pittsburg sand proving more productive than expected. But the geology remains especially tricky, with wildcat success ratio at about 1 in 10.

Coos basin off Oregon's southwestern coast may contain very good prospects for commercial deposits of oil and gas (Fig. 9–11). Most of the industry's exploratory interest in Oregon has been on the Mist area and in Linn County, both in the northwestern corner of the state. But that young gas play could be on trend with possible deep hydrocarbon prospects on the Outer Continental Shelf off Coos Bay, more than 100 miles south. Prospects have been drilled on and off portions of Coos basin and adjacent areas since the turn of the century, usually resulting in noncommercial oil and gas shows. But the state's opposition to petroleum development off its coast has prevented development of the more promising prospects since the 1960s. Two offshore wells then found good hydrocarbon shows off Oregon's southwest coast. Seismic profiles and well data analyses show a large structure with perhaps 20,000 ft of Tertiary sediments southeast of Heceta Bank off the town of Reedsport. This would probably be the best spot to look for oil and gas.

Subcommercial production was established in Washington in 1957 when Sunshine Mining completed 1 Medina in Grays Harbor County for 180 b/d of oil. Production declined rapidly, and the oil saga in Washington died with it. However, wildcatters have returned to the area.

Among the major exploratory disappointments of 1982 was Shell's significant but fruitless effort to put

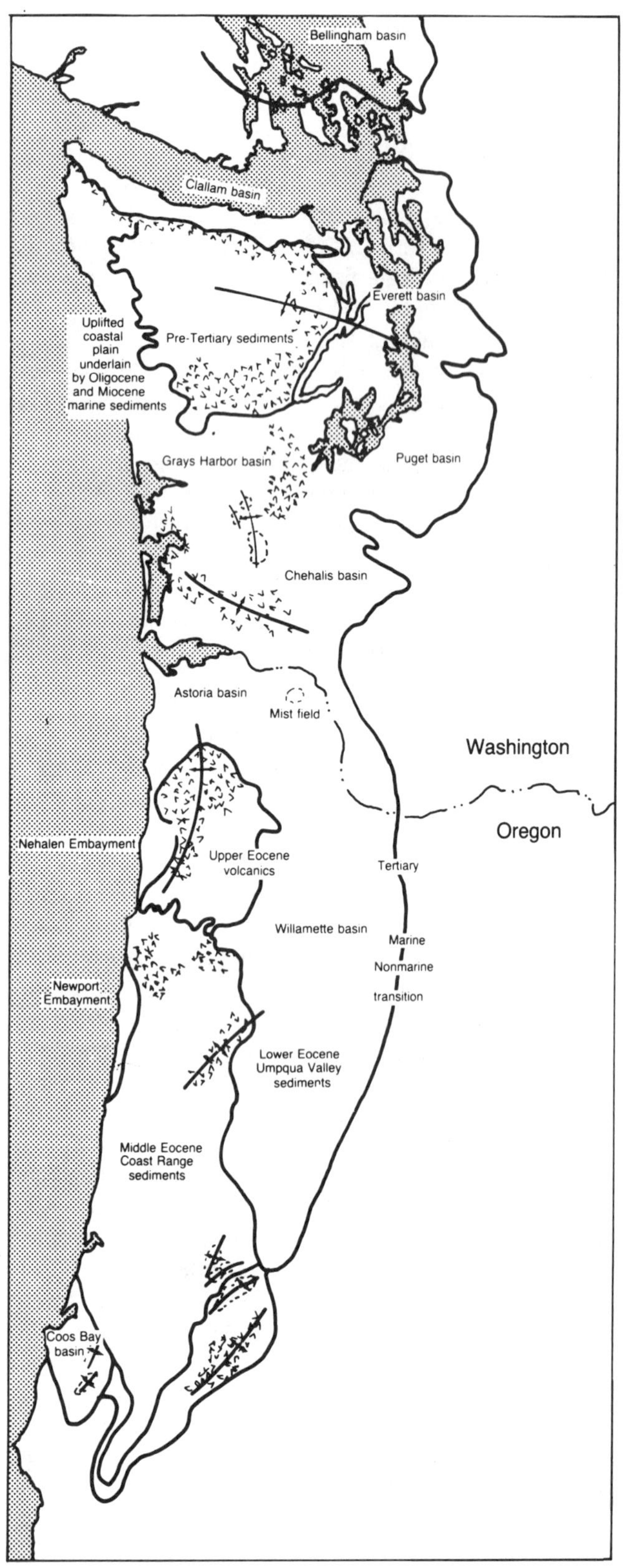

Fig. 9–10 Oregon production areas (courtesy *OGJ*, March 26, 1962)

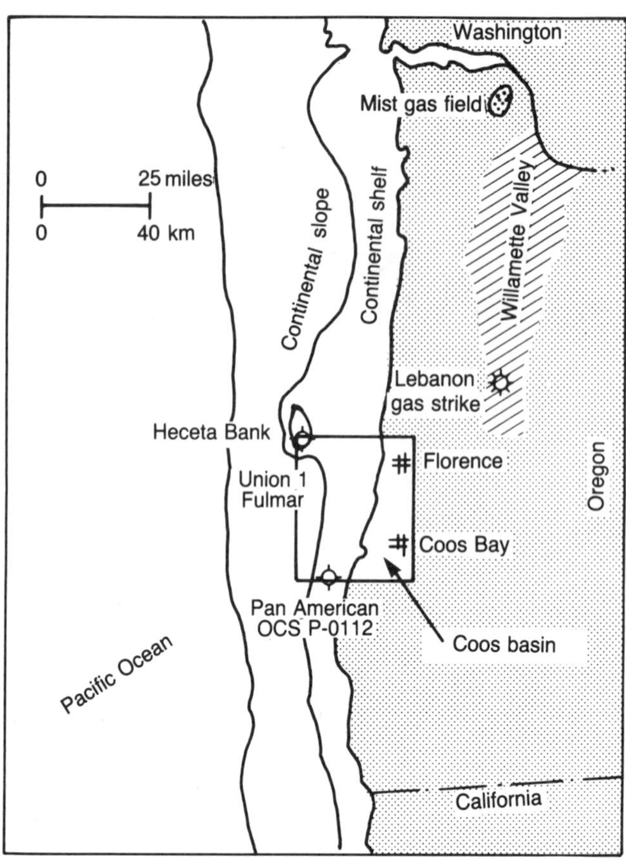

Fig. 9–11 Coos basin offshore prospects (courtesy *OGJ*, November 16, 1981)

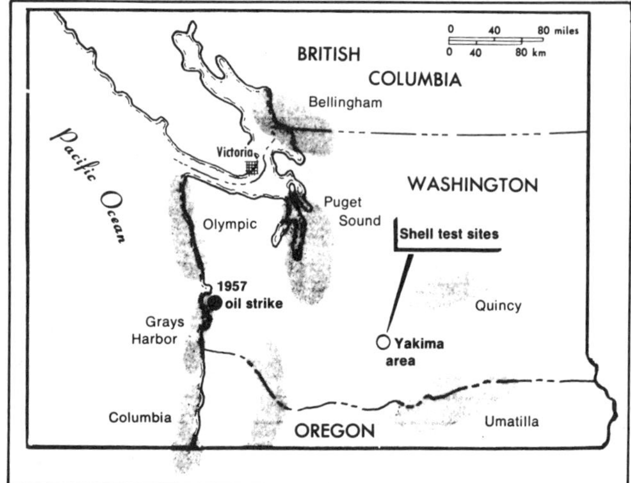

Fig. 9–12 Wildcats return to Northwest U.S. (courtesy *OGJ*, June 9, 1980)

Washington in the column of producing states. The firm tested in five zones of its intensely watched 1-33 Yakima Minerals, but even the combined rate was not commercial. A sixth test (to 16,199 ft) failed due to mechanical problems.

Shell also left its twin, the 2-33 Yakima Minerals. This 5,604-ft wildcat turned up a prospective zone found in the first well but could not be tested through three casing strings. Shell was testing on its 1-29 Bissa in 1982 in the Whiskey Dick Mountain area of Kittitas County to 20,000 ft (Fig. 9–12).

Meanwhile, Texaco scheduled an exploration campaign on a big acreage spread in Washington and Oregon with Boise Cascade. The agreement grants Texaco exploration rights for 2 years on about 500,000 acres. Texaco expects to concentrate its efforts on 200,000 acres in the Yakima-Ellensburg area of central Washington. This entire region will continue to be an intriguing theater of operations for at least another 8 years.

Washington is divided into six physiographic provinces. These reflect three different terrains: plateau lava, igneous and metamorphic, and sedimentary beds. The Okanogan Highlands and the Cascade Mountains are mostly igneous rocks, including eroded batholiths and volcanic cones and metamorphics.

Shell's wildcats in Washington, though unsuccessful so far, have certainly stirred the exploratory imagination in the area. The advent of oil and gas exploration in the Columbia Plateau has stimulated renewed interest in an old geological question—what lies beneath the Columbia Plateau? The basalt that forms the plateau is of Miocene age. It is estimated as locally reaching a thickness of about 1 mile and covers an areal extent of 50,000 sq miles in parts of the Northwest (Fig. 9–13).[13] Rocks cropping out around the edges of the plateau range in age from Precambrian to Oligocene and include nearly every possible rock type. State geologists feel neither the Eocene arkosic units, the Swauk/

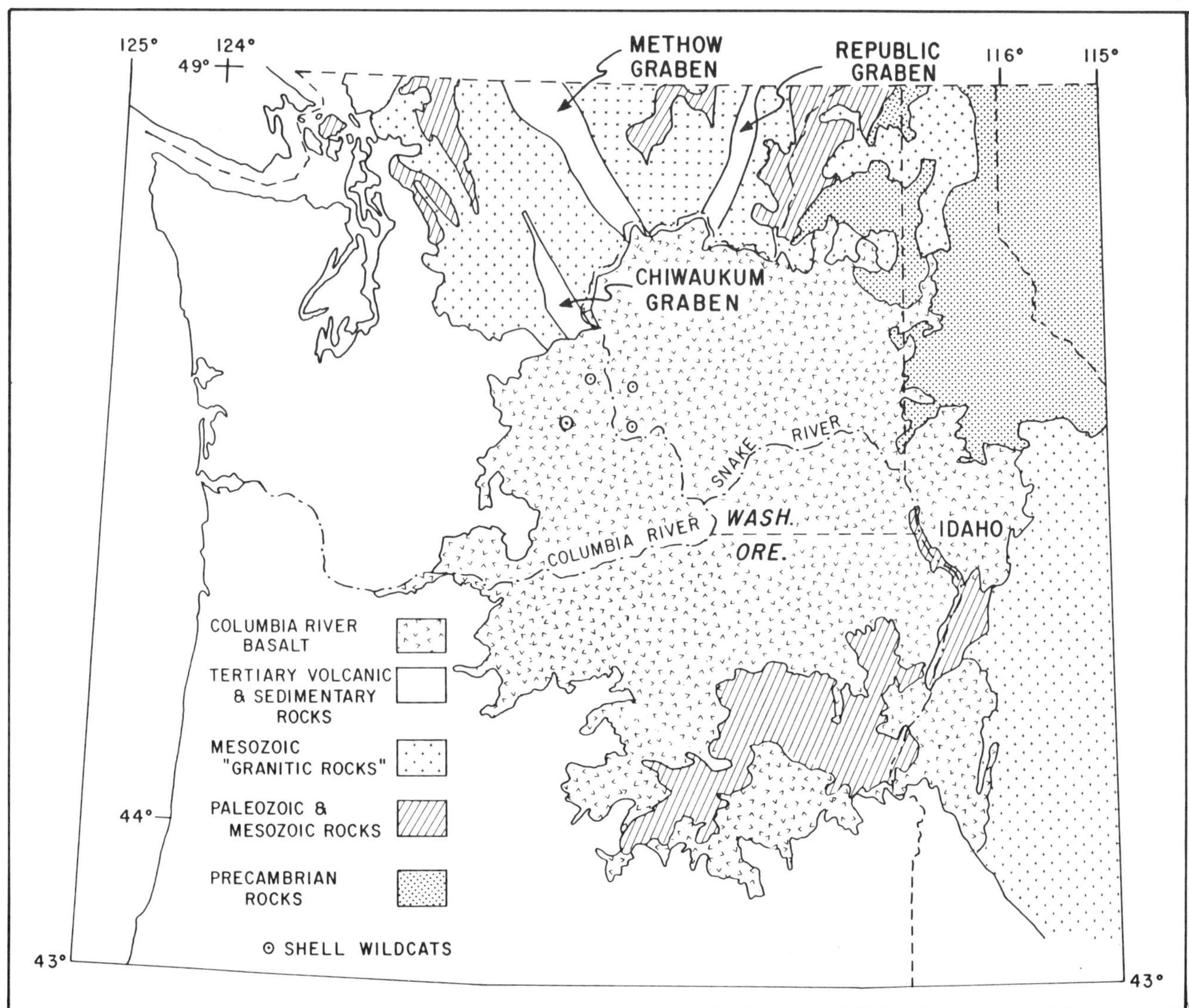

Fig. 9–13 Geology of the Columbia Plateau

Chuckanut, nor Chumstick/Roslyn have much oil potential, but they contain enough organic material to have gas prospects. More Swauk formation probably exists to the southeast of the Shell area, but its position depends on the amount of right-lateral movement on the Entiat fault, for which there is currently no geologic control. Because of the regional plunge, Washington University geologists believe the Chumstick and Roslyn formations probably merge to the south beneath the plateau and may spread out to the east below the plateau, thickening southward. They feel the Swauk and Chumstick are potential source rocks for gas to the south and east of the present exposures.

The best reservoir rock is the porous, nearly unconsolidated quartz sand of the Wenatchee formation, which overlies the Swauk and Chumstick formations with angular unconformity. Most explorationists think the late-Tertiary basins of the continental margins offer the greatest hopes for oil production in the Pacific Northwest. But the lack of drilling throughout those areas and over the rest of the region makes it hard to predict the future. At this time, it is almost impossible to say that one area has more promise than another.

Only wildcatting will reveal secrets to help unfold the story. Today's giant strides in geophysical know-how and geological boldness plus companies with wildcat fever will surely find something of intense interest and importance in the Pacific Northwest.

REFERENCES

1. M. Kamen-Kaye, *Oil & Gas Journal* (January 2, 1978), p. 116.
2. Ibid.
3. Ibid., p. 117.
4. Ibid.
5. Ibid.
6. Dolly Foster, *Oil & Gas Journal* (May 5, 1980), p. 303.
7. L. McFarland and R. Greutert, *Oil & Gas Journal* (August 16, 1971), p. 111.
8. J.C. Mahjer, R.D. Carter, and R.J. Lantz, *Petroleum Geology of Naval Petroleum Reserve No. 1,* USGS Paper 912 (1975).
9. K. Hall and J. Curran, *Oil & Gas Journal* (September 2, 1974), p. 90.
10. Ibid., p. 92.
11. Ibid., p. 93.
12. Ibid., p. 93.
13. R. Gresens and R. Stewart, *Oil & Gas Journal* (August 3, 1981), p. 157.

10

The Rocky Mountains and the Overthrust Belt

It was common knowledge in 1953 that for a generation or more the Rocky Mountain region had held promise of becoming one of the great oil-producing areas of the world. But only during the years immediately before 1953 had the region seen exploration activity equal to its vast potential. Many of the sedimentary basins are a geologist's dream because the structures and formations can be determined at the surface.

Why then the delay in exploration? The main reason was a lack of markets. A market for the regional oil did not develop until World War II, thus stimulating activity in the Rockies. Crude lines were built; basins were opened; and giant, intermediate, and small fields were discovered. Pioneering always has been a way of life in the Rocky Mountains. The vast region began to cash in on its many years of oil pioneering in the 1950s, and this spirit has never waned.

It was believed in the 1950s that the intermontane basins of the Rocky Mountains and those basins of the Great Plains just east of the Rockies, particularly the Williston basin, formed the greatest undeveloped oil region in the U.S. This is still true today, even with the addition of Alaska and the country's offshore potential.

This tremendous oil province still has plenty of room in which to search for oil and gas, although production has been known here since the late 1800s. The Rocky Mountain region includes nearly one-quarter of the estimated volume of sedimentary rocks above 20,000 ft of burial in oil-producing states (except Alaska). However, in 1953 it yielded less than 6% of the total crude oil output of the country. Today's output in the Rockies is 598,000 b/d of oil, and more than 8 billion bbl of oil have been produced since the early days of the Canon City and Florence fields in the Canon City embayment of Colorado. North Dakota, not included in the total, produces 126,000 b/d of oil. Wyoming is the leader in the Rockies with 361,000 b/d of oil.

Several oil giants are located in the Rockies, most of them in Wyoming. Colorado has one at Rangely, discovered in 1933. Annual output is 15.3 million bbl of oil at this Piceance basin field. Cumulative production stands at 649.2 million bbl of oil with 74.6 million bbl of oil remaining to be produced.

Three oil giants are located in Montana—Bell Creek, discovered in 1967, Cut Bank in 1926, and Pine in 1951. All three are in different basins or provinces of the state. Bell Creek is a big Powder River basin field, Cut Bank is in the

Sweetgrass arch portion of the state, and Pine is a Williston basin success (one of the first in the basin). The largest field is Cut Bank with 151.3 million bbl of oil produced and 44.6 million left. Bell Creek has produced 117 million bbl of oil and has 36 million remaining. Hundreds of dry wildcats were drilled before the initial discovery of Bell Creek. The find set off an intensive search for new oil in the southeastern portion of the Powder River basin.

North Dakota claims six giant oil fields, most of them rather recent finds. Beaver Lodge was the first Williston basin discovery in the U.S. It has produced 99.7 million bbl of oil and has 22.3 million remaining. The other giants are on the Billings Nose where exploratory activity has uncovered vast new oil reserves for the basin. Utah has five giant oil fields, two of them on the Overthrust Belt. Largest field is now East Anschutz Ranch on the Overthrust Belt with estimated remaining reserves of 800 million bbl of condensate. Greater Aneth in the Paradox basin is the next largest field with 313 million bbl of oil produced and 55 million left to produce. Altamont-Bluebell and Greater Red Wash are the two large Uinta basin fields in northeastern Utah. Pineview, where the discovery well for the Overthrust Belt was drilled in 1975, has produced 20 million bbl of oil and has 125 million or more remaining.

THE WYOMING BASINS

More basins are located in Wyoming than in any other oil-producing state. All of these basins have some production, either oil or gas, and they all have exploratory histories. Some of the state's basins are prolific oil producers, while others have huge gas deposits in addition to the oil reserves. None of the basins has been completely drilled, evidenced by recent discoveries in the Wind River and Big Horn basins.

The Wind River basin

The Wind River lies between the Big Horn and Green River basins (Fig. 10–1). It is well-defined, except on the east side, either by mountain ranges or prominent erosional escarpments. The basin is bounded on the north by the southern portion of the Absaroka Range and the Owl Creek-Bridger Mountains, on the east by a line of folding and faulting between the Laramie and Big Horn Mountains (which separates the Wind River and Powder River basins), on the south by the Sweetwater uplift, and on the southwest and west by the Wind River Mountains.

The Wind River basin is a major asymmetrical syncline with the basin axis near the northern limits of the structural depression. The basin is bordered by a series of sharp anticlines. Many of these anticlines, unlike those in most of the

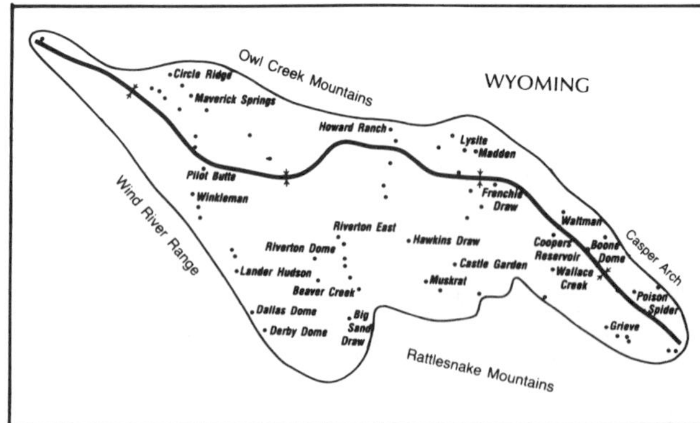

Fig. 10–1 The Wind River basin (courtesy *OGJ*, October 12, 1981)

intermontane basins of the Rocky Mountains, do not parallel the basin margins.

Oil-producing structures found distributed almost in line along the western border of the basin. The surface expressions of these productive anticlines do not parallel the Wind River Mountains but trend more to the north. Other oil fields are found along similar-trending anticlines farther to the east near the southern margin of the basin. Oil fields are found on the east side on small separate closures superimposed on a major asymmetrical anticlinal fold trending northwest-southeast just to the west of the line of folding and faulting between the Laramie and Big Horn Mountains. Deep exploration in recent years has uncovered gas fields in the Upper Cretaceous rocks on the highly faulted, complex, and folded north flank. This exploration continues today.

In the area of greatest downfolding of the Wind River basin, there is an estimated 15,000–20,000 ft sedimentary rock section. All ages of rock from Precambrian are represented in the basin, except the Silurian. Devonian rocks are restricted to the western side.

This area was not drastically disturbed until the close of Upper Cretaceous time when the bordering mountains and most of the present structures were formed. The pre-Cretaceous section in the northwestern part of the basin is not unlike that found in the Big Horn basin across the Owl Creek Mountains to the north.

Tertiary rocks cover 75% of the basin and constitute about half of the total volume of sediments in Wind River. Oil and gas production comes from rocks ranging in age from Tertiary and Paleocene Fort Union sands and others to the Madison limestone in the Mississippian. The largest producer is the Pennsylvanian Tensleep. Other big gas pays include the Permian Phosphoria Embar and the Cretaceous Lakota, Dakota, and Frontier sands. Most Tertiary rocks remain almost undrilled despite new discoveries in recent years on the north side of the basin. Major unconformities exist in and below the Tertiary and may have trapped oil and gas in great quantities. These rocks offer great potential for future explorers in both stratigraphic and structural traps. Great volumes of Cretaceous rock also are still relatively virgin under the Tertiary.

First exploration in the Wind River basin was in the southeastern part. A well drilled on Dallas Dome in 1886 had production in the Permian Phosphoria and Pennsylvanian Tensleep. When the railroad came into Lander at the turn of the century, Dallas Dome was developed further. In 1923 the Maverick Springs and Circle Ridge fields were discovered. Pilot Butte was found in 1916. In the 1940s, Pilot Butte got extra pay zones in the Phosphoria and Tensleep. Winkleman Dome was opened in 1917 with Lakota and Jurassic Morrison production. Many other large fields were discovered in the 1940s. During the 1950s, more oil was found at Big Sand Draw in the Mesaverde, Frontier, Muddy, Tensleep, Lakota, and Madison. Later, the Riverton Dome and East Riverton Dome fields were discovered with production from the Frontier down to the Tensleep. The east side saw the Poison Spider field's birth in 1917 with Sundance oil.

The really deep play in the basin began at West Poison Spider in 1948 in the Tertiary Fort Union. Other discoveries included Bull Frog with Morrison, Sundance, Lakota, and Muddy sand production. Another new find of significance to the basin's future is Teepee Flats, a subthrust structure on the northeastern flank near Waltman.

The Madden Unit was discovered in 1968 in the Fort Union (Fig. 10–2). Drilling here continues with reserves set at 2–3 trillion cu ft of gas. In 1978, Dome Petroleum completed an important well at 6-1 Arapahoe-Shoshone-Tribal in Fremont County, flowing 2 MMcfd of gas from Lance Upper Cretaceous at 13,805–14,026 ft. Location is 12 miles southeast of the Muddy Ridge gas field and 35 miles northwest of the Madden field.

An important discovery opening the 1980s was Chaparral Resources 1-29 Moneta Hills in Fremont County. Flow was 4.7 MMcfd of gas from the Cretaceous Shannon at 20,604–20,700 ft. This field is in the Madden area, 3½ miles south of nearest production in the field. The Madden field has produced more than 124 billion cu ft of gas from Tertiary Fort Union and several Upper Cretaceous zones since 1969. The Wind River basin's future lies in its deep zones yet to be tested adequately.

The Green River basin

The Green River basin is a structural and topographic basin in southwestern Wyoming (Fig. 10–3). It is bounded on the east by the Rawlins uplift, on the south by the overthrust Uinta Mountains and southeastward extension, on the west by the rugged zone of folding and overthrusting now known as the Overthrust Belt, and on the north and northeast by the Wind River Mountains and the Sweetwater uplift.

The Washakie, Red Desert, and Bridger basins are included in the Greater Green River basin. Sediments range in age from Eocene through Cambrian. There is in excess of 30,000 ft of sedimentary rocks near the margins of the basin. Production has been found in the Cretaceous, Triassic, Jurassic, Eocene, Paleocene, and Paleozoic sections. Important fields in the basin, not including those in the Overthrust Belt, are Baxter Basin, Church Buttes, Hiawatha, Lost Soldier, Wertz Dome, Crooks Gap, and Happy Springs.

Large structures, numerous known pay zones, strat traps, and untapped deep zones all combine to make the

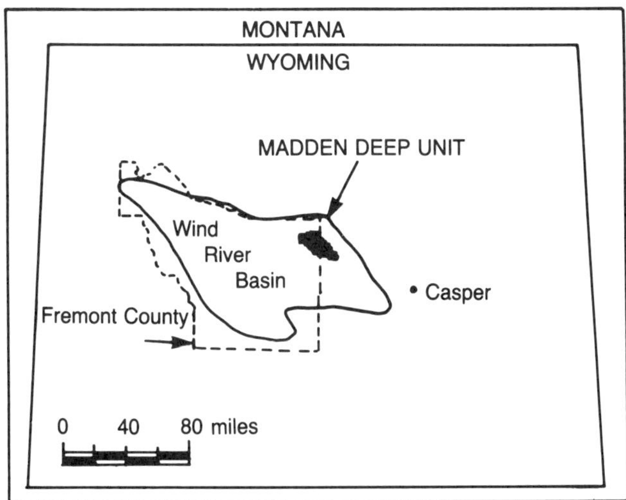

Fig. 10–2 The Madden Deep Unit of the Wind River basin

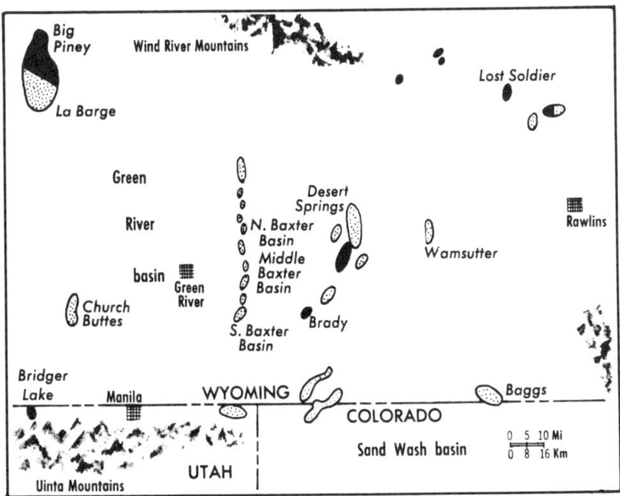

Fig. 10–3 The Green River basin play (courtesy *OGJ*, March 13, 1978)

Green River basin empire an attractive exploration area. In 1971, Mountain Fuel Supply, one of the most active operators in this basin, set out to tap one of these structures, discovering the Butcher Knife Springs field in Uinta County. That discovery opened the deepest production in the Rocky Mountain province at that time at 18,200–18,280 ft in the Pennsylvanian Morgan. Flow was 5,700 Mcfd of gas and 300 b/d of condensate. The field was one of three Morgan pools opened on the Moxa arch at that time. The other finds are 9 miles north in the Church Buttes field and in Moxa, farther up the line 26 miles. Church Buttes is one of the largest fields in this region, producing great amounts of gas and condensate from the Cretaceous Dakota.

Through the years the basin has rewarded the explorer with some of the industry's best oil and gas discoveries. The year 1978 was an especially good one for the basin. Several multipay discoveries were made in the area; one of these was the Dines Unit with five pays.

The Greater Green River basin has everything it needs to be a major oil and gas province. Along the north rim of the Uinta Mountains are undoubtedly more untapped fields like the flush one at Bridger Lake where 15,000-ft Dakota oil was found. The western side of the basin used to scare the explorer, but the birth of the Overthrust Belt dispersed this fear. The industry has long been interested in the gas and oil prospects of the Greater Green River basin. This area now has more interest factors as a result of Overthrust Belt play.

Credit for old Church Buttes itself goes entirely to the seismograph. The big structure is completely masked by a Tertiary blanket of great thickness. Tertiary rocks cover most of the basin except along the rims where ancient rocks crop out. Along the east side, the sprawling Rock Springs uplift brings Cretaceous rocks to the surface.

Some recent developments in the basin show the way for future work. Texaco made a new pay discovery at the Table Rock field in Sweetwater County at 37 Unit. Flow was 4.45 Mcfd of gas from the Pennsylvanian Weber formation at 17,095–17,240 ft. This pay lies between the Madison and Nugget formations. The Madison, with production at 18,300 ft, is the deepest pay in the field. The Nugget zone is at 15,200–15,700 ft. There are three more pays in the field above the Nugget. Texaco opened the field in 1946, and it now covers 12,700 acres. About 270 billion cu ft of gas and 1.4 million bbl of oil have been produced in this area since discovery. Average daily output is 200 b/d of oil and 100 MMcf of gas. This discovery proves there are more pays to be found in and around the existing fields in the Green River basin. Depth seems to be the key to future Green River success. The industry is busy developing the vast area, spurred on by the tremendous success in the Green River basin's neighbor to the west, the Overthrust Belt.

The Powder River basin

The Powder River basin, one of the world's largest in area, lies mostly in northeastern Wyoming but includes parts of Montana and South Dakota (Fig. 10–4). The basin is a busy one, offering many kinds of traps and favorable sands. Though Cretaceous sand exploration has dominated the basin's drilling, there is interest in the pre-Cretaceous, too.

The first oil discovery in the Powder River basin was in 1889, when, on the basis of seeps, a well was drilled just north of the Salt Creek structure. Following geologic work, the structure itself was drilled in 1906, opening Wyoming's vast Salt Creek field. This field was the largest in the Rockies, until Rangely came along in Colorado.

After the discovery at Salt Creek, other surface-expressed anticlines on the basin's rims were tested as productive. Geophysics entered the scene and was credited with a discovery in 1938. Meanwhile, in 1920, oil was found at structureless Osage on the east side of the basin. After Osage, interest on the eastern side of the basin died until several big wells were completed at Mush Creek in 1947. During the next 3 years, the spotty, stratigraphic-trapped pool of the Mush Creek, Skull Creek and Fiddler Creek areas was developed rapidly. In 1952, the discovery of oil in the Cretaceous Shannon at Ash Creek, 110 miles north of the Meadow Creek area, gave new life to the search along

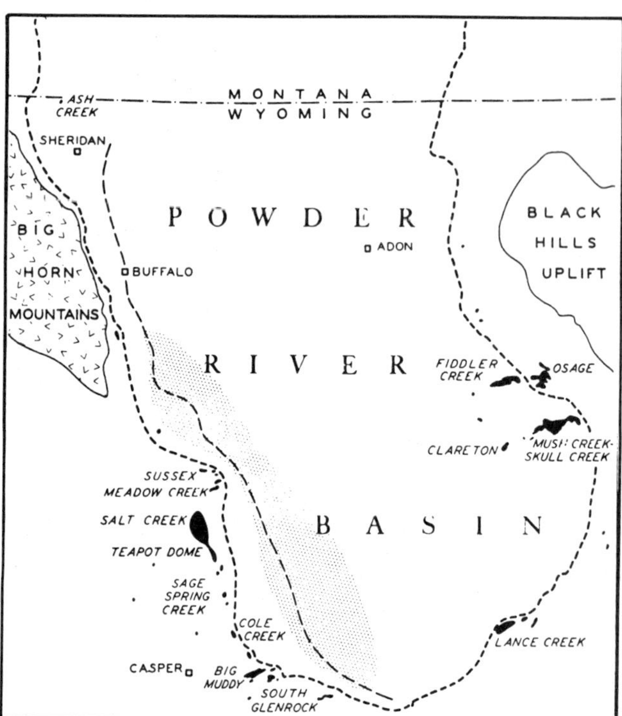

Fig. 10–4 The Powder River basin area (courtesy *OGJ*)

the western edge of the Powder River basin and increased interest in the productionless northern sweep of the basin.

Geologically, the Powder River basin is bounded by pronounced positive tectonic elements except on the north. Here a dividing line between this basin and the Williston basin is without a definite topographic or structural break. On the east side of the basin lie the Black Hills, on the southeast the Hartville uplift, on the south the Laramie Range, on the southwest the line of folding and faulting between the Laramie and Big Horn ranges, and on the west the Big Horn Mountains. The basin is a broad syncline with marginal zones of relatively steep folding and faulting but apparently very low folding in its interior. It is estimated the sedimentary section is more than 16,000 ft.

Strata in this basin range from Middle Cambrian to Eocene. Lower Cambrian, Silurian, and Devonian strata are absent. Middle Cambrian strata occur only in the western portion. Ordovician rocks are restricted to the north.

The Powder River basin contains some of the most important fields in the Rocky Mountains. Production is from the Cretaceous, Pennsylvanian, and Jurassic. The important producing areas in the basin today are shown in Fig. 10–5.

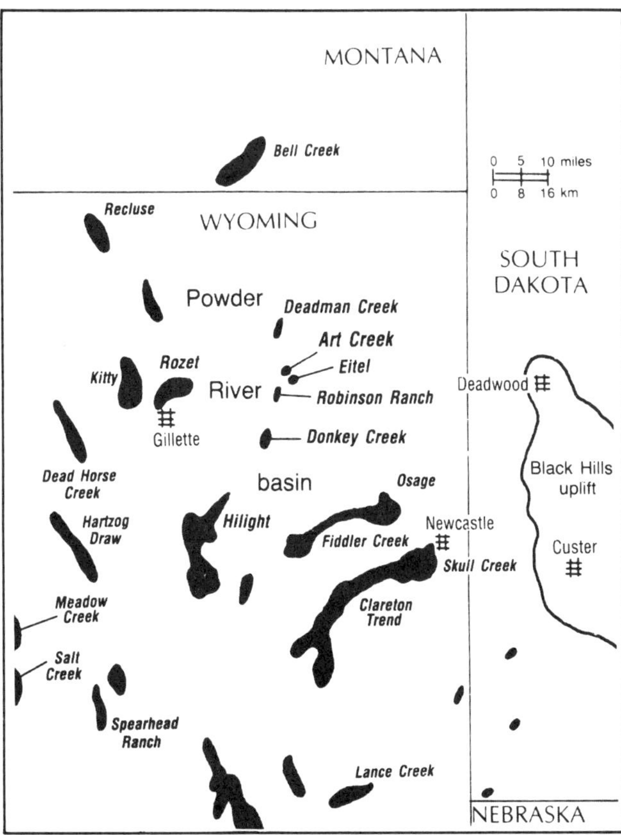

Fig. 10–5 Powder River basin today (courtesy *OGJ*, December 14, 1981)

Exploratory and development work in this big basin always has been busy. Not a year has gone by without a major oil find in the basin. Most of the obvious structures on the southwestern and southeastern rims of the basin had been drilled by World War II. But drilling really peaked in the 1950s when there was an avalanche of Cretaceous Newcastle-Muddy, Fall River, and Parkman discoveries. These finds were followed by a wave of Muddy discoveries during the 1960s and 70s when giants like Hilight field were developed. This search continues today, although at a slower pace.

The first upper Minnelusa success was made in 1957 in the Powder River basin. But Minnelusa drilling really did not get swinging until the early 1960s. Then, after 1967, another wave of Muddy-Newcastle exploration swept the Powder River basin scene. But the interest in Minnelusa prospects has not floundered. There still will be more fields located in the basin within this pay zone.

The Cretaceous system, of course, has accounted for the lion's share of Powder River basin oil and still has the greatest potential for future discoveries of size. Most of these large Cretaceous fields in later years have been found in stratigraphic traps on the gently dipping eastern flank of the basin.

There always will be oil to be found in the Powder River basin. Interest in the basin has shifted even into extreme southwestern South Dakota in Fall River County where Pennsylvanian Leo sands have been found to be quite productive.

The Big Horn basin

Though this basin is only 3,000 sq miles, it is the biggest producing basin in the Rocky Mountains. This was true in 1953, as shown in Fig. 10–6, and is still true today.

The Big Horn basin, in its earlier days, saw many advances in exploratory techniques. Structures with shallow Cretaceous fields were drilled deeper to productive Permian, Pennsylvanian, and Mississippian rocks and were found to show increasing closure with depth. The discovery of Bonanza field in 1950 proved surface closure is not needed for closed traps at depth. Sage Creek-Tensleep, discovered in 1952, proved that the tilted-water-table theory of exploration has practical application.

Geologically, the Big Horn basin is a well-defined topographic, structural depression in northwestern Wyoming (Fig. 10–7). It is bounded on the northeast by the Big Horn Mountains, on the south by the Owl Creek-Bridger Mountains, on the west by the Absaroka Range, and on the northwest by the Bearstooth Mountains. There is no topographic break northward into Montana through a thin neck of the basin bounded on the east by the Pryor uplift.

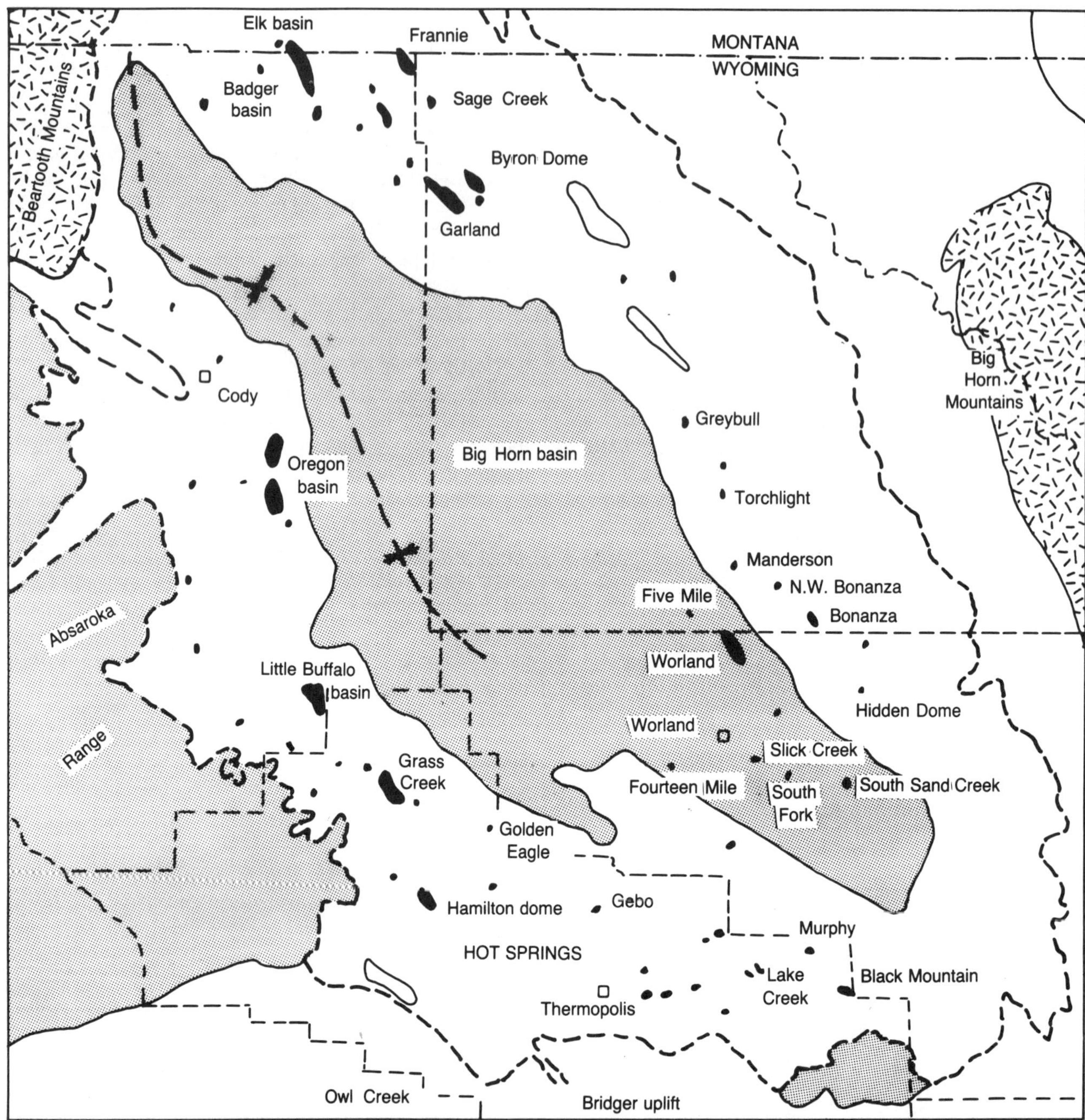

Fig. 10–6 The Big Horn basin area

In an east-west direction across the Big Horn basin, the strata dip rather gently from the Big Horn Mountains toward the axis of the basin and rise comparatively steeply onto the volcanic mountains to the west. There are two prominent lines of surface-expressed folding in the basin, which include well-defined anticlines roughly paralleling the bordering mountain ranges. With few exceptions, the fields are restricted to the edges of the basin on these series of satellite anticlines. In recent years, fields have been found by seismic means off these anticlines. This is indeed where most of the new oil will be found in the years ahead in the Big Horn basin.

Sediments in the basin range in age from Recent through Cambrian with the maximum sedimentary section much thicker than orginally assumed. First estimates were about 17,000 ft of sediments. No doubt there will be more

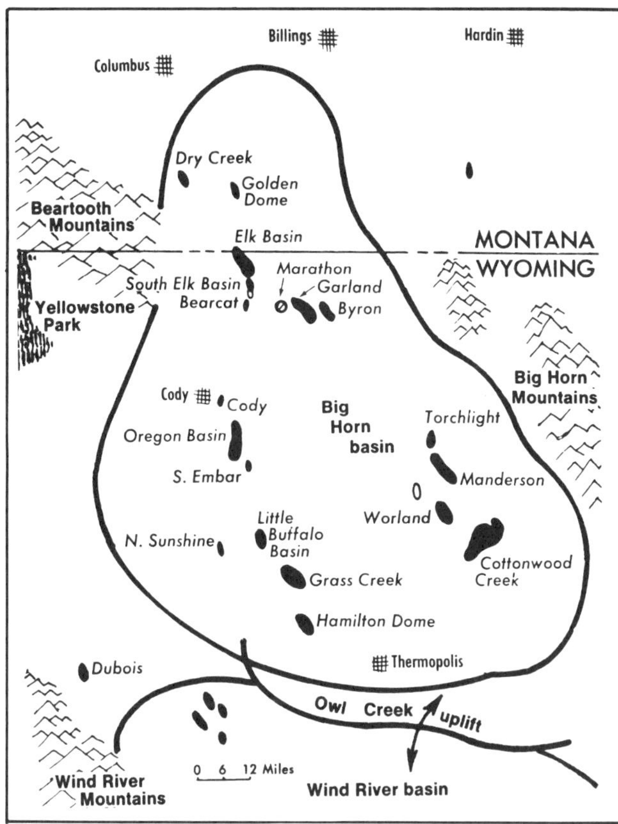

Fig. 10–7 The active Big Horn basin (courtesy *OGJ*, December 11, 1978)

structural accumulations found under the masking pyroclastic rocks and thrust sheets on the west side of the basin. New geophysical work and methods will find some oil-bearing structures under the Tertiary rocks of the basin interior. Each new deep hole in this small but prolific Wyoming basin is an important step toward keeping the Big Horn the largest producer in the Rockies.

Many large fields—Elk Basin, Frannie, Byron, Worland, Bonanza, Grass Creek, Little Buffalo Basin, and Oregon Basin—contribute toward making the Big Horn such a prolific oil basin. Another important field is Cottonwood Creek, the first major discovery in Wyoming drilled specifically as a stratigraphic-trap prospect. The discovery well was completed in 1953 by Stanolind Oil & Gas on a structural nose near a dolomite-to-shale facies change. Geophysics, deeper drilling, and determined exploratory work will uncover some more large oil and gas fields in the Big Horn basin.

Other Wyoming basins

The Laramie, Hanna, and Shirley basins of southeastern Wyoming are sleepers that need more work and

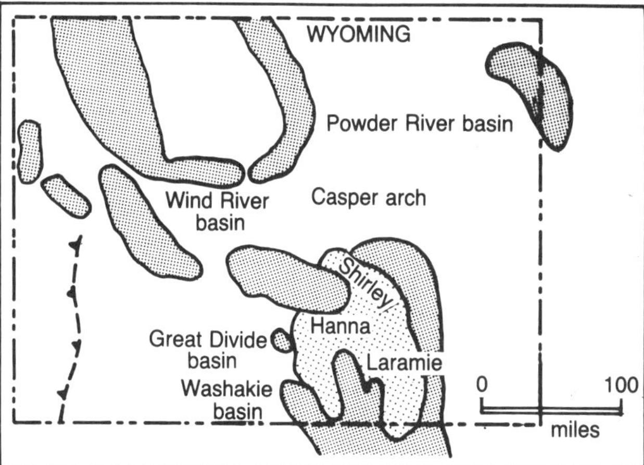

Fig. 10–8 Wyoming basins (courtesy *OGJ*, October 1, 1979)

examination. The three basins are small contiguous intermontane basins covering about 5,000 sq miles in parts of Carbon, Albany, and Natrona counties (Fig. 10–8).[1] Although the rocks in these three basins are equivalent to units producing in active basins to the west, northwest, and east, the last big exploration campaign here took place in 1960.

The first field in the area was found in 1918 at Rock River in the Laramie basin. This oil field until 1979 had cumulative production of 45 million bbl of oil from the Cretaceous Frontier, Muddy, Dakota, Lakota, and Jurassic Sundance. There are 31 fields in the three-basin area with a cumulative production of more than 80 million bbl of oil and 12 billion cu ft of gas.[2]

Most of the oil comes from the Early Cretaceous in the Muddy, Dakota, and Lakota sands. All production in the three basins is from either structural or combination traps. The potential does exist in the Laramie, Hanna, and Shirley basins, and many new oil fields should be found in the region. Three types of traps are present: structural, stratigraphic, and combination.

THE DENVER BASIN

If the industry sits still long enough to write a book on its successful exploration in the U.S., the Denver basin will certainly merit a chapter all its own (Fig. 10–9). One of the best success stories in the business has been written on the plains of northeastern Colorado, southwestern Nebraska, and southeastern Wyoming. Denver basin is a busy, extremely successful oil and gas province. From 1949 when it all began at Gurley field in Nebraska to the mid-1960s, more than 17,000 wells have been drilled in the three-state basin. This exploratory campaign scored highly with a total of about 800 million bbl of ultimately recoverable oil reserves.

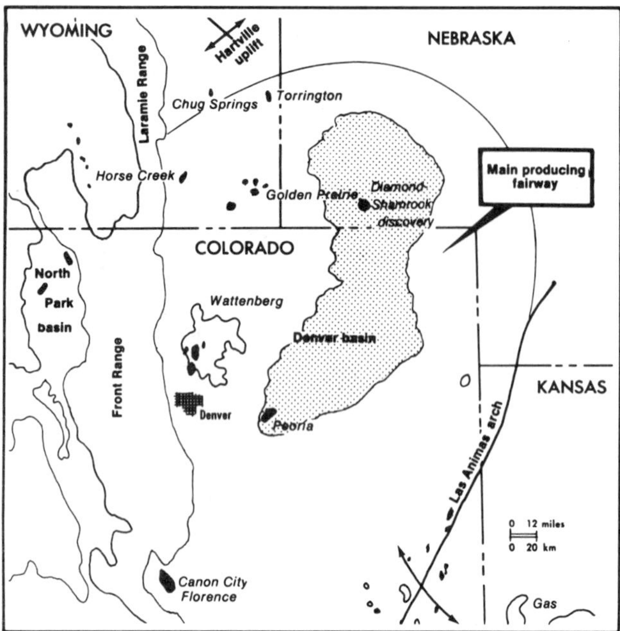

Fig. 10–9 The Denver basin (courtesy *OGJ*, January 19, 1981)

After the mid-1960s the basin activity simmered a bit, but this slowdown was short-lived. Drilling east of Denver and along the Union Pacific Railroad in the early and mid-1970s poured new reserves into the till. Drilling and exploration work fanned out in all directions into nearly every corner of the basin, as thousands more wells were completed.

The Denver basin is about 360 miles long and stretches about 20 miles at its greatest width. Though the basin sits directly in front of the Front Range of the Rocky Mountains, it is considered to be in the Great Plains physiographic province. It is limited on the west by the Front Range and Laramie uplift, on the northwest by the Hartville uplift, and on the north by the Black Hills.

Topographically, the basin has no other boundaries, but geologically it is bounded on the northeast by the Chadron arch and on the southeast by the Las Animas arch. The basin is asymmetrical with a very gentle east flank and a very steep west flank where the rocks climb abruptly, almost vertically onto, or are overridden by, the Front Range uplift.

Most Denver basin production is from Cretaceous rocks. Major oil and gas production comes from the Dakota group series of sands. Production appears along definite trends based on both structure and permeability changes. Production also is found on east-west-trending structures and westward-nosing structures where proper sand conditions are present. Many fields are located on terraces.

Denver basin production goes back many years to 1862 when the Florence field was found north of Canon City, Colorado. That Pierre shale Cretaceous pool came just 3 years after the discovery of oil in Pennsylvania by Colonel Drake at Titusville. But years passed before new oil was found in other parts of the Denver basin like the Permian Lyons pools on the west flank near the Front Range. Then came Gurley, and hundreds of Cretaceous pools popped up.

Another recent target in the basin is the Niobrara, also of Cretaceous age. But this sand was considered uneconomical to produce, until gas prices forced a new look at the tight nature of the sand. Several fields have been discovered in the past few years with Niobrara pay, and the search continues.

Very few pre-Cretaceous wells have been drilled in the Denver basin. In 1980, Diamond Shamrock drilled an 8,000-ft wildcat in Nebraska, 15 miles west of Sidney. The well found oil in the Paleozoic at 7,055 ft. The nearest similar oil was at Pierce, Black Hollow, and New Windsor fields in western Weld County, Colorado, near the Front Range.

Some Paleozoic oil exists in the Wyoming part of the basin. Deeper exploration will no doubt find more such pools in the Denver basin in the future. Meanwhile, more Cretaceous oil and gas fields will be found with no end in sight.

A definite Mississippian oil play is shaping up in southeastern Colorado in the midst of an already established Pennsylvanian program (Fig. 10–10). Several companies are developing new oil and gas pools in Pennsylvanian and

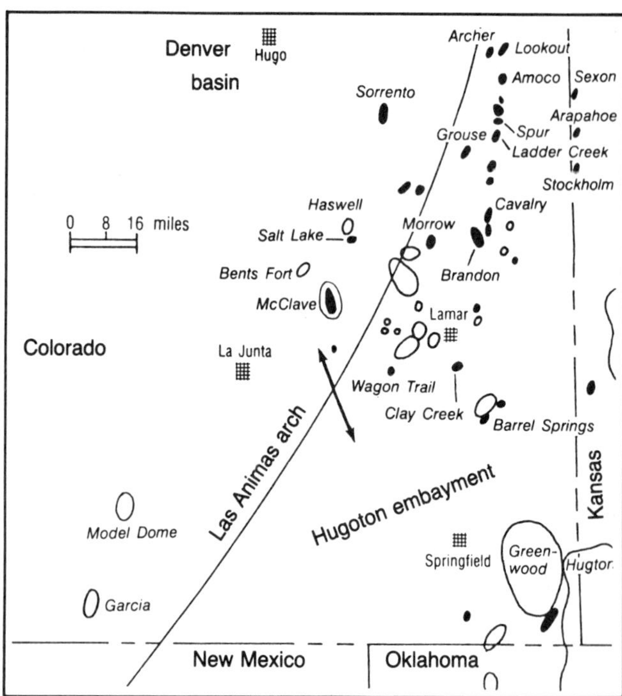

Fig. 10–10 Southeast Colorado exploration areas (courtesy *OGJ*)

Mississippian reservoirs. This seems to be one of the busiest periods of drilling in southeastern Colorado. There have been 11 new Morrow oil and gas fields discovered in the region since early 1980, making this a significant period of history for the Las Animas arch sector. One of these new strikes flowed at an initial rate of 700 b/d of oil and 750 Mcfd of gas from the Morrow sandstone. The trapping mechanism here, as in most new pools in the area, is combined stratigraphic and structural, with structure taking a secondary role. The depositional history of Morrowan sandstones in southeastern Colorado was strongly influenced by regional tilt into the Anadarko basin established at the close of Mississippian time.[3]

Mississippian rocks were subject to truncation in the area of the Las Animas arch and development of Karst topography on the erosion surface. Later transgression of the sea from the southeast in Morrowan time resulted in the deposition of a sandstone and shale sequence.

Distribution of elongate sandstone bodies is generally erratic and meandering in the area, leaving many areas unexplored in southeastern Colorado. Recent Morrow discoveries force the explorationist to take another look at this part of the upper reaches of the Anadarko basin and along the Las Animas arch. The arch is the main structural feature in southeastern Colorado. This section of the state has been explored for many years. There was a big Mississippian oil play in the 1960s as a result of the Brandon field discovery. Then there was the McClave field with its Pennsylvanian Morrow production. In October 1981, cumulative Morrow production was 37,001,474 Mcf of gas and 112,458 bbl of oil at McClave. Brandon is on the Brandon axis and McClave on the Eads axis, which also includes Wagon Trail and Beta fields.

New discoveries since early 1980 along the arch area are Sorrento, Grouse, Stockholm, Haswell, and Clyde fields. Stockholm is in Greeley County, Kansas. In May 1982, Pintail Petroleum confirmed Spur field in eastern Cheyenne County at 1 Champlin-Cobb, NE SE 7-14s-44w. This well pumped 217 b/d of oil from open hole in the Spergen at 5,453–5,480 ft. The new well offsets to the northwest the discovery well of the field, 1 Price in 8-14s-44w, completed in 1981 on pump for 35 b/d of oil in the Spergen at 5,453–5,462 ft. This well is located 1 mile southwest of Mississippian oil at Cheyenne Wells field.

In March 1982, Texas Oil and Gas completed a confirmation to Salt Lake field in Kiowa County at 1 Frazee-B in 4-19s-50w. The well pumped 90 b/d of oil from the Morrow at 5,167–5,176 ft. This well lies ¾ mile northwest of the discovery well at 1 Shotton in 10-19s-50w, completed in 1981, flowing 617 Mcfd of gas from the Morrow at 5,131–5,142 ft. Location is also 3½ miles southeast of Haswell field, which produces from the Morrow.

In December 1981, Champlin had a good well at 1 Pelton 41-31 in 31-12s-44w, Archer field. It flowed 1,533 Mcfd of gas and 1 bbl of oil from the Morrow at 5,314–5,318 ft. To the northwest of the mapped area in Elbert County, Samuel Gary Oil Producer had a small discovery in late 1981 at 13-16 Amoco in SE SE 13-6s-64w, 14 miles east of Parker. The well pumped 7 b/d of 38° oil from the Fort Hays at 7,810–7,826 ft, a Cretaceous zone and not connected with the current Paleozoic interest in southeastern Colorado. However, it is indeed a remotely productive area of eastern Colorado.

The search for Mississippian oil certainly is not new to the plains country of southeastern Colorado. Oil was first found in the Mississippian here at the Comanche field in 1964, but the Brandon discovery in 1965 was the spark that set things moving. In 1968 it was the fourth largest field in daily output in all of Colorado. Drilling action that year netted five indicated discoveries: three in the Mississippian rocks, two in the Pennsylvanian zones.

The major tectonic element is the Las Animas arch. This arch has relatively low relief. It is an Upper Paleozoic structural link between the Sierra Grande uplift to the south and the Sioux uplift to the northeast. The absence of Mississippian rocks on these uplifts, largely by truncation, is a reflection of the magnitude of the uplifts.[4] Mississippian rocks also are absent over the Central Kansas uplift because of similar pre-Pennsylvanian tectonics. Until 1968 only about 400 wildcats had penetrated the entire Pennsylvanian section in this area of southeastern Colorado and western Kansas. The McClave field, discovered in 1951, was the first Morrow field in the area. It is stratigraphically entrapped in a lenticular sandstone in the Morrow shale.

The newest round of exploration in this interesting area of the Las Animas arch country of southeastern Colorado should uncover some more oil and gas pools in the Mississippian and in the Pennsylvanian. There are bound to be more Brandon fields along this arch; only drilling will find them.

THE UINTA BASIN

The Uinta basin of northeastern Utah—harboring major oil reserves in two giant fields, Altamont-Bluebell and Greater Red Wash—was the last of the big basins in the Rockies to become an oil producer. The search for oil here began in 1944, even though seeps had been known for many years. One reason the remote basin was bypassed earlier was the idea that oil was to be found only in marine sediments. In fact, this idea was in vogue until the completion of the discovery well of the first field in Uinta basin at Roosevelt in 1949. Production was found in a fractured zone of the Eocene-Green River, a laminated, impure carbonate of lacustrine origin. This discovery rapidly dispelled the marine theory.

The Uinta basin is a structural, topographic basin in northeastern Utah, about 100 miles north-south and 130 miles east-west (Fig. 10–11). It is bounded on the north by the east-west-trending Uinta Mountains, on the east by the Douglas Creek arch, which separates the Uinta and Piceance Creek basins, on the southeast by the Uncompahgre uplift, on the southwest by the San Rafael swell, and on the west by the north-south-trending Wasatch Mountains. The basin is asymmetrical. Its axis runs east-west just south of the Uinta uplift, making the major portion of the basin a northward-dipping monocline.

The most important formations in the basin are the rocks of Tertiary-Eocene age. These sediments were deposited in a lake that existed from early Paleocene time through the Eocene and into early Oligocene. During Green River time, this lake reached its maximum development and deposited highly organic sediments that were to become the much-publicized oil shales of the area and were the source of the abundant Gilsonite veins in the eastern and central parts of the Uinta basin. The Uinta and Wasatch formations, overlying and underlying the Green River, respectively, are mostly nonlacustrine throughout most of the basin area.

Sediments in the basin range from Oligocene through Cambrian in age. In the axis of the basin, these sediments may reach a thickness of 25,000–30,000 ft. Pre-Tertiary rocks crop out around the edge of the basin, and in this area the Pennsylvanian Weber sandstone has been oil-bearing at Ashley Valley and big Rangely field. Oil and gas fields in the Uinta basin produce from Tertiary rocks of Eocene and Paleocene age, with some Cretaceous-age gas found in the southern part of the basin.

The discovery well at Roosevelt was a Carter Oil and Stanolind Oil and Gas strike in 1949 on a seismic anomaly. It flowed 1,633 b/d of 35° gravity oil from the basal Green River formation at 9,400 ft. The oil is dark green to black and has a pour point of 90° F.

The second big discovery in the basin was the Red Wash field. Red Wash was opened by the California Com-

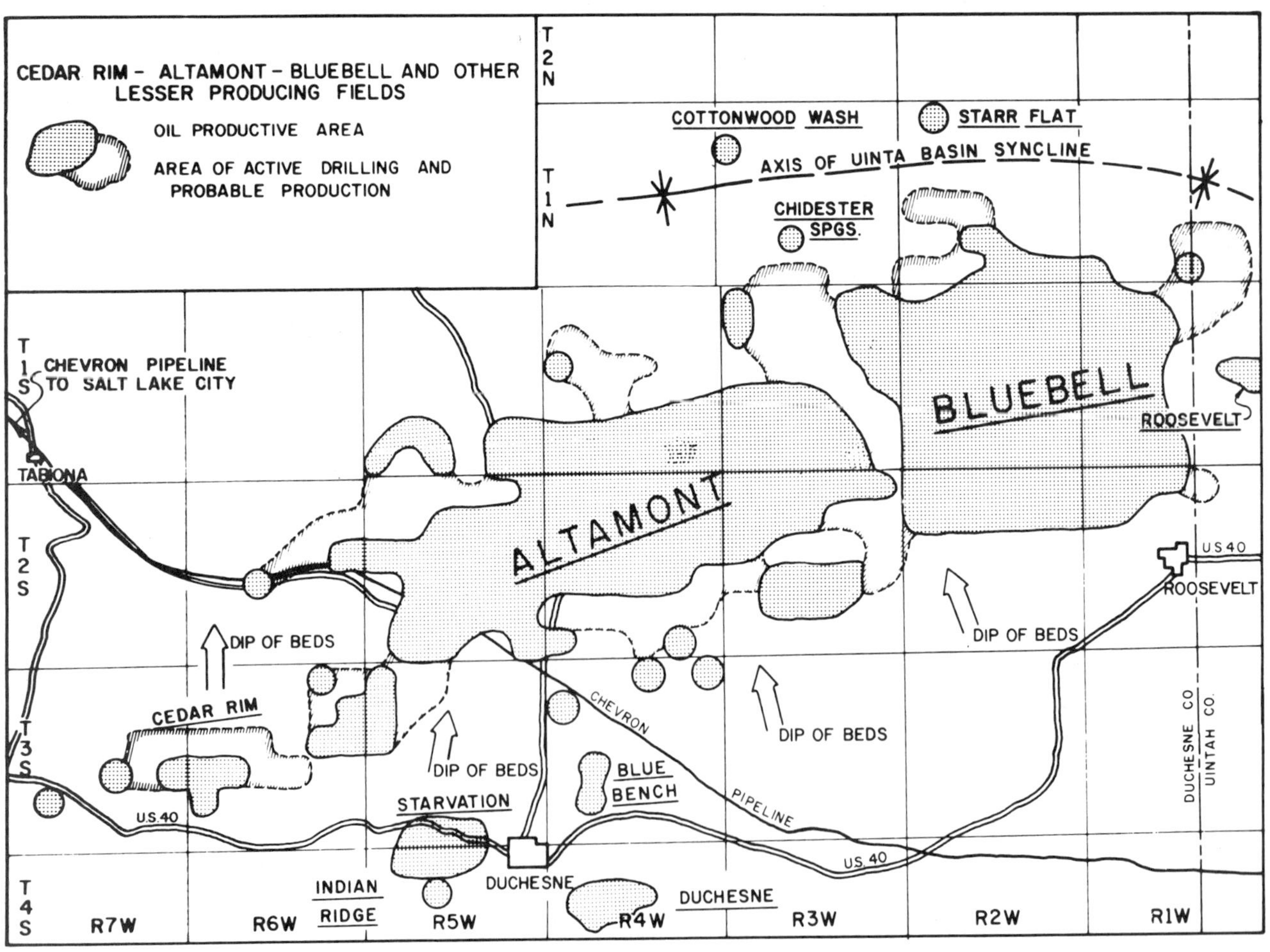

Fig. 10–11 Utah's lesser-producing fields (courtesy Utah Geological and Mineral Survey *Quarterly Review*, February 1974)

pany in 1951. The field produces from a series of thin sandstones in the basal Green River formation at about 5,500 ft. Crude is similar to that of Roosevelt, and like Roosevelt, the field is on a nose. This nose, however, is better developed than the one at Roosevelt and plunges westward.

In the early 1970s, fast-moving developments took place in the Uinta basin. The result was the Altamont-Bluebell complex (Fig. 10–11). Three fields—Bluebell, Altamont, and Cedar Rim—began to grow together, resulting in the big field known as Altamont-Bluebell. Some call the area Greater Altamont after the central field. The structure of the area is simple.[5] At the surface, formations dip gently at 1½–3° to the north and northwest toward the axis of the Uinta basin. Because the formations thicken from south to north as the basin is approached, the dip increases gradually with depth but probably does not exceed 6–8° at 15,000–18,000 ft. The total sedimentary section of Tertiary, Mesozoic, and Paleozoic rocks in the deepest part of the basin probably is 30,000–32,000 ft.

Shell, discoverer of Altamont field in 1970, and Chevron—discovery well operators at Bluebell in 1967 and 1971, respectively—were pacesetters in 1973. At the beginning of 1973, 90 wells were producing 9,000 b/d of oil. In 1974 the field made 1.3 million bbl/month. The field now makes 7.5 million bbl/year, has produced 104.4 million bbl of oil since its opening, and has 260 million bbl of oil yet to be produced. Uinta basin exploratory work has slowed in recent years, but the industry still needs to look for new prospects along the edges of the basin, particularly in the southern and southeastern sectors.

THE PICEANCE BASIN

The Piceance basin of northwestern Colorado, a neighbor of the Uinta basin, has a long history of slow development, although its gas potential has been recognized for many years. The basin covers about 7,000 sq miles (Fig. 10–12). The boundary of the Piceance basin, home of the one giant at Rangely, is defined as the outcropping base of the Mesaverde group, except for the western margin where the Colorado state line provides a boundary. Although the Douglas Creek arch may be separated as a distinct tectonic unit, it can be considered as a part of the Piceance basin.[6] Producing reservoirs, entrapping mechanisms, and hydrocarbon production are similar and correlative.

Significant developments in the basin did not really take place until the discovery of the Rangely field in 1933. Then in 1955, the Pacific Northwest Gas Pipeline was completed into this area, providing market for new gas reserves.

There are rocks from late Tertiary age to Cambrian in the basin, but the Jurassic Morrison reservoirs and those of

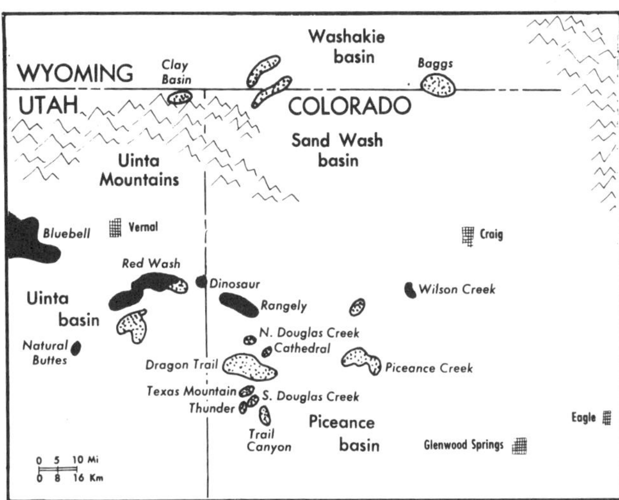

Fig. 10–12 The Uinta-Piceance basin (courtesy *OGJ*, April 3, 1978)

younger ages have proven to be the only producers. Oil from the Jurassic Entrada has been found just below the Morrison in structural traps around the edge of the Piceance basin.

Potential production depths range from 2,500–15,000 ft. Drilling has been confined primarily to the shallow portions of the basin in areas more accessible or nearer existing pipelines. The basin has a good future, since no more than 10% of the area has been tested. This region needs an extensive wildcat program, provided by improved economic incentives and technological advances in drilling and completion methods.

The Kaiparowits basin

There's a small basin tucked away in the southern part of Utah where wildcats are really wildcats (Fig. 10–13). This basin is Kaiparowits. It first drew industry attention in 1962 when Tenneco Oil found Permian Kaibab oil at 2 Upper Valley unit in Garfield County, climaxing a long history of dry holes in the region. The discovery well rated 300 b/d of oil. Its success triggered a leasing campaign that gobbled up 750,000 acres in a five-county area in some of the wildest country in the union. Nearly all the known anticlinal features, previously drilled or not, were leased. Tenneco had completed two more wells at Upper Valley (1965) to establish the basin as a new oil province. But little work has been done in the region since.

The basin had a tedious entry into the exploratory scene. The California Company had the first oil show in the basin in 1948 when it found oil in three formations in the Upper Valley anticline's first test. Oil was reported in the Timpoweap member of the Triassic Moenkopi formation

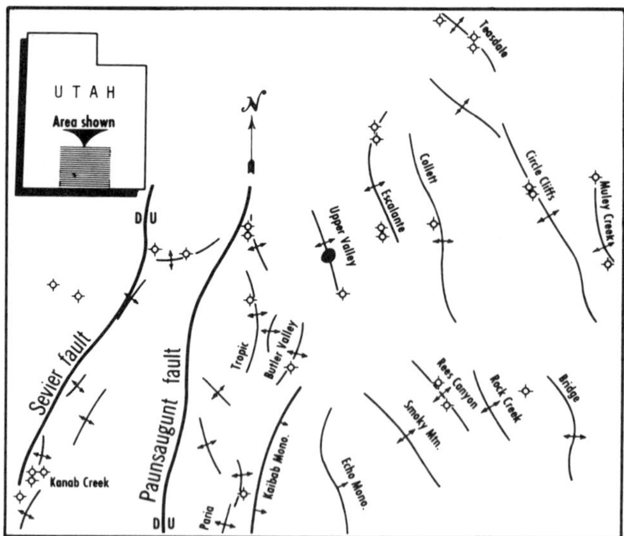

Fig. 10–13 The Kaiparowits basin

and from the Mississippian Redwall limestone. These shows led to other tests to Mississippian on the Johns Valley, Teasdale, Muley Creek, Circle Cliffs, Rees Canyon, Kanab Creek, Collett, Bryce Canyon, Kaibab, and Escalante anticlines. No fields were found, but shows were good enough to make the hunt more interesting.

The industry needs to take another look at this vast area of rugged canyons, eroded wonders, and rough country. Very few tests have been drilled in the 7,500-sq-mile basin characterized by large open folds in a desolate and sparsely populated area east of Bryce Canyon National Park. There has been some production at Virgin and Medicine Hat for many years. But the Tenneco discovery at Upper Valley gave exploration people a new basin to think about. This was also the first important Permian oil production in Utah. The nearest big oil is 125 miles east in giant Greater Aneth field in the Paradox basin.

The Kaiparowits basin occupies the central part of Kane and Garfield counties and is separated from the Paradox basin by the Circle Cliffs upwarp. It is contiguous with Arizona's Black Mesa basin. In fact, the two are quite similar in many respects. Geographically, they are separated by the canyon of the mighty Colorado River.

Geologically, the stratigraphy and structure of the basin are encouraging for oil and gas exploration. The Kaibab is a natural to look for, but the Timpoweap is usually oil stained and is a member of the Moenkopi where oil and gas are bound to be found. And the Mississippian Redwall dolomite should not be overlooked in future exploratory plays here. Because of its vuggy to cavernous porosity and its production at Big Flat in the Paradox basin, the Redwall has a good future.

THE PARADOX BASIN

The Shell discovery in 1954 in the Desert Creek area of the Blanding-Sage Plain structural basin substantiated the concept that the large marine Paradox evaporite basin held promise of major oil fields.[7] This strike in the Paradox basin was followed by other San Juan County, Utah, discoveries. Extensions of the Desert Creek field, the discovery of the East Boundary Butte field in 1954, North Boundary Butte in 1955, Texaco's Aneth in 1956, plus the Hunav field strike by Humble in the same year set the stage for an aggressive wildcat and development program that swept the southeastern Utah area for some time. Several large fields resulted, but the big one was Greater Aneth. This field produces 6.7 million bbl/year of oil. Cumulative production is 313.4 million bbl of oil with about 55 million left to be produced. One wonders why it took so long for interest to grow in the prospects of the Pennsylvanian rocks of this basin. Large amounts of oil were found in 1929 at the Rattlesnake field in San Juan County, New Mexico.

The Paradox is one of the three large basins in the Four Corners region (Fig. 10–14). The Paradox geosyncline and its associated shelves cover more than 40,000 sq miles of territory in Utah, Colorado, Arizona, and New Mexico.

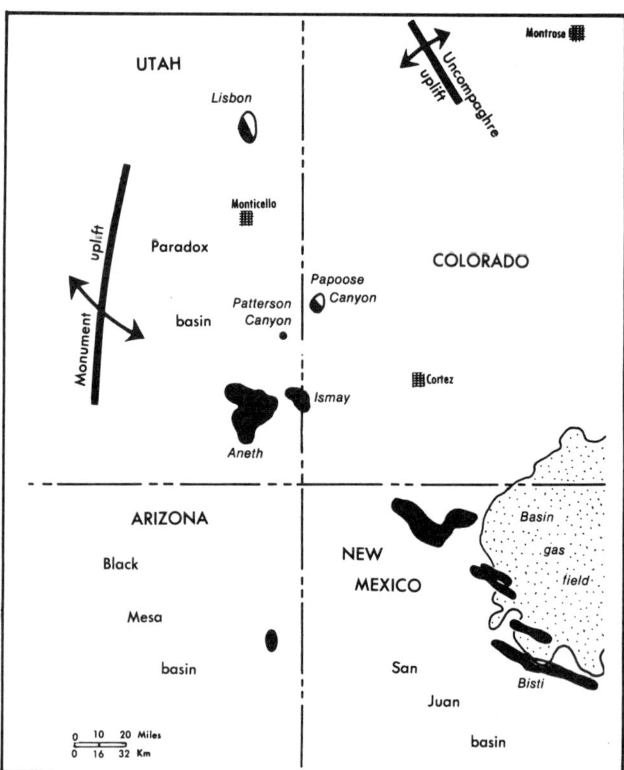

Fig. 10–14 The Four Corners (courtesy *OGJ*, October 15, 1979)

This basin is genetically limited by the Uncompahgre-San Luis-Penasco uplifts on the west and southwest.[8] Other basins in the region are the Black Mesa and San Juan.

Aneth always has been the big oil field in the Paradox country. Geologists feel there are more fields of this class yet to be found, though perhaps not quite as large. To date, the Pennsylvanian sequence in the basin contains the most prolific reservoirs, which are porous, mound-like algal limestone.

Large reserves are seen for the basin with some estimates as high as 30 billion bbl in place and unfound. Oil will be found in the Devonian, Mississippian, Pennsylvanian, and Permian sediments. Stratigraphic traps provide most of the present reserves.

The basin contains many paleostructures and associated unconformities. Numerous facies changes also are in the Paleozoic section. These may harbor home for numerous reservoirs and trapping mechanisms. A large Mississippian reservoir also is in the Paradox basin at the Lisbon, Utah, field. This is the Leadville formation, found at Lisbon in 1960. Lisbon has produced more than 40 million bbl of oil and large quantities of gas. The discovery of Mississippian production at Lisbon set off a flurry of exploratory action in the northern part of the basin, but little came of it. But the Leadyille can certainly be considered a promising target for future exploration. It is difficult to explore for this formation due to the thick Pennsylvanian salt section that overlies it. Further studies and new techniques may open the Mississippian reservoirs.

Other parts of the Paradox basin need to be explored. Over a hundred townships in Emery, Grand, Wayne, Garfield, and San Juan counties in Utah may be harbingers of hydrocarbons. This western Paradox basin area lies in a structural saddle between the San Rafael swell on the northwest, the Monument upwarp on the southeast, the Paradox fault-fold belt of salt anticlines on the northeast, and the Henry Mountains on the southwest.[9] This region is considered a good place to drill for oil and gas because of its stratigraphic section, tectonic history, and location straddling a zone of diverse marine regional tilts throughout geologic time.[10]

THE OVERTHRUST BELT

The December 1974 discovery of the Pineview oil field in northeastern Utah's Summit County set off an oil and gas play that belittles other modern-day exploratory campaigns in the U.S. At least 18 oil and gas fields have been discovered along a 50-mile productive fairway in northeastern Utah and southwestern Wyoming (Fig. 10–15). This region is becoming a legend before its time. In 1981 just 6 years after the Pineview discovery in Utah, the trend's potential was being compared with that of Prudhoe Bay in Alaska. But this vast region, stretching from Canada to Mexico, remains 95% unexplored. One of the reasons for so little drilling is that the U.S. government has been slow to permit drilling in wilderness areas or in areas that may be so designated.

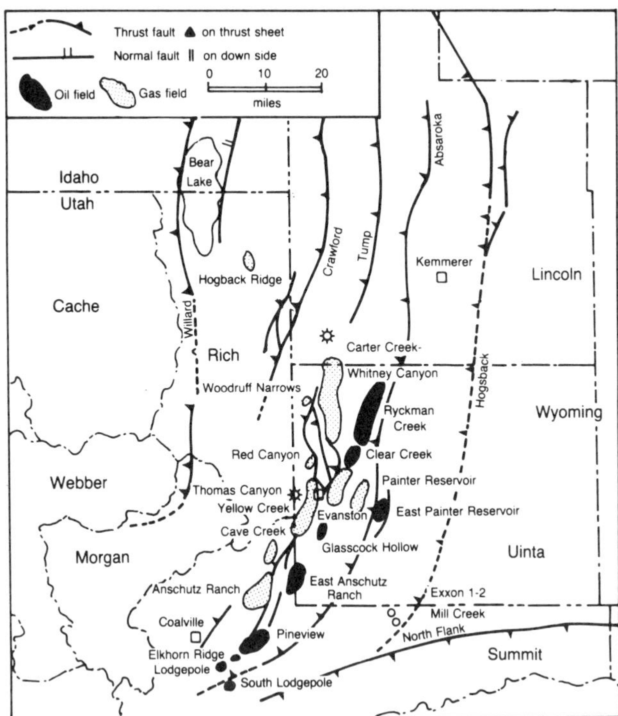

Fig. 10–15 The developing Overthrust Belt fairway (from *IPE*, courtesy Utah Geological and Mineral Survey)

The thrust belt is one of the most challenging exploration theaters in the world today. The complicated geology that provides so many traps for oil and gas also puzzles seismic workers and plays havoc with drilling operations.

The Overthrust Belt runs from Alaska to Central America. It is believed to be a result of a collision of the Pacific and North American plates. Sandstone and limestone layers were thrust up and over or down and under other strata. Many geologists think the main movement was down and under and prefer to call it the thrust belt. The northern part of the belt is called the Disturbed Belt. South of the fairway fields in Utah, the Overthrust Belt joins a separate geologic feature known as the Hingeline, a region of few wildcats but plenty of interest.

The complicated geology makes exploration a tough and expensive job. Rough terrain pushes seismic costs in some cases to $20,000–25,000/line mile. Using copters can double the cost. Three-dimensional coverage costs as much as $180,000/sq mile. Hard rocks, steeply dipping formations, and Jurassic salt snarl drilling operations. But this tremendous effort has paid off handsomely with fields established along the fairway and two in Montana. In 1981, the USGS estimated resources of the Idaho-Wyoming-Utah segment of the thrust belt at 2.7–13.3 billion bbl of oil and 29.1–105.4 trillion cu ft of gas. Mean resource estimate was 6.7 billion bbl of oil and 58.4 trillion cu ft of gas.

The fields

At least six of the thrust-belt fields are giants with reserves of more than 100 million bbl of oil or 1 trillion cu ft of gas. Fairway fields are still in various stages of development. These fall generally into two anticlinal trends running northeast-southwest. The eastern trend is oil; the western has gas.

A late 1980 study by Amoco, one of the main operators in the region, divided 13 fairway fields into five categories, estimating discovered potential reserves in those fields at 10 trillion cu ft of gas and 900 million bbl of liquids. It defined the Pineview, Ryckman Creek, Painter Reservoir, and Clear Creek fields as Jurassic Nugget oil fields, almost fully developed at the time of the study.

The Yellow Creek and Anschutz Ranch fields produce mainly sweet gas from the Jurassic Twin Creek. Yellow Creek includes what was formerly called Evanston field. The big Carter Creek-Whitney Canyon complex is a Paleozoic sour-gas field. The Cave Creek field and the Amoco 1 Urroz Working Interest Unit deeper pool discovery in the Yellow Creek field produce sour gas from deep formations. The East Painter Reservoir, East Anschutz Ranch, and Glasscock Hollow fields produce retrograde condensate. Production along the fairway is still limited by lack of outlets and processing equipment.

Wildcatting continues in the Idaho portion of the Overthrust Belt, but no discoveries have been made despite a series of expensive and deep holes. The thrust belt runs along the state's eastern border and contains a number of large, seismically defined structures. Geologists believe this region has all it needs for oil and gas discoveries.

Oil and gas discoveries in Montana's Disturbed Belt set off a flurry of wildcat action in the western part of the state a few years ago, and this activity continues today. The Montana segment of the Overthrust involves the Lewis overthrust in the north and the Eldorado and Lombard overthrusts in the south. The thrusts brought Precambrian belt supergroup strata east onto Paleozoic and Mesozoic rocks during the Laramide orogeny of late Cretaceous and early Tertiary time.

The Disturbed Belt, a zone of thrust-faulted and folded stata ranging in age from Precambrian to early Tertiary, lies east of and in the lower plate of the overthrusts. The best reservoir and source rocks probably are in the Cretaceous, but there are potential Paleozoic traps, too. Thrusts in the western part of the belt have displacements of several miles and may conceal major structures beneath the thrusts. There will be more discoveries in the Disturbed Belt like Blackleaf Canyon and Two Medicine Creek.

In 1979, Williams Exploration found gas in the 1-19 Two Medicine in Glacier County, Montana. The well flowed 6.6 MMcfd of gas and 90 b/d of condensate from the Mississippian Sun River at 8,934–8,980 ft. Two offsets did not do so well. In 1980, Williams found the first modern oil discovery in the Montana thrust belt at 1-11 Two Medicine. This well pumped a modest 145 b/d of oil from the Cretaceous Greenhorn at 7,142 ft. The 1-3 well made 131 b/d of oil, confirming East Glacier field. In Teton County, Williams discovered Backleaf Canyon field at 1-8 Blackleaf-Federal. The flow was 5.1 MMcfd of gas from the Sun River at 5,256–5,412 ft. Other wells will follow. Meanwhile, wildcatting continues to probe the Overthrust Belt sector of western and southwestern Montana. A discovery would open up a huge new link in the thrust belt.

History

The first North American production on the Overthrust Belt was found in Alberta in 1918 at Turner Valley field. This gas discovery set off an extensive exploratory effort resulting in 32 fields with thrust-faulted reservoirs, containing 9.3 trillion cu ft of initially recoverable reserves, 143 million bbl of natural gas liquids, and 132 million bbl of oil.[11] Extensive seismic surveys by Shell Canada led to the 1957 discovery at Waterton Parks, 135 miles southwest of Calgary. Reservoir rocks of the Alberta Foothills Belt are Paleozoic carbonates of Upper Devonian and Mississippian age. They contain about 8% and 87%, respectively, of the belt's estimated reserves.

Geology

Because of common characteristics, the Canadian thrust belt is considered to be a geologic cousin to the Overthrust Belt in the U.S. Similarities include general structural configuration, trap types, reservoirs, stratigraphy, timing of migration of hydrocarbons, depth of burial, and age of tectonic transport.[12]

More than 500 dry holes were drilled in the Overthrust Belt before the discovery in 1975 at Pineview in Summit County, Utah. Amoco opened Ryckman Creek field 40 miles north in 1976 in Wyoming. Yellow Creek, another Amoco find, was opened the same year. This discovery indicated the possibility of a whole series of fields between Pineview and Ryckman Creek. Four fields had been opened by 1977, and the heat was on in the Overthrust Belt. By

1980 more than 60 MMcf of gas and 29,000 b/d of oil were being produced in the field.

From Canada, the belt extends into northwestern Montana where it curves to the west around the south side of the Boulder batholith and continues south-southwest into Idaho. Geologically, the Idaho-Wyoming-Utah portion of the Overthrust Belt is defined as the region south of the Snake River Plain and north of the Uinta Mountains.

All but one of the fields on this fabulous new fairway produce from reservoirs occurring in anticlinal closures on the folded, leading edges of the Absaroka or Tunp thrusts. The exception is in northern Utah in an anticlinal trap on the older Crawford thrust. All of the production to date has been found on the upper thrust sheets; however; geologists feel there is great hydrocarbon potential present in deeper, repeated thrust sheets.

There are two definite anticlinal trends defined: the Pineview-Ryckman Creek (where oil production is limited to reservoirs in rocks of Mesozoic age) and the Whitney Canyon-Carter Creek Paleozoic gas trend. Production has been found in many pays, including the Jurassic Twin Creek limestone, the Jurassic Nugget sandstone, the Triassic Thaynes limestone, and the Dinwoody and Woodside formations. Production also is in the Ordovician Big Horn and in the Cretaceous Kelvin limestone.

The Hingeline of Utah is closely associated geologically with the Overthrust Belt. Howard Ritzma of the Utah Geological and Mineral Survey notes that the Hingeline began to take shape more than 500,000 million years ago with subsidence of the earth's crust west of a narrow zone of weakness running northeast to southwest across Utah. Thick sediments built up on the subsided area, now eastern Nevada and western Utah. As a result, some Precambrian through middle Paleozoic intervals are 6–12 times thicker on the west side of the hinge boundary than they are on the east.[13]

The future

The potential seems endless for the Overthrust Belt. The Utah and Wyoming discoveries on the south and the Montana strikes on the north are only the beginning. Stirred on by the great successes made in the belt in 1981 and the approaching expirations of many primary lease terms, operators are certain to keep the brisk pace going in Overthrust Belt drilling beyond 1983.

More wildcatting is needed in southwestern and western Montana and along the Disturbed Belt in that state. Continued wildcatting should eventually tap a reservoir in eastern Idaho, most geologists think. The key there has not yet been found, but the industry is working on it. More discoveries are expected beyond the northern end of the fairway in southwestern Wyoming. There will be more Utah discoveries, both on the belt and on the Hingeline as time wears on.

REFERENCES

1. L. Porter, *Oil & Gas Journal* (October 1, 1979), p. 118.

2. Ibid.

3. Loren Avis and Donald Boothby, *Oil & Gas Journal* (May 31, 1982), p. 163.

4. H. Ritzma, *Quarterly Review* (Utah Geological & Mineral Survey, 1974).

5. Roger Matson and R. Schneider, *Oil & Gas Journal* (November 18, 1968), p. 204.

6. Fellows, Pendley, and Rinehart, *Oil & Gas Journal* (November 11, 1973), p. 212.

7. S. Wengerd, *Oil & Gas Journal* (September 24, 1956), p. 106.

8. Ibid.

9. S. Wengerd, *Oil & Gas Journal* (January 26, 1970), p. 172.

10. S. Wengerd, *Oil & Gas Journal* (February 9, 1970), p. 97.

11. *Oil & Gas Journal* (January 4, 1982), p. 152.

12. Ibid.

13. Bob Tippee, *International Petroleum Encylopedia* (1982), p. 290.

11 Alaska, the Arctic, and Canada

Alaska harbors the largest field in North America—Prudhoe Bay. It also will soon have the second largest field in the U.S. at Kuparuk, and the state continues to be the hope for future U.S. giant oil and gas fields.

Only a small part of this huge state has been tested adequately for potential oil and gas reserves. Future lease sales may open the vast untouched and expected oil and gas coffers in offshore basins and in the state's remaining undrilled onshore regions. Operators at Prudhoe Bay are progressing with a 5-year, $10.5-billion program to maintain the giant's output of 1.5 million b/d of oil and to ensure ultimate recovery of about 40% of the oil in place.

ARCO Alaska and Sohio Alaska Petroleum (Prudhoe operators) have drilled about 450 wells in the field. Of that total, about 75% are capable of producing, with each operator having about the same number of wells.

Closer well spacing is expected to help boost recovery by about 4% of the original oil in place. Operators began shifting to 160-acre spacing from 320 acres in 1979. Infill drilling began on 80-acre spacing in August 1981. Under the initial development plan, about 500 wells will ultimately be tapping the Sadlerochit oil at Prudhoe Bay (Fig. 11–1). The field should begin its decline during 1985–1987.

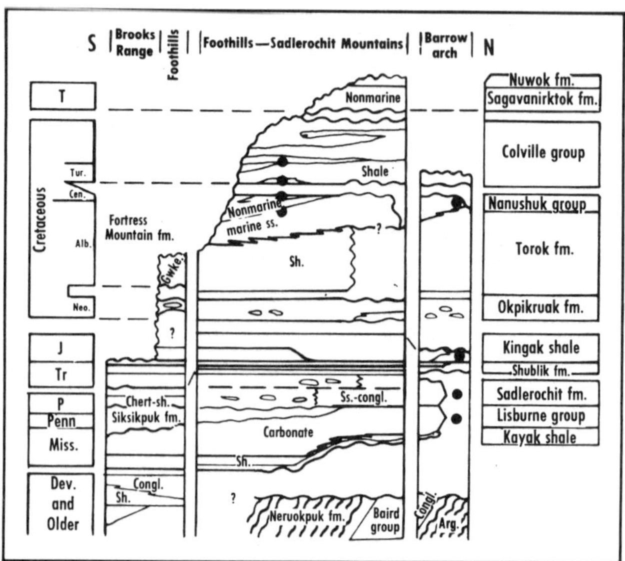

Fig. 11–1 North Slope formations and pays (from USGS, courtesy *OGJ*, April 20, 1970)

The Kuparuk field, also on the North Slope of Alaska, promises to become the second largest field in the U.S. during the 1980s. ARCO Alaska plans to spend about $7.2 billion (inflated dollars) to develop Kuparuk and to produce about 250,000 b/d of oil. This field was found in 1969; however, at that time it was uneconomic to develop. But rising oil prices and cost cutting turned the field into a good proposition. ARCO Alaska brought the field on stream in late 1981 with 50,000 b/d of oil from 24 wells. Production rose to 80,000 b/d by 1982. Current production is from 49 wells at five drill sites. There were to be eight drill sites and about 60 producing wells by the end of 1982. A second phase will add 181 drill sites and about 150 producing wells in 1983. The third phase will cover development of the unitized area to the northwest. Sealift for CPF-3 will be in 1985 or 1986, with startup of as much as 100,000 b/d of oil from 110 producers on 18–20 drill sites by 1986 or 1987 (Fig. 11–2).

The reservoir

The Kuparuk field produces from the early Cretaceous Kuparuk River, a fine-grained marine sand. Its thickness is about 40–50 ft. Wells produce 1,500–2,000 b/d of oil on the average. The development area of the field covers more than 200 sq miles. The estimated original oil in place is 4–5 billion bbl, with total recoverable reserves tagged at about 1.25 billion bbl from primary and secondary recovery.

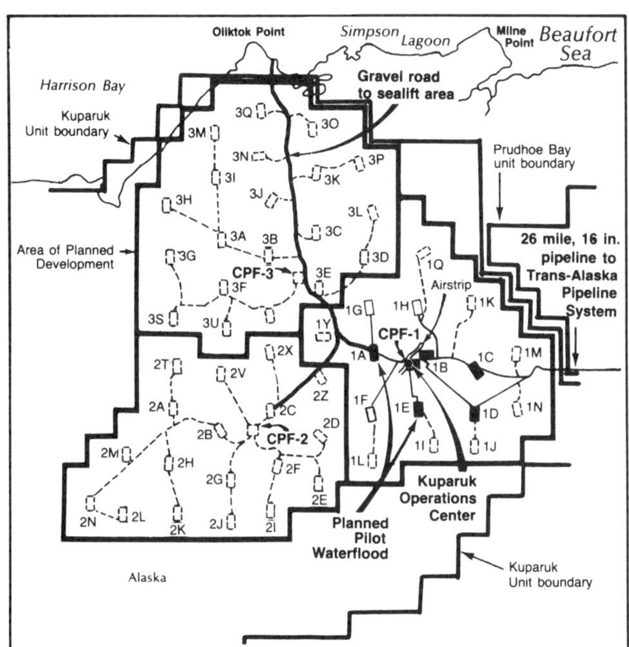

Fig. 11–2 Planned development for Kuparuk field (courtesy *OGJ*, July 12, 1982)

ARCO hopes to keep production at the 90,000-b/d level for the next few years, boosting it to about 180,000 b/d of oil during the mid-1980s. This will be after CPF-2 goes on stream. Production should peak at about 250,000 b/d of oil during the last part of the decade and then should enter into a normal waterflood decline of about 10–12%/year.

Alaska's resources

Alaska's other large oil fields are Granite Point with 87.6 million bbl produced and 32 million left and McArthur River with 458.2 million bbl produced and 66 million remaining. Swanson River (the field that started the Alaskan oil play in 1968) has produced 180.8 million bbl of oil, with 21 million remaining. Industry may spend more than $300 billion to develop Alaska's oil and gas resources during the next half century. But such a capital outlay could be cut sharply if the state adopts tough tax and regulatory policies.

The state's huge untapped resource base promises a big share of industry spending will be earmarked for exploration/development in the state into the next century. Studies estimate U.S. Arctic petroleum potential stands at 33–44 billion bbl of oil equivalent. Undiscovered oil and gas resources in the Arctic could amount to as much as 40% of the U.S. total. For oil alone, the Arctic could hold 50% of the U.S. total.[1]

If those estimates are accurate and the reserves are found, industry could spend almost $300 billion on development, transportation, and operating expenses during the next 40–50 years. This staggering figure does not include lease bonuses and dry-hole costs. Alaskan petroleum development so far, as well as upcoming projects, will account for more than $50 billion of industry spending.

THE BEAUFORT SEA

The icy Beaufort Sea is where the offshore wildcatter of Alaska is headed today. Nine wildcats were drilled there during the 1981–82 winter season. At least some are thought to be potential producers. These wildcats were drilled to evaluate prospects that drew $1.056 billion at the joint federal-state sale in Fairbanks in December 1979.

Sohio drilled three holes in 1981–82, including two from the same island. Exxon drilled four, and Amoco and Chevron each put down one exploratory test. Sohio drilled two holes from Endeavor Island (Fig. 11–3) to assess an apparent new field discovery further. The island is in the Sagavanirktok River delta area, 3 miles offshore and 8 miles northeast of the Prudhoe Bay field. Both wells are thought to be producers.

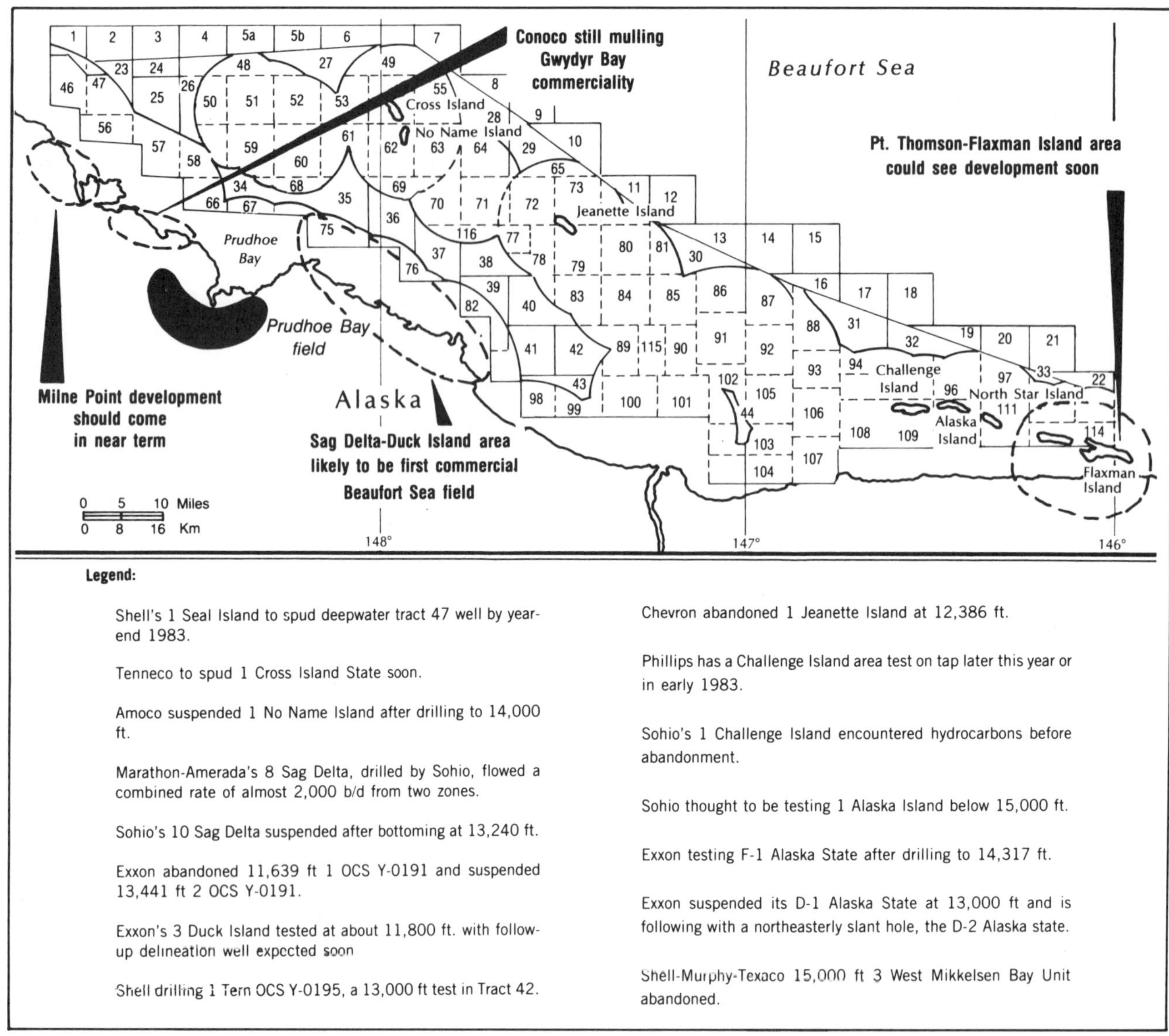
Conoco still mulling Gwydyr Bay commerciality
Beaufort Sea
Pt. Thomson-Flaxman Island area could see development soon
Cross Island
No Name Island
Jeanette Island
Prudhoe Bay
Prudhoe Bay field
Challenge Island
North Star Island
Alaska Island
Flaxman Island
Alaska
Milne Point development should come in near term
Sag Delta-Duck Island area likely to be first commercial Beaufort Sea field
0 5 10 Miles
0 8 16 Km
148°
147°
146°
Legend:
Shell's 1 Seal Island to spud deepwater tract 47 well by year-end 1983.
Tenneco to spud 1 Cross Island State soon.
Amoco suspended 1 No Name Island after drilling to 14,000 ft.
Marathon-Amerada's 8 Sag Delta, drilled by Sohio, flowed a combined rate of almost 2,000 b/d from two zones.
Sohio's 10 Sag Delta suspended after bottoming at 13,240 ft.
Exxon abandoned 11,639 ft 1 OCS Y-0191 and suspended 13,441 ft 2 OCS Y-0191.
Exxon's 3 Duck Island tested at about 11,800 ft. with follow-up delineation well expected soon.
Shell drilling 1 Tern OCS Y-0195, a 13,000 ft test in Tract 42.
Chevron abandoned 1 Jeanette Island at 12,386 ft.
Phillips has a Challenge Island area test on tap later this year or in early 1983.
Sohio's 1 Challenge Island encountered hydrocarbons before abandonment.
Sohio thought to be testing 1 Alaska Island below 15,000 ft.
Exxon testing F-1 Alaska State after drilling to 14,317 ft.
Exxon suspended its D-1 Alaska State at 13,000 ft and is following with a northeasterly slant hole, the D-2 Alaska state.
Shell-Murphy-Texaco 15,000 ft 3 West Mikkelsen Bay Unit abandoned.

Fig. 11–3 Beaufort Sea action (courtesy *OGJ*, July 12, 1982)

No one believes the Beaufort Sea will be a bad deal. The hopes are higher than ever that the sea will yield giant oil fields. In fact, one such probable giant already may have been found at Duck Island Unit and Sag Delta east of Prudhoe Bay. As much as 100,000 b/d of oil could be produced by the end of this decade if it is proven commercial. A decision was likely early in 1983 by Exxon, Sohio, and partners on whether to proceed with the first commercial development of an oil structure in the Beaufort Sea.

Another likely candidate for big production is the Flaxman Island-Point Thomson area at the southeast tip of the 1979 Beaufort Sea lease sale area. More drilling and seismic work are needed here, however, before any positive statements can be made. Exxon and Sohio have led the parade in the far eastern part of the Beaufort Sea play, finding hydrocarbons in several wells just west of the highly prospective Arctic National Wildlife Range. The range is believed to be on trend with large Beaufort strikes off Canada.

Any fields found in the Beaufort Sea must be in the megagiant class to warrant commercial development because of the huge costs involved. It is the major, thick sediments that will determine the most efficient Beaufort Sea technology. Even the largest accumulations will not justify scattering structures around a wide area to develop a thin pay section. For example Sag Delta-Duck Island, while not a large structure by North Slope standards, offers a thick, quality sandstone, perhaps justifying developments of a "marginally" giant reservoir in the Arctic.

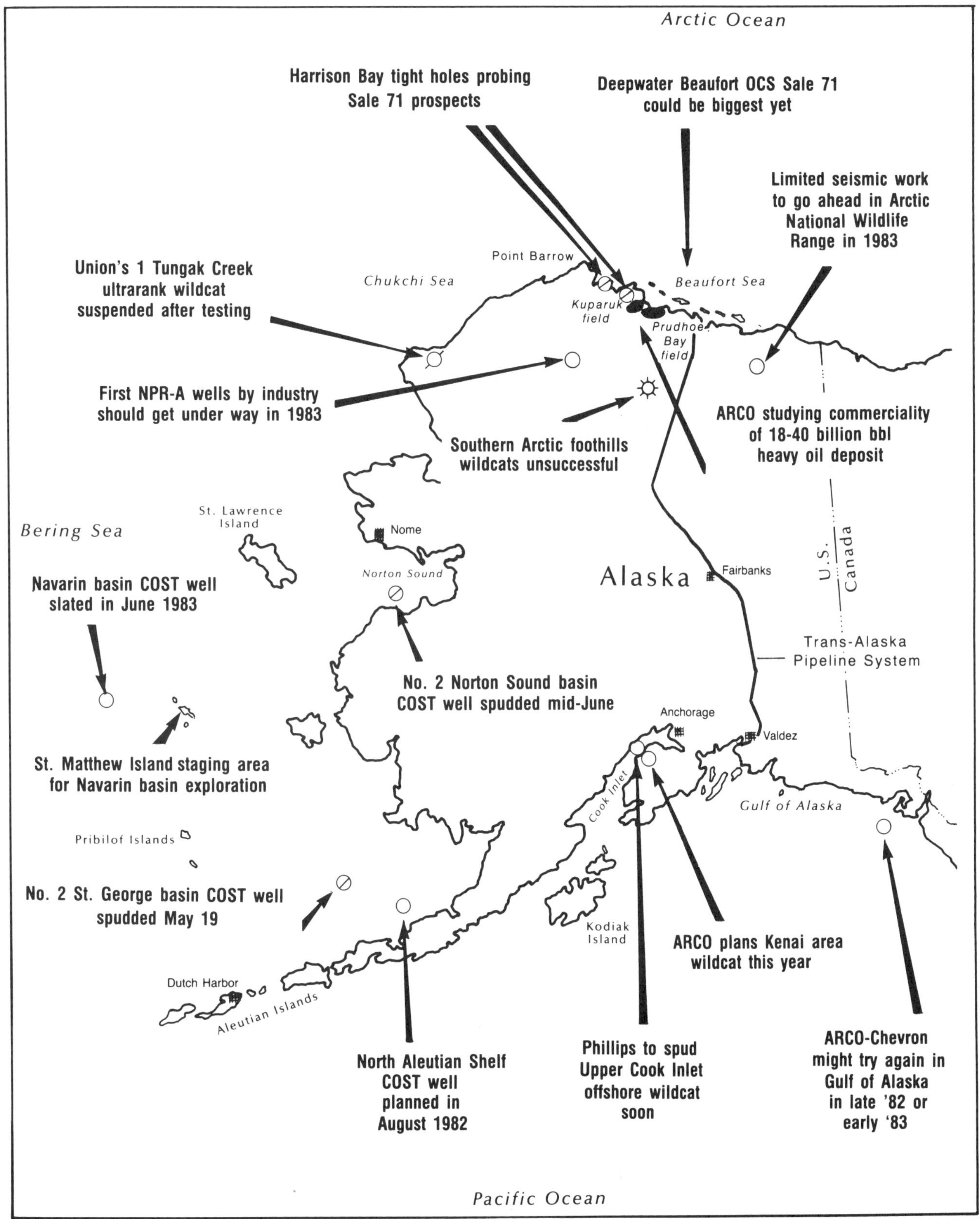

Fig. 11–4 Alaska's exploration action focuses offshore (courtesy *OGJ*, July 12, 1982)

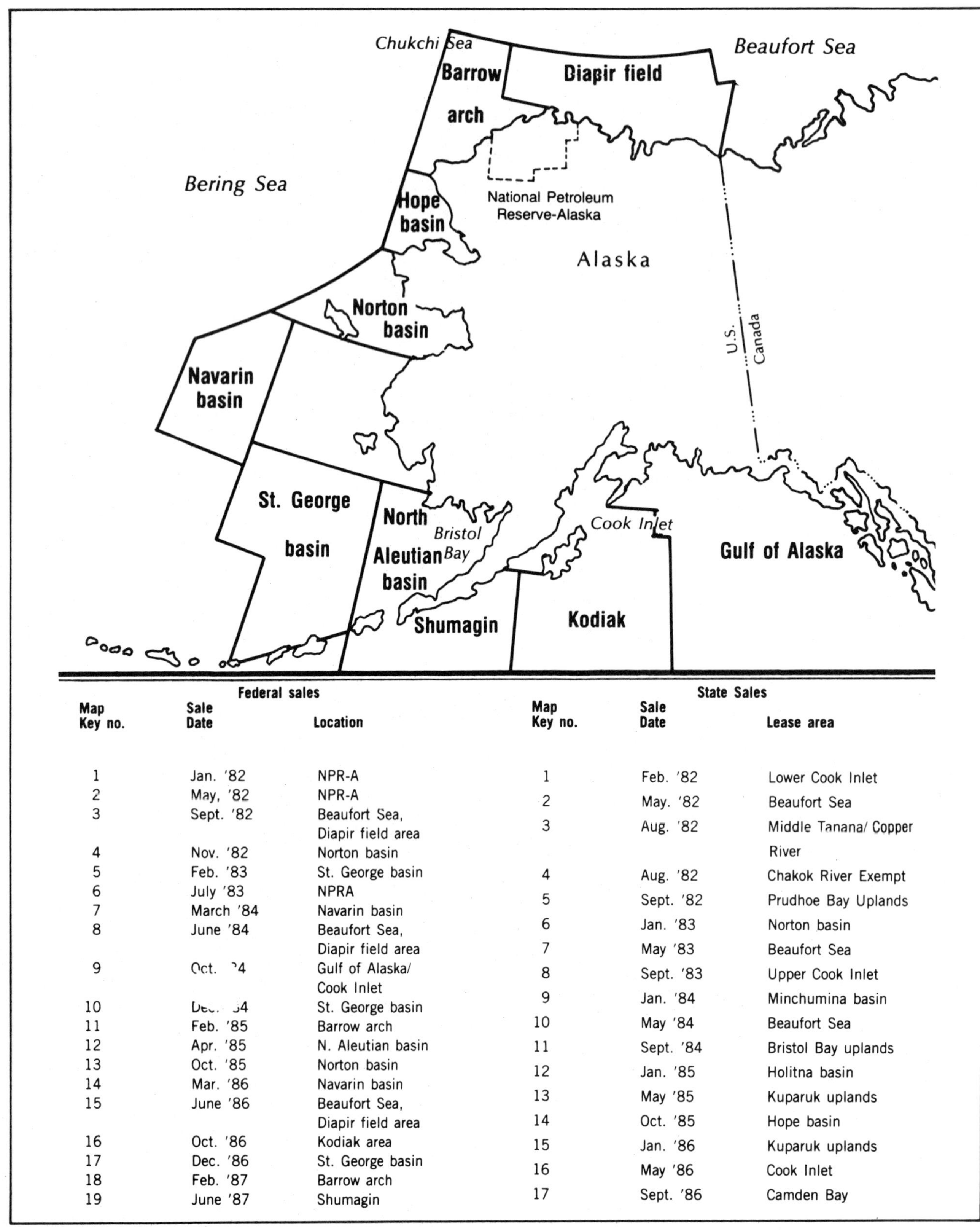

Federal sales

Map Key no.	Sale Date	Location
1	Jan. '82	NPR-A
2	May, '82	NPR-A
3	Sept. '82	Beaufort Sea, Diapir field area
4	Nov. '82	Norton basin
5	Feb. '83	St. George basin
6	July '83	NPRA
7	March '84	Navarin basin
8	June '84	Beaufort Sea, Diapir field area
9	Oct. '84	Gulf of Alaska/ Cook Inlet
10	Dec. '84	St. George basin
11	Feb. '85	Barrow arch
12	Apr. '85	N. Aleutian basin
13	Oct. '85	Norton basin
14	Mar. '86	Navarin basin
15	June '86	Beaufort Sea, Diapir field area
16	Oct. '86	Kodiak area
17	Dec. '86	St. George basin
18	Feb. '87	Barrow arch
19	June '87	Shumagin

State Sales

Map Key no.	Sale Date	Lease area
1	Feb. '82	Lower Cook Inlet
2	May. '82	Beaufort Sea
3	Aug. '82	Middle Tanana/ Copper River
4	Aug. '82	Chakok River Exempt
5	Sept. '82	Prudhoe Bay Uplands
6	Jan. '83	Norton basin
7	May '83	Beaufort Sea
8	Sept. '83	Upper Cook Inlet
9	Jan. '84	Minchumina basin
10	May '84	Beaufort Sea
11	Sept. '84	Bristol Bay uplands
12	Jan. '85	Holitna basin
13	May '85	Kuparuk uplands
14	Oct. '85	Hope basin
15	Jan. '86	Kuparuk uplands
16	May '86	Cook Inlet
17	Sept. '86	Camden Bay

Fig. 11–5 Alaskan leasing scenario (courtesy *OGJ*, July 12, 1982)

OTHER OCS AREAS

Thick, high-quality pay zones also are necessary in any giant fields the industry might discover in other Alaskan OCS waters (Fig. 11–4). Very little is known about these prospective, yet forbidding, frontiers. Seismic work has identified large structures in the western OCS, but only two wells have been drilled. These were the joint-venture Continental Offshore Stratigraphic Test (COST) wells operated by Atlantic Richfield in Norton Sound and St. George basin. A three-well program costs about $200 million. A second single-well program will cost about $35 million. ARCO is partner in both ventures, with 16 partners in the multiwell program and 18 in the latter.

Drilling is under way at 2 St. George basin, 115 miles northwest of Cold Bay. The 2 Norton basin is active, 48 miles north of Kotlik. The 1 North Aleutian Shelf was to fire up in August 1982, 80 miles north of Cold Bay. In the summer of 1983, the 1 Navarin basin will drill 290 miles southwest of Gambell.

A rundown on the leasing outlook in Alaska is shown in Fig. 11–5. State officials are pursuing a stronger leasing policy in Alaska, an effort underscored recently when lagging crude prices slashed the state's budget by more than half.

Navarin basin

The Navarin basin is perhaps the most attractive of the offshore basins in Alaskan waters. This basin may present the greatest challenge to the oil industry. Weather is terrible, and the basin is virtually out of reach by all but the longest-range helicopters, making operating and transportation expenditures horrendous. Alaskan operators are thinking about using St. Matthew Island as a possible staging area for Navarin and other western OCS operations.

The Navarin basin province is a relatively unexplored frontier area beneath the northwestern shelf of the Bering Sea shelf (Fig. 11–6). The entire shelf covers more than 400,000 km^2, an area larger than California. The Navarin province itself covers an area of at least 45,000 km^2 (11 million acres). Water depth ranges from 100–200 ft. The province contains at least three basins, each filled with 10–15 km of sedimentary strata.[2]

No wells have been drilled in the Navarin basin to date, but the USSR has drilled about 25 offshore wells in the Anadyr basin, adjacent to the Navarin basin. Gas was found in Miocene sandstones, and oil shows were reported in the lower Tertiary strata. A single well drilled in Khatyrka basin, southwest of Cape Navarin, flowed commercial amounts of gas of as much as 1.06 MMcfd from three Oligocene sandstone units. Geophysical data indicate Anadyr and the Navarin basins are juxtaposed features and probably share a

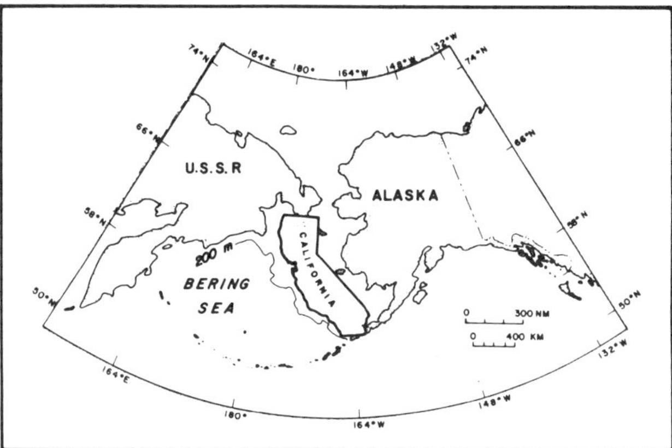

Fig. 11–6 The Navarin basin regional picture (courtesy *OGJ*, October 29, 1979)

common geologic history. The Anadyr basin is also known as the Northeast Siberia province.

In 1976 and 1977 surveys, 800 km of 24-channel seismic reflection data were collected, using a sound source of five air guns totaling 1,326 cu in. The multichannel data revealed the Navarin basin province comprises a series of northwest-trending basins and ridges. The sedimentary section is 10–15 km thick. Although Mesozoic beds may form part of the basin fill, most of the section is probably of Cenozoic age.[3] Sedimentary beds in the southern two basins are underformed except along the basin flanks where reflectors are cut by normal faults.

The sedimentary section in the northern basin is folded into large anticlines 10–15 km wide that may have diapiric cores. Little is known about possible source beds in the province's sedimentary section, but the province should be counted as a significant hydrocarbon-bearing region for future exploration.

The Norton basin

The Norton basin, adjacent to the Yukon Delta, lies beneath the east central part of the northern Bering Sea. The Norton basin is an extensional basin that most likely developed on a basement composed of Precambrian and Paleozoic rocks like those exposed on the Seward Peninsula and on St. Lawrence Island.[4] Norton basin fill is as thick as 6.5 km. These rocks are of probable Cenozoic age, although some uppermost Cretaceous sediments could be present locally at depth. The basin formed as a rift. Klemme has said that rifted basins tend to have high geothermal gradients and large traps for hydrocarbons, and that 35% of rifted basins contain giant oil fields. These general characteristics of rifted basins, plus the thickness and probable age of the rocks in the Norton basin, suggest this basin could have hydrocarbon reserves.[5]

The United States Geological Survey says the basin could contain 0.2 billion bbl of oil and 1 trillion cu ft of gas. Only three wells from which public data can be obtained have been drilled near the Norton basin. Two lie north of the Seward Peninsula in Hope basin. The third penetrated Cretaceous rocks east of the Norton basin.

Potential traps in the Norton basin could be of four main types: traps within horsts, traps in drape or compaction structures, traps in structures that develop along faults, and stratigraphic traps along the flanks of horsts. USGS geologists think the drape or compaction structures are the most likely to contain hydrocarbons in the Norton basin. These traps formed primarily in Neocene time. Oil began to be generated in deep rocks before these structures began to form. No bright or flat spots are evident in seismic data collected over any of these structures.[6]

OTHER ALASKA EVENTS

At the opposite end of the North Slope is the Arctic National Wildlife Range. This region has tempted geologists for many years but has always been off limits to the drillers. Some seismic work on a limited scale will be allowed in the reserve in 1983. But there is no timetable or other provision to allow any followup drilling in the range after industry hands over the seismic results to the federal government. Geologists are unhappy about these limitations, particularly since the real interest is in only a small strip of prospective lands along and off the coast of the range.

ARCO will spud an onshore wildcat in the Cook Inlet area during the 1980s and participate in more seismic work on native corporation leases near the Swanson River oil field. Union Oil is delineating the gas play in the Cannery Loop area of Kenai and Beaver Creek.

Offshore Cook Inlet exploration was to involve Chevron in seismic surveys in the summer of 1982. This company won all the interesting tracts at OCS Sale 60. Phillips also plans two wildcats in the inlet. There may be new interest in the disappointing Gulf of Alaska if ARCO and Chevron put together a planned joint venture for a near-shore test off Yakutat.

Exploratory growth in Alaska should continue to grow well into the next century. Estimates of Alaska's total unfound oil and gas account for 40% of the U.S. total. These figures alone stir the exploratory enthusiasm of a new oil-hungry industry.

CANADA

The Canadian oil industry planned to spend $1.2 billion in 1982 on frontier exploration and development programs, a slight increase from 1981. Subject to the resolution of some political problems, the spending pace may speed up in 1983.

In the Beaufort Sea, the major operators—Dome Petroleum, Gulf Canada, and Esso Resources—will try to prove the commercial value of several very promising oil discoveries. These include East Tarsiut N-44, a step-out well for which Gulf is operator, and Esso's Issungnak oil strike. More wildcats will be drilled in the sea, and all three major Beaufort operators are working on new generation drilling systems.

On the east coast in Atlantic Ocean waters, exploration should continue at the same level in the Grand Banks region off Newfoundland, site of the Hibernia oil field. The Sable Island sector of the Atlantic off Nova Scotia, over which an energy agreement was signed in 1982 between the federal government and the province, should see some brisk action.[7] Five rigs were working off Nova Scotia during the winter of 1982–83. These will either be new wells or delineation wells of the Venture gas field. Production from Labrador is not expected until at least the mid-1990s.

Beaufort Sea

Dome Petroleum still hopes to produce the first oil from this Arctic province as early as 1985. This forecast depends on proving up commercial reserves at Gulf et al. Tarsiut. Dome says the N-44 is probably the best well drilled in the sea to date. Dome has drilled 15 wells. This includes four oil discoveries and three gas strikes.

Gulf Canada, operator at Tarsiut, says that the N-44 well should have a sustained productive capacity of 3,500 b/d of oil, with flows of as much as 1,972 b/d of oil from several sands. Two more wells were being drilled in 1982. The level of continuing and planned activity in the Beaufort Sea by major operators is a measure of the industry's conviction of the sea's potential.[8]

Structures identified by Dome in the sea to date are shown in Fig. 11–7. Estimates of Beaufort Sea potential vary widely from 6.9–34 billion bbl of oil and more than 100 trillion cu ft of gas. Dome's estimates are in the higher range. The company expects to confirm 4.5–7.5 billion bbl of oil in the Beaufort Sea by 1990. The Geological Survey of Canada has estimated recoverable reserves of 9 billion bbl of oil and 112 trillion cu ft of natural gas.[9]

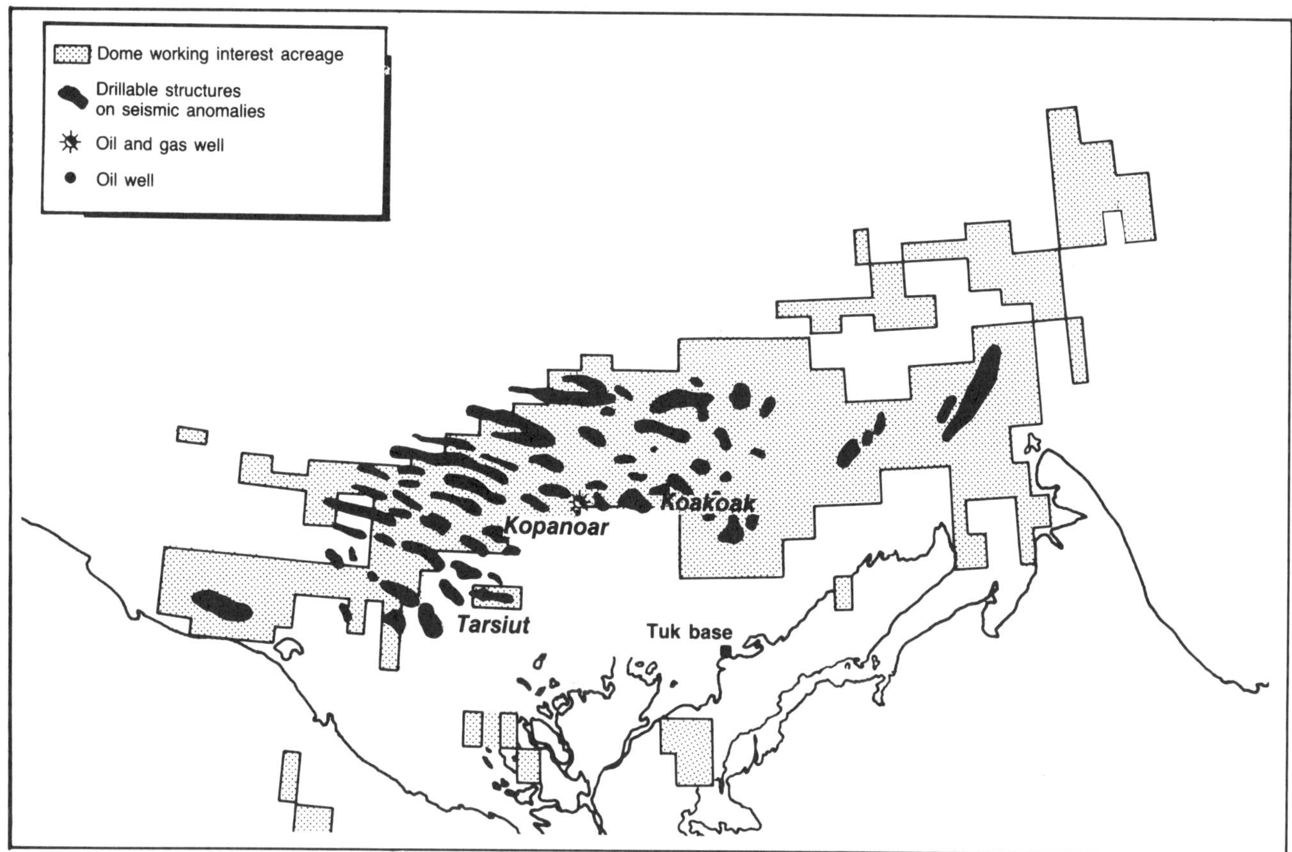

Fig. 11–7 Beaufort Sea structures identified by Dome Petroleum (courtesy *OGJ*, July 12, 1982)

Arctic Islands

Panarctic scored several important oil and gas strikes in the Arctic Islands during the 1981–82 winter drilling season. But this hunt is expected to slow down due to cash-flow problems among participants and the need to renegotiate exploration permits with the Canadian government.

One of two deep wildcats programmed for the Sabine Peninsula of Melville Island has been put off until 1983, and Panarctic trimmed its 1982 exploration budget from $240 million to about $150 million. One well will be drilled during the 1982–83 season to 20,000–24,000 ft on the Sabine Pensinsula.

Panarctic expects to spend most of its time and money at the Cisco oil discovery off Lougheed Island, the big discovery of the 1982 drilling season. Cisco C-42, a 5-mile step-out from the 1981 Cisco B-66, 10 miles west of Lougheed Island in 950 ft of water, tested 9.7 MMcfd of gas and 1,691 b/d of oil from one zone and 1,465 b/d of oil from another.

The Cisco step-out is a positive confirmation of Panarctic's contention that the islands do indeed contain substantial oil reserves as well as the 18.3 trillion cu ft of natural gas which the company says has been proven. Oil was found at Bent Horn on Cameron Island in 1975, but step-outs were all dry.

The Cisco C-42 confirmed oil-bearing zones in the Zwingak formation extend south-southeast from the original discovery well at least 5 miles. One test had 944 b/d of oil, and 900 b/d was tested from a second zone. Panarctic has not declared the Cisco structure commercial but does estimate it could contain more than 1 billion bbl of oil.

Skate B-80, 33 miles north of Cisco, flowed 7.3 MMcfd of gas and 775 b/d of oil. The MacLean I-72 wildcat, also near Lougheed Island, flowed 12 MMcfd of gas and 1,800 b/d of oil. Other new wells in the islands include Panarctic et al. Sculpin E-08, 11 miles off the Noice Peninsula of Ellef Ringnes Island. It flowed 7.5 MMcfd of gas at 4,915–4,979 ft in the King Christian sands.

Panarctic owns more than 50% by Petro-Canada and is operator for the Arctic Islands Exploration Group. It has been exploring in the Arctic Islands since 1967. The firm's score is 12 gas fields with estimated reserves of 18.3 trillion cu ft of gas with total reserves of oil set at 750

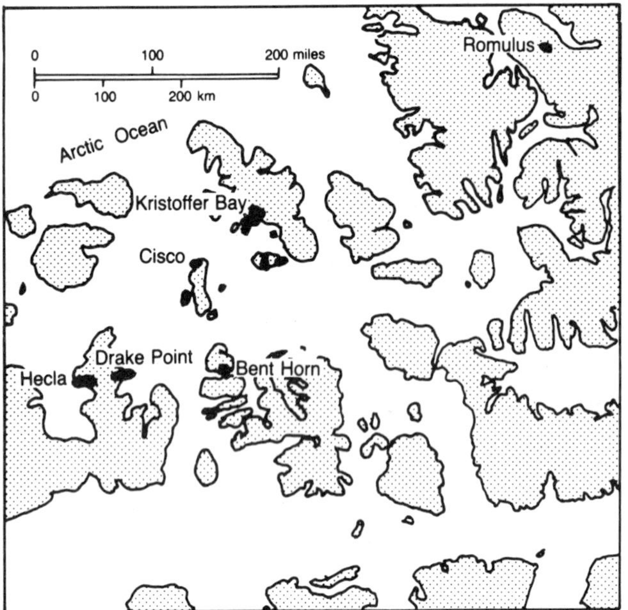

Fig. 11–8 Fields in the Arctic Islands

million bbl. The oil and gas fields of the Arctic Islands are shown in Fig. 11–8.

The area

Canada's Arctic Islands comprise the fragmented land north of the mainland, covering 285,000 sq miles. The area, onshore and offshore, is about 665,000 sq miles.

A large portion of the islands contains possible hydrocarbon-bearing sedimentary rocks. The major sedimentary basins, the Franklinian and the Sverdrup, have large oil and gas reserves. The entire region is the largest known undeveloped sedimentary area in the Western Hemisphere.

The sedimentary area in the Arctic Islands is composed of four major geologic features: the Arctic Lowlands, the fold belt, the Sverdrup basin, and the Arctic Coastal Plain. Most geologists believe the Arctic Lowlands are the lowest member of the group in hydrocarbon prospects. The fold belt covers an area of 90,000 sq miles and contains 250,000 cu miles of prospective sediments. The Sverdrup basin has the best prospects of becoming a major oil and gas province. Good prospects also exist on the Arctic Coastal Plain.

East Coast

Exploration action off Nova Scotia near Sable Island was brisk in 1982 as a result of a new energy agreement between the federal government and the province. But the outlook was cloudy for the Grand Banks region off Newfoundland where the province and the federal government were still locked in a bitter feud over ownership and jurisdiction.

Mobil was busy in 1982 completing two tests. One was the Olympia A-12, 8 miles west of Venture B-43. The other was the South Venture O-59, a third appraisal well to the 1979 Venture D-23 gas discovery, 10 miles east of Sable Island (Fig. 11–9).

The first appraisal well, B-13 Venture, 2 miles east of the discovery well, tested gas in three separate zones at 13.7, 18.6, and 15.3 MMcfd. One zone also had 300 b/d of oil, and two others had 384 b/d and 204 b/d of condensate. An earlier test flowed 18.8 MMcfd of gas, and three of the four tests were in sands corresponding to those of the D-23 discovery hole.

Mobil said the second appraisal well flowed up to 15.8 MMcfd of gas. The company thinks recent drilling on the Venture structure has confirmed enough gas to justify commercial development. Mobil hopes to begin building its production facilities on the Venture structure in 1984. Previous estimates have placed 3 trillion cu ft as the minimum requirement to justify development of the structure. Development could take until late 1987 or early 1988 and cost $3–4 billion.

On the Grand Banks off Newfoundland, the Canadian Petroleum Association gave its first official recognition to the East Coast in its reserves estimates by giving 1.1 billion bbl to the Hibernia field. Other estimates run as high as 1.8 billion bbl of oil at this Atlantic Ocean prize. Initial estimates by the Newfoundland Petroleum Directorate and a private consulting firm say Hibernia reserves are 1.5 billion bbl of oil. This study shows 1.037 billion bbl of oil in the Avalon zone and 519 million in the Hibernia zone. The study says the Avalon could sustain production of as much as 135,000 b/d of oil and 130,000 b/d of oil from the Hibernia zone. This report says production could begin at any time from 1985–1989. Gulf Oil Canada, a major leaseholder with 25% interest, says unsolved political disputes over jurisdiction will push production from Hibernia back to 1987 at the earliest.

The structure is a large north-northeast-trending rollover anticline bounded on the west by a major listric growth fault. It is divided into a number of separate blocks by transverse faults. A major mid-Cretaceous unconformity truncates the crest of the structure. The most productive reservoirs are Lower Cretaceous sandstones with porosities ranging from 17% to more than 30%.

Hibernia lies 200 miles east of St. John's, Newfoundland, in about 260 ft of water. It is near the southeast end of the Jeanne d'Arc subbasin, where thicknesses of Mesozoic and Cenozoic sediments range to more than 39,000 ft.

There were six wells drilled on the Grand Banks in 1981. This total included three more delineation wells at

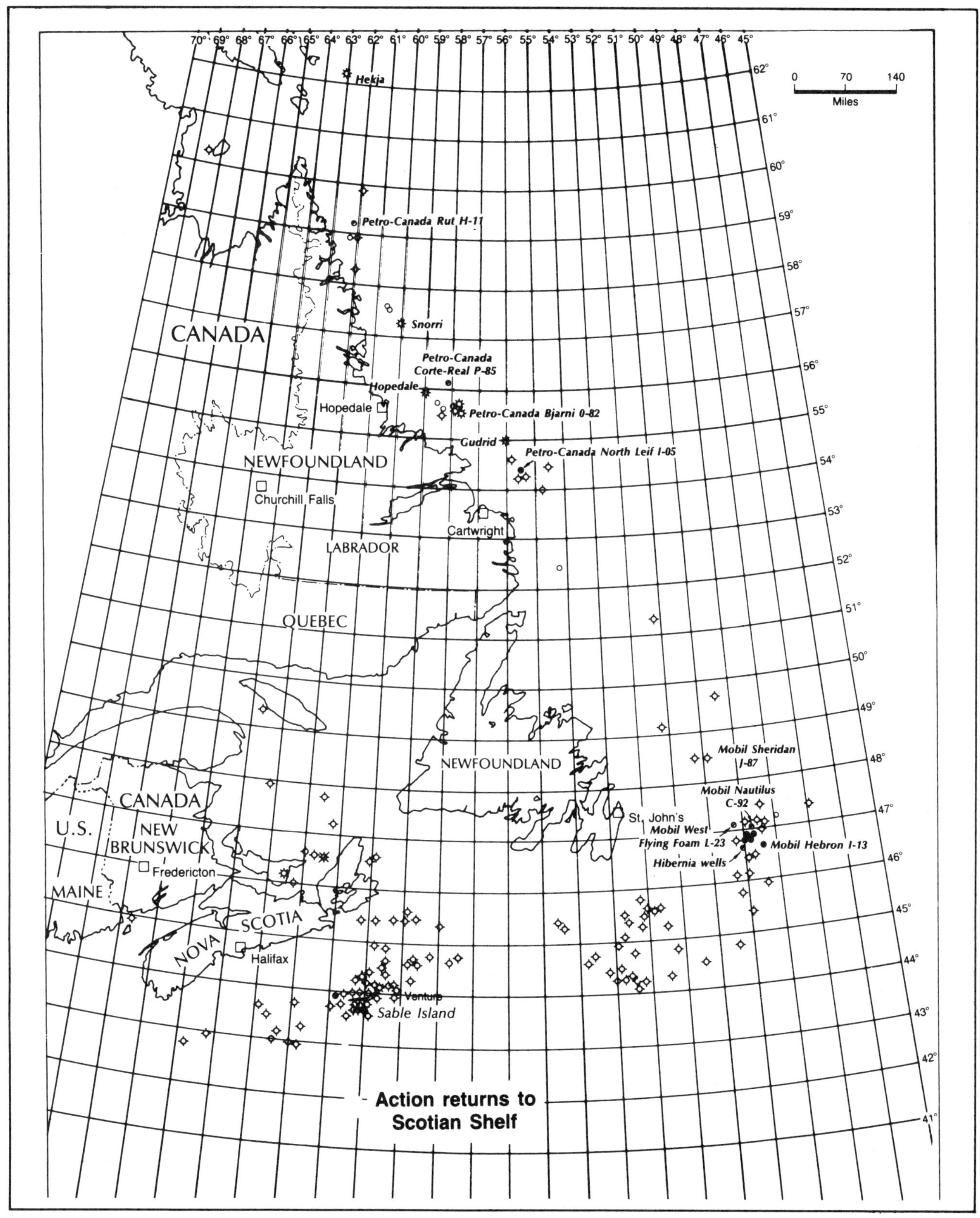

Fig. 11–9 Offshore action—East Canada (from *IPE*, PennWell)

Hibernia itself. The K-18 well flowed 4,642 b/d of oil and 6.68 MMcfd of gas from nine zones.

The Mobil Oil Canada group has found more oil and gas off the east coast at C-92 Nautilus, a wildcat 9 miles north of the Hibernia oil field. It flowed from two of five zones tested. It is on a separate structure. The well flowed 2,630 b/d of 31° gravity oil and 2.4 MMcfd of gas on a ½-in. choke from a 36-ft interval below 10,909 ft.

Some geologists feel that the East Newfoundland basin has an indicated potential similar to the North Sea. But development of Hibernia and other giant discoveries hinges on how well industry overcomes many obstacles.

Six gas discoveries have been made in the Baffin/Labrador area off eastern Canada. These have been found on the Labrador shelf. There is a definite potential for oil here, too. Oil was recovered on a drillstem test of the North Leif I-05 well, and oil is considered present in a thin interval of a core in the Roberval K-92 well (Fig. 11–10). Hydrocarbons

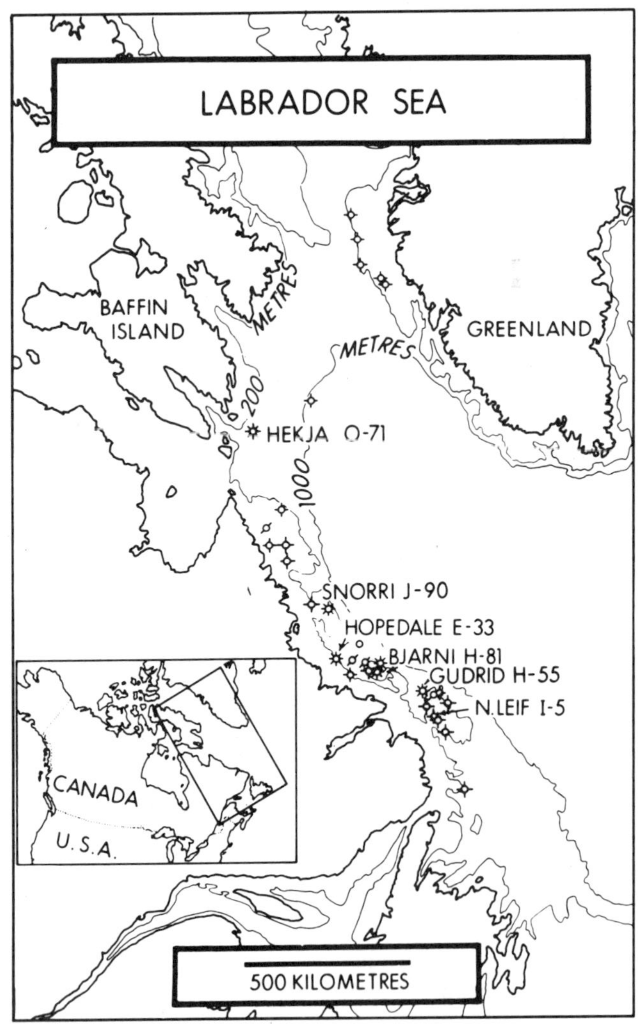

Fig. 11–10 The Labrador Sea region (courtesy *OGJ*)

are present off Labrador from Hekja on the north to Leif North on the south, a distance of about 950 km. It is difficult to estimate reserves due to lack of delineation wells.

Information obtained during the 1950s and 60s reveals marine sediments of Tertiary to Cretaceous age occur offshore, and structures of substantial size are present. A well was drilled in 1971 to the Eocene at 1,088 m on the Leif structure. It was not until 1973, when dynamically positioned drillships became available, that drilling began in earnest.[10] The Bjarni gas-condensate well in 1973 was the first of five discoveries to be made in the area. To date, 17 prospects have been tested.

Sediments range in age from Tertiary to Paleozoic and are mostly marine except for the Lower Cretaceous which is mostly fluvial in origin. Three known types of traps are recognized thus far in the Baffin shelf area. One involves structure related to major wrench faulting at the south end of Baffin Island, such as the Hekja and Gjoa structures. The main targets are Eocene and Paleocene sands. The other two types of traps involve the possible presence of Mesozoic and Paleozoic sediments in the area.

Since 1971, 21 wells have been drilled on the Labrador shelf. Prospects on the Baffin/Labrador shelf are various. Four of the 12 classes of prospects have yielded gas condensate. More are bound to follow as exploratory work continues. More drilling prospects probably will be uncovered in the near future. Fig. 11–11 reviews the various estimates of potential recoverable reserves (resources) which have been made to date by different agencies and workers.[11]

Onshore Canada

Efforts will continue through the years to uncover new oil and gas reserves in the western provinces of Canada. In spite of the large areas of the Alberta Foothills Belt in which exploration is restricted, it is estimated that 6–14 trillion cu ft of gas may be found yet.

Estimated proven and probable ultimate reserves of marketable natural gas in Alberta in 1980 were 80.5 trillion cu ft, of which about 10.5 trillion cu ft are in Paleozoic carbonate reservoirs that have been involved in thrust faulting in the Foothills Belt. This belt is the easternmost of four major physiographic and structural divisions of the southern Canadian Rocky Mountains. The eastern boundary of the Foothills Belt is marked by a zone of underthrusting. The western boundary is defined by the surface trace of major thrusts which bring Paleozoic or older strata to the surface.[12]

The Foothills Belt contains at least 18 significant gas-bearing structures. Until 1980, about 5 trillion cu ft of reserves of gas had been found. Geologists feel exploration of

	PROBABILITY	OIL B. BARRELS	GAS TCF
Mc MILLAN 1973 1979 2000 m W.D.	OPTIMISTIC	10.0	63
GEOLOGICAL SURVEY OF CANADA 1977 MAXIMUM 450 m W.D.	10%	4.5	45
	50%	2.6	27
	90%	1.7	18
NEWFOUNDLAND GOVERNMENT 1980	10%	NOT ESTIMATED	270
	50%		60
	90%		20
AQUITAINE (KLEMME METHOD) 1981 1000 m W.D.	RICH	12.0	72
	AVERAGE	7.0	30
	POOR	1.0	9
AQUITAINE (GEOCHEMICAL METHOD) 1981 1000 m W.D. [ASSUME 1/250 HYDROCARBON IS TRAPPED]	AVERAGE	9.0	64

Fig. 11–11 Labrador reserve estimates (courtesy *OGJ*)

the Foothills Belt, though not new, has only begun. Large volumes of gas are still to be found.

In 1979 production began at the Elmworth gas field, part of the vast Deep Basin area of northwestern Alberta. This basin is estimated to hold potential reserves of 400 trillion cu ft of gas.

The Deep Basin was first identified by Canadian Hunter Exploration in 1976. It stretches about 500 miles northwest from Grand Prairie in northwest Alberta into northern British Columbia, covering about 16.3 million acres.

The Elmworth field, located in the Elmworth-Fahler trend, forms a small part of the overall Deep Basin area. Details of the Deep Basin area released in 1978 said the Mesozoic rock underlying the area was saturated with gas below 3,500 ft and had huge potential reserves in storage.

Alberta enjoyed the West Pembina oil play in 1978. Observers felt the area would ultimately yield 1 billion bbl of oil or more. This play involved a series of Devonian pinnacle reef discoveries mixed with a few Mississippian strikes. Chevron Standard opened the play with its Nairb Pembina zone discovery in the Devonian in 1977. Production from the big Pembina field nearby is from the Cretaceous Cardium sand at 5,100 ft.

The Cardium, Upper Devonian Nisku below 9,000 ft, and in some places, the Mississippian gave drillers multizone prospects to produce. The play in 1977 covered a 900-sq-mile area. At one time 36 rigs were working and almost as many seismic crews were busy looking for new prospects. Drilling has since slowed considerably, but there are still more prospects in this large area west and northwest of Pembina. Operators say that when they found the prolific Cretaceous oil at Pembina in the early 1950s, they largely overlooked the deeper prospects.

REFERENCES

1. *Oil & Gas Journal* (April 12, 1982), p. 182.
2. M. Marlow, *Oil & Gas Journal* (October 29, 1979), p. 191.
3. Ibid., p. 192.
4. M. Fisher, W. Patton Jr., and M. Holmes, *Oil & Gas Journal* (May 3, 1982), p. 352.
5. Ibid., p. 365.
6. Ibid., p. 371.
7. Bob Williams, *Oil & Gas Journal* (July 12, 1982), p. 84.
8. Ibid., p. 89.
9. Ibid., p. 89.
10. M. Hriskevich and N. McMillan, *Oil & Gas Journal* (July 12, 1982), p. 145.
11. Ibid., p. 148.
12. P. Gordy, *Oil & Gas Journal* (June 30, 1980), p. 145.

12 Mexico and Latin America

Latin America has been considered for years as the most exciting region of the world for the future of oil exploration (Fig. 12–1). The oil and related industries have risen to astronomical heights in a few short years. New exploration and development thrives, and there are hopes of new virgin oil and gas production from the Rio Grande to the Falkland Islands off southeastern South America.

New and important exploratory work will uncover reserves in the San Jorge offshore basin of Argentina. It remains to be seen whether the huge delta of the mighty Amazon River of Brazil will pan out as an untapped oil and gas reserve. Geologists still have hopes for the Chaco basin of Paraguay and Bolivia. Trinidad and Tobago waters, on the

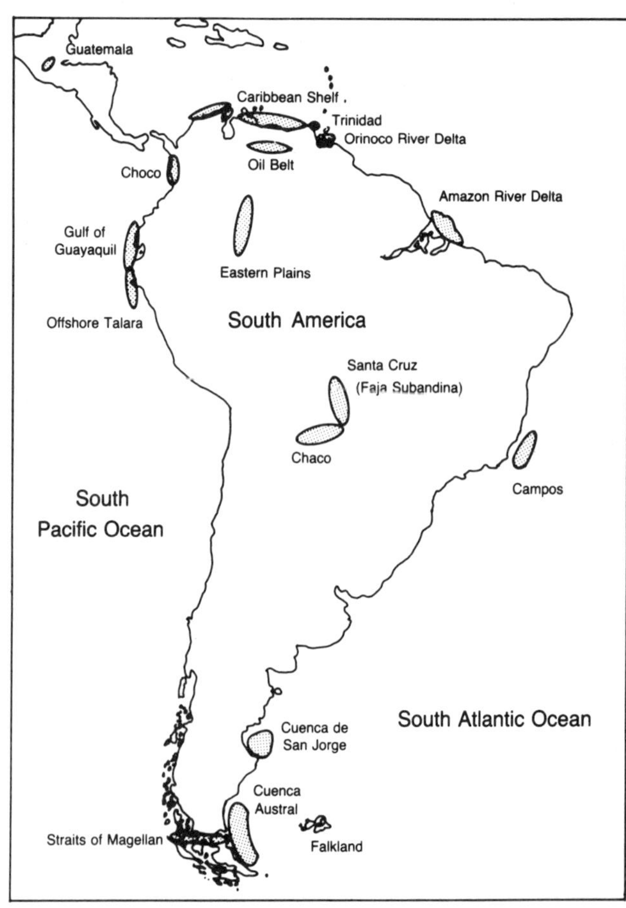

Fig. 12–1 Future prospects in Latin America (from *IPE*, PennWell)

north and on the southeast side of the country, are expected to be areas of huge new oil discoveries. The continental shelf off Venezuela can be expected to become an exciting new oil and gas province.

Mexico is still quite confident its Baja California Peninsula contains sizable reserves of hydrocarbons. Some geologists feel large portions of Baja are structurally similar to the Santa Barbara region of southern California. Some say the marine extensions of the Baja basins may be more attractive than the onshore portions, both on the Pacific Coast and on the east side (the Gulf of Cortez, California).

Mexico's oil boom continues after several years of tremendous exploratory success—namely, rapid fire Cretaceous and Jurassic discoveries in the Reforma region of Chiapas and Tabasco states. Recent discoveries in Colombia have opened a new gas-bearing province in a strategic location. This new area is onshore and offshore in northeastern Colombia near the city of Riohacha. Reserves have been estimated to be at least 2.4 trillion cu ft of gas. Geologists for years have rated the northern coastal plain and the adjacent continental shelf as prime prospects. The Guajira basin, which covers some 15,000 sq km, also has been considered a possible extension of the prolific Maracaibo basin to the south.

The large fields

Some of the world's largest oil fields are in Latin America, mostly in Venezuela and Mexico. There are nine oil fields in Venezuela that already have produced more than 500 million bbl of oil. Guara, discovered in 1946, has produced 538 million bbl of oil from pays at 5,000–10,000 ft; Quiriquire, opened in 1928, has produced 749 million bbl of oil from pays at 7,000–7,200 ft; Boscan, 1946, has produced 607 million bbl of oil; and Bachaquero, 1930, has produced 5.7 billion bbl of oil. Cabimas, 1917, has produced 1.5 billion bbl of oil from pays at 2,000 ft; Lagunillas, 1926, has produced 9.8 billion bbl of oil at 3,000 ft; Lama, 1937, has produced 2.2 billion bbl of oil from pay at 8,320 ft; Lamar, 1958, has produced 951 million bbl of oil from pay at 13,000 ft; and Mene Grande, 1914, has produced 612 million bbl of oil from pay at 4,130 ft.

Mexico has four supergiant oil fields with others indicated in the Gulf of Campeche. Ebano-Panuco, found at the turn of the century, has produced 934 million bbl of oil from pay at 1,450 ft. Then there is Naranjos field with 1.2 billion bbl of oil produced already from pay at 1,800 ft. Famous Poza Rica field, found in 1930, has produced 842 million bbl of oil from pay at 7,090 ft. Cretaceous Samaria field, discovered in 1973, has produced 647 million bbl of oil. The largest offshore field is in the Gulf of Campeche at Cantarel. It was discovered in 1976 and has produced 432 million bbl of oil from pays at 8,528 ft. In 1981 the field was producing more than 1 million b/d of oil.

Reserves

Mexico now leads the U.S. with the most oil reserves of Western Hemisphere countries. The country's oil reserves stand at 56.99 billion bbl with 75.4 trillion cu ft reserves of gas. Venezuela's reserves stand at 20.3 billion bbl of oil and 47 trillion cu ft of gas. This compares with only 29 billion bbl in the U.S. Gas reserves in the U.S. are 198 trillion cu ft and 89 trillion cu ft in Canada.

Mexico has 3,593 producing oil wells, Venezuela 13,610. Oil production in Mexico runs at 2.4 million b/d of oil, Venezuela 2.1 million b/d, the U.S. 8.6 million b/d.

MEXICO

Mexico is really a newcomer on the list of major oil and gas producing nations. Its vast area of sedimentary basins and offshore shelves and basins barely has been scratched. Exploration activities—seismic, geochemistry, and wildcatting—have increased by about 25%/year since 1977.

In 1980, Pemex drilled 64 wildcats, 35 successful. That compares with 58 wildcats, 30 productive in 1979; 67 wildcats, 28 productive in 1978; and 66 wildcats, 32 productive in 1977. Discoveries in 1980 included 6 gas and condensate fields in the Northeast Zone, 3 oil and gas fields in the South Zone, and 6 oil and gas fields in the offshore South Zone in the Sound of Campeche.

Pemex also drilled 20 new productive wells in the blanket sands of Chicontepec. In addition, the national oil company made an important strike in the Gulf of California off the mouth of the Colorado River.

The state company says key exploration areas at present are the Sound of Campeche, Mesozoic areas of Chiapas and Tabasco and Sierra de Chiapas in the South Zone, the Gulf of Sabinas basin and Gulf of California in the Northeast Zone, and the Plataforma de Cordoba in the Poza Rica Zone. Good prospects also await the explorer in Chihuahua, Baja California, and off Mazatlan. Long-term objectives for future exploration have been identified by Pemex in Michoacan, Guerrero, Taxiaco (Oaxaca), Zongolica (Veracruz), and what Pemex calls the "neovulanic axis."

Recent discoveries

In the Gulf of California, the 1 Extremeno flowed 5.7 MMcfd of gas and some condensate through a ¼-in. choke from pay at 13,510–13,543 ft. Onshore discoveries in the

Northeast Zone were 1 Bario, 1 Catarrink, 1 Centinela, 1 Llanero, 1 Paso de la Lomz, and 1 Patricia. This zone includes the Gulf of Sabinas basin, which many geologists say is associated with the Overthrust Belt of the western U.S.

New discoveries in the South Zone include 101 Cardenas, 1 Carmito, and 2-A Jugo. These found production in the Middle Cretaceous and Late Jurassic limestones and dolomites.

The Sound of Campeche has become Mexico's largest producing area, and it continues to yield major discoveries (Table 12–1). Production reached 1.308 million b/d in late 1980. In 1981, production was 1,319 b/d of oil and 627 MMcfd of gas, most of which is flared. The Campeche fields are in a 270 sq mile area about 50 miles off the shore of Campeche state. Most production comes from fields in the Cantarel complex, which flowed its first oil in 1979. Output is 1.3 million b/d of oil. The Abkatun field has made about 83,000 b/d of oil since it came on stream in 1980.

Five giant fields have been found in the Isthmus Saline basin of Mexico (Fig. 12–2). The north end of the basin is the Gulf of Mexico marine platform, the Chiapas uplift is the southern limit, the Chiapas-Tabasco Mesozoic area is to the east, and the Isthmus depression lies to the west.

Oil and gas production comes from the Tertiary sandstones, mostly of Encanto Lower Miocene age. The five giants are Tonala-El Burro with a cumulative production of more than 95 million bbl, El Plan with 147 million, Cinco Presidentes with 227 million bbl, Oraggio with 119 million, and Magallanes with 126 million. The largest dome is the Magallanes, which is about 18 km long by 7 km wide.

Table 12–1

HISTORY OF MEXICAN PETROLEUM ACTIVITY

1869	First oil well drilled near Vera Cruz.
1884	Mining Law allows private ownership of subsurface minerals.
1904	First commercial production established.
1906	President Porofiro Diaz encourages foreign investment in mineral rights.
1910	Golden Lane development begins.
1917	New constitution deeds all subsurface resources to government.
	Supreme Court states government ownership not retroactive.
1921	Mexico producing 530,000 b/d, second largest in world.
	Salt water begins to invade Golden Lane wells.
1924	First wells drilled on wooden platforms in Tamiahua Lagoon.
1930	Production drops sharply as government enacts stiff tax on production.
	Foreign operators begin worker layoffs.
	Shell discovers Poza Rica field.
1935	Petroleum Workers union formed.
1937	Union calls strike against 17 foreign oil companies.
	Golden Lane reservoirs exhausted.
1938	President Lazaro Cardenas nationalizes oil company assets on March 18.
1942	Mexico makes $180 million in reparations to oil companies.
1958	Offshore exploration begins off Coatzacoalcos.
1959	Santa Ana field discovered off Tabasco state.
1971	Mexico begins importing oil for domestic needs.
1972	Chiapas-Tabasco discoveries made.
	60 structures mapped in Campeche Sound.
1973	Foreign investment and technology import laws enacted.
	Production returns to 500,000 b/d plus level.
1974	Mexico resumes crude exports.
1975	First Campeche well, Chac, comes in as oil producer.
	Five Reforma wells developed.
1976	Mexican currency devalued.
1977	Akal and Bocab discoveries made.
	First Baja California discovery made in Sebastian Vizcaino area.
	117 Reforma wells producing.
1978	Chac 2 discovery made.
	Chicontepec reassessed—120 rigs needed for 16,000 wells.
	178 drilling units manned by 2,200 drilling crews in operation.
1979	Cantarel complex production system on stream.

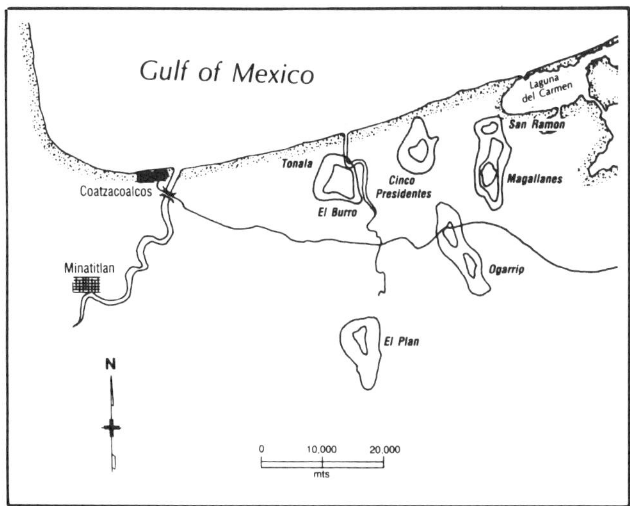

Fig. 12–2 Giant fields in the Saline basin, Mexico (courtesy *OGJ*, July 20, 1981)

Macuspana basin

This basin is between the Chiapas-Tabasco Mesozoic fields and the Yucatan platform. Giant fields discovered to date are José Colomo, Chilapilla, and Hormiguero. All three are giant gas and condensate producers. Production is from anticlinal structures with pays in the upper Amate Middle Miocene sandstones.

Chiapas-Tabasco Mesozoic area

This region covers about 7,500 km^2. It is between the Isthmus Saline basin and the Macuspana basin (Fig. 12–3). Sitio Grande field was discovered in 1972 in the Middle Cretaceous at 4,197 m. The structure is a dome-type anticline 10 km long and 6 km wide. Cumulative production here is more than 200 million bbl, and at least 600 million bbl of reserves are yet to be recovered.

Cactus field was discovered in the same period. Location is on a dome 10 km long and 7 km wide. Production is from Cretaceous rocks. Cumulative production exceeds 258 million bbl of oil. Expected reserves are 1,500 million bbl (primary or secondary).

An extensive exploration campaign followed the Cactus and Sitio Grande discoveries in 1972. The Samaria well was drilled in 1973 and completed for 3,416 b/d of oil. Other large fields were opened, forming what is known as the Antonio J. Bermudez complex. This wide structure is 16 km long and 14 km wide. Cumulative production was more than 200 million bbl of oil at last count. Expected recovery is more than 5,000 million bbl of oil. Cumulative gas production is more than 3 trillion cu ft.

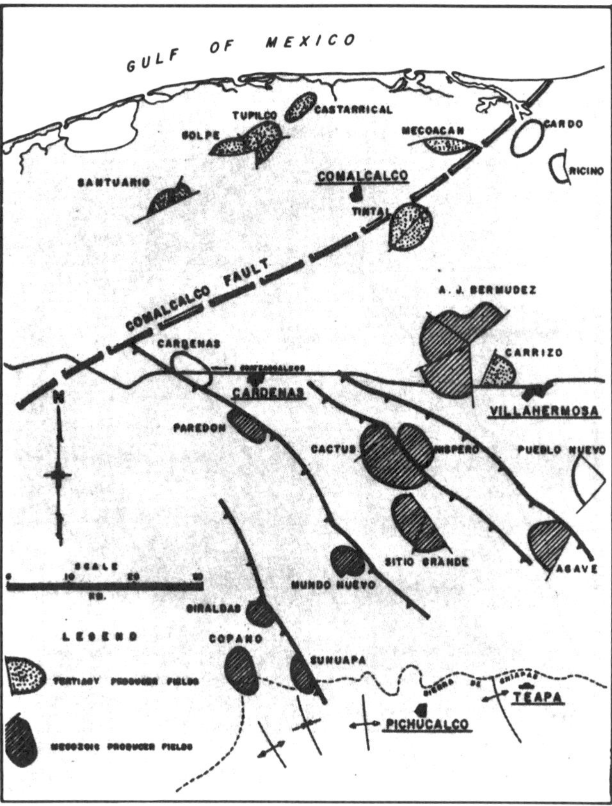

Fig. 12–3 Chiapas-Tabasco Mesozoic area (courtesy *OGJ*, July 20, 1981)

Iris-Giraldas complex

This giant field is 11 km long and 8 km wide. It was discovered in 1977 with oil and gas in the mid-Cretaceous. Another well was added at 1 Iris in 1979. This complex can be rated as a supergiant (more than 500 million bbl of oil) and a giant gas field (more than 1 trillion cu ft). Proved reserves at Giraldas-Iris are 1,500 million bbl of oil.[1]

Campeche marine area

The area covers 8,000 km^2 and was explored by reflection seismology. A large number of anticlinal structures were identified, and new fields have been opened. More than 46 wildcats have been drilled in the Bay of Campeche, with 82% successful. Total proven hydrocarbon reserves as of December 1981 stood at 72 billion bbl. Production in March 1982 was 2.55 million b/d of oil. A dozen producing fields have been found to date.

The discovery well was the 1 Chac in 1976, from the Lower Paleocene dolomite breccia. The Akal field is on the top of the structural anticline; the Nohoch field on the south. This complex is called Cantarel. Proved reserves are more

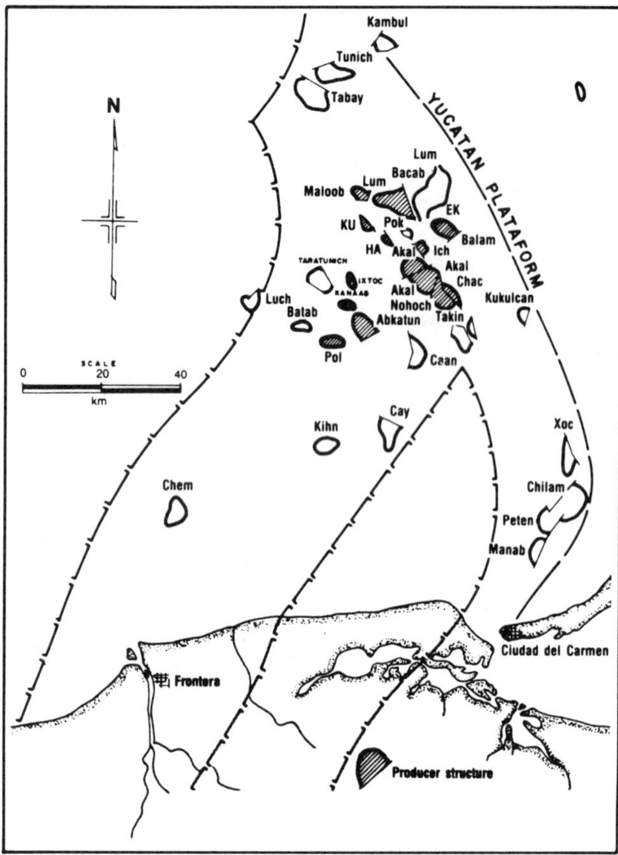

Fig. 12–4 Campeche marine platform (courtesy *OGJ*, July 20, 1981)

than 8 billion bbl of oil with output at more than 600,000 b/d of oil. This is the most productive area of Mexico.

Geologically, the Reforma-Campeche shelf trend is stratigraphically analogous with the equivalent aged Golden Lane of the Tampico-Nautia embayment to the north (Fig. 12–4). The Reforma-Campeche shelf carbonate talus trend extends northwestward toward the Golden Lane and has been found productive at several small fields southwest of Veracruz.[2] Eventually, the whole bank-edge talus trend shown on Fig. 12–4 should prove productive. The main fields produce from strongly faulted salt pillows.

In the Reforma area, the main fields are on the so-called Villahermosa horst, a complex uplifted zone with numerous north-south-striking horsts and grabens. These horsts and grabens are very ancient features, predating the Late Jurassic producing section. In the offshore Campeche shelf trend, horst and graben structures also formed. West of the Jurassic-Cretaceous bank edge, the overall fault pattern is one of concave, down-to-the-gulf faults. The producing structures parallel the faults.

Giant Chicontepec

The biggest single contribution to Mexico's sharply higher proved crude oil reserves in 1979 came from a complex of skinny pays in an old producing area that drillers had passed off as noncommercial for many years. Pemex then claimed Chicontepec oil reserves of 10.96 billion bbl. Modern production methods will improve permeabilities and flow rates from the field's tight pays.

The Chicontepec basin covers 11,300 km^2, about the same as the state of Queretaro. It is located inside the coast between Tampico and Posa Rica Tecolutla.

Other Mexican theaters

Mexico is confident its Baja California Peninsula may have hidden hydrocarbon reserves of substantial size. Several onshore and offshore prospects exist in this area. One of these is the Altar Desert in the northwestern part of the country.

The basin lies in the physiographic province of Salton and is an elongation of the San Luis basin in the southwestern corner of Arizona. Geologists think a sedimentary marine basin lies under the Altar Desert and under the updip waters of the Sea of Cortez (California). The presence of a basin has been delineated by seismic studies.

The discovery of gas recently in the Sea of Cortez by Pemex and other reported successes in Baja California plus some in the state of Chihuahua indicate exploration is needed in northwestern Mexico.

CENTRAL AMERICA

Most of the petroleum potential of Central America is in the Late Jurassic through middle Cretaceous trend of Reforma in southern Mexico. The second largest area of potential production of hydrocarbons is the Cretaceous and Late Jurassic section of the Chapayal basin in western Guatemala and southeastern Mexico.[3] The early Tertiary volcanic basement of the Nicaragua Rise may overlie deeper Cretaceous carbonate objectives, particularly in the southern part of the rise. Tertiary reefs also may be present. The potential of this large province has not been tested adequately (Fig. 12–5).

Of the Tertiary basins of the southern Central American orogeny, only the Limon-Bocas del Toro basin appears to have potential for commercial production. However, geologists believe this potential is small.

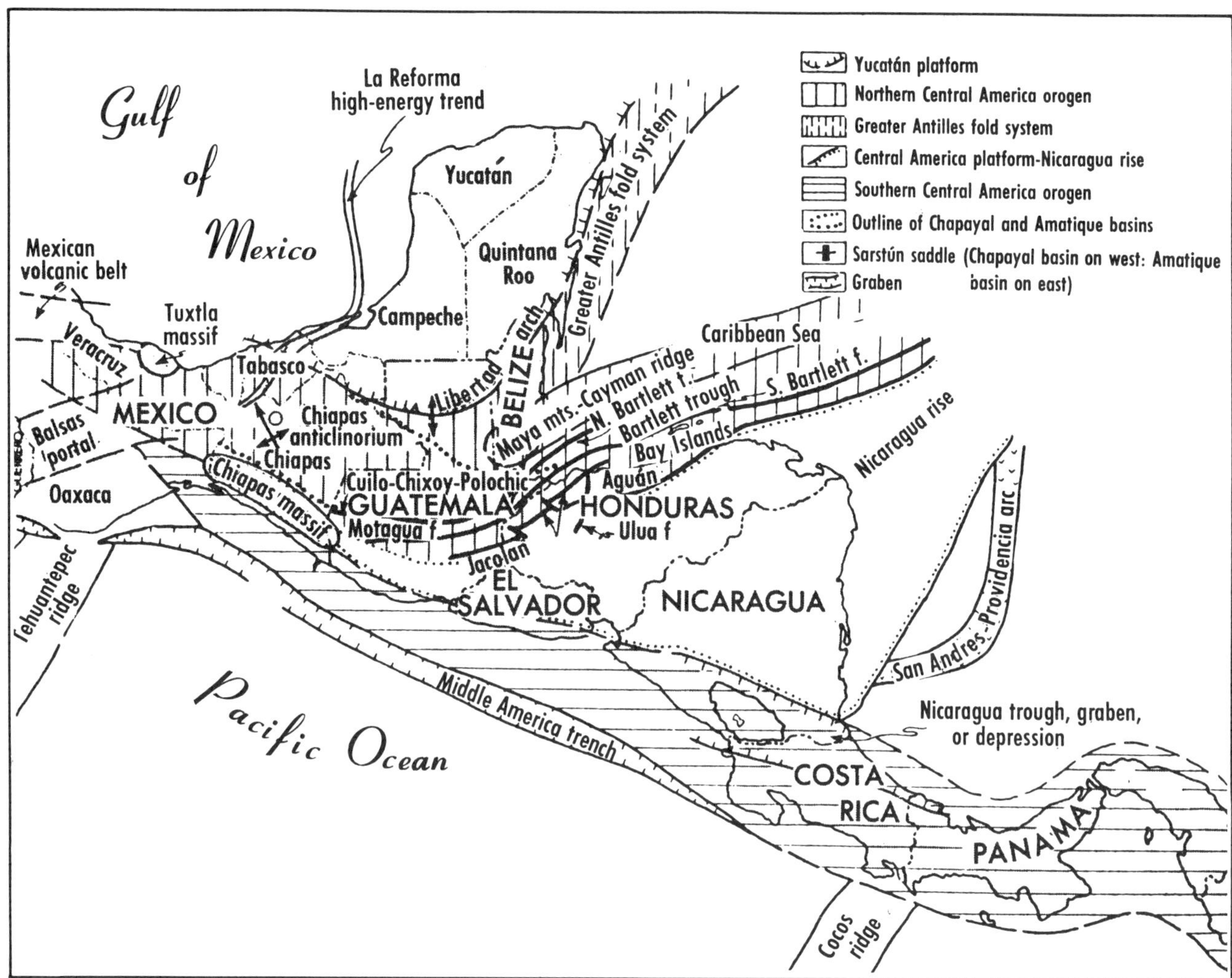

Fig. 12–5 Principal tectonic-stratigraphic provinces and tectonic features (courtesy *OGJ*, October 17, 1977)

OFFSHORE LATIN AMERICA

Brazil is the leader in offshore work in South American waters. Petrobras expected to produce 265,000 b/d of oil by the end of 1982 and 500,000 b/d of oil by 1985. Oil consumption is now about 1 million b/d.

Further development of the Campos area, step-outs to new strikes in the Amazon and Bahia, opening the remaining offshore risk blocks, and a trend away from Petrobras ownership of rigs has improved the exploratory outlook here. The country had a new drilling record in 1981 with 687 wells drilled, of which 121 were offshore.

Brazil has been in the offshore drilling business for 13 years. During this period, more than 500 wildcats and 200 development wells have been drilled on the continental shelves. A total of 330,000 km of offshore seismic lines have been run since 1968. At least 20 commercial fields have been discovered. Offshore wells average 125,000 b/d of oil, or more than half of Brazil's petroleum demand.[4]

There are 12 subsea sedimentary basins of more than 1 million km^2. Large areas of these basins have not been explored and should be thought of as potential oil-bearing basins. Offshore wildcatting will definitely increase in Brazil through 1985. A total of 147 wells were to be drilled in 1982, 166 in 1983, 179 in 1984, and 193 in 1985.

Brazil's proven oil reserves rose in the first 2 months of 1982 to 1.53 billion bbl. New discoveries in the Campos area pushed this figure up. One of the new Campos basin fields, Geroupinha, is a new field and not a step-out to Garoupa.

Petrobras has stepped up its exploration operations on the south bank of the mouth of the Amazon. Location of a new wildcat is 60 km southwest of the first discovery in the region, the 1 PAS-9, which tested 500 b/d of oil in 1980.

The 1-PAS-11 well, the second oil strike in the region, flowed 817 b/d of oil. Petrobras did a rerun of seismic data in this area, concluding there are isolated but potentially productive structures. Pecten-Chevron-Union found oil at 1-BAS-64 off southern Bahia in 1981. Flow was 1,000 b/d of oil at 2,370–2,379 m.

Argentina

A number of discoveries have been reported off Argentina. The most recent was the Total-Dimenex-Bridas group find off Tierra del Fuego. The Arie-X-1 wildcat had several thousand barrels of oil per day and gas from three zones. This was the fourth strike in this area.

Esso's Argentine affiliate found oil and gas in 1982 at 2 Salmon, off the southern tip of the nation in the eastern part of Tierra del Fuego Block 2. The well flowed 20 Mcfd of gas and 600 b/d of gas liquids at 2,625 m. In 1981, Esso had 3,100 b/d of oil in Tierra del Fuego East Block 1.

Caribbean

Mobil is the only oil firm active in Barbados, which has 600 b/d of oil production. Several new wells in shallow zones should increase this oil flow soon. New gas deposits also have been found off Trinidad and Tobago, but crude output is falling. National Gas of Trinidad-Tobago Ltd. was to begin recovering and marketing gas from two oil fields off Trinidad late in 1982. Production is expected to be 75 MMcfd at first, peaking to 100 MMcfd. Teak field is about 25 miles off Galeota Point, and Poui field is 10 miles off the point. These fields are operated by Amoco Trinidad Oil.

Exploration adjacent to the Tobago Trough has resulted in Miocene gas discoveries. Continued exploration between Trinidad, Tobago, and La Blanquila at the southern end of the end of the Aves Ridge, should prove rewarding.[5]

Ecuador

The Corp. Estatal Petrolera Ecuatoriana (CEPE) discovery of the Amistad gas field in the Gulf of Guayaquil has shown the potential of that area. A production platform is being built for production and transportation to shore. CEPE has identified seven prospective structures in the Gulf of Guayaquil concession and hopes to drill all of them during the next three years.

Venezuela

Petroven, the national company, planned to drill 208 wildcats and 1,100 development wells and work over another 2,020 wells in 1982. Lagoven, a subsidary of Petroven, slated 600 wildcats and workovers on 975 other wells both onshore and offshore in Lake Maracaibo in 1982. Lagoven is very happy with its new discovery in the Caribbean Sea an plans further exploration at Caribe One. This discovery averaged 3,526 b/d of oil on tests.

The Gulf of Venezuela, between Lake Maracaibo and the open Caribbean Sea, has not been explored, but it should be a highly prospective area. The onshore part of this basin in Falcón state has been a disappointment, but the offshore gulf area should be underlain by cleaner sandstone reservoirs than those onshore, which are too close to the sediment source.[6]

The Gulf of La Vela has had four oil discoveries in the Tertiary carbonates, but lack of funds has prevented their development. Other offshore basins look promising. The Cariaco basin should be productive, as may the flanks of the Bonaire and Los Roques Trenches. Any oil to be found here will probably be in Miocene reservoirs, mostly sandstone.

The Gulf of Paria, because of its great structural complexity and stratigraphic facies changes, has been rather unrewarding. However, the Orinoco delta is being explored.[7]

ONSHORE SOUTH AMERICA

Argentina has begun an ambitious effort to become energy self-sufficient by the end of this century. The nation hopes to increase its oil reserves to 6.3 billion bbl by 2000, compared to 2.5 billion in 1982.

One of the brightest spots in the country's campaign to boost production lies in the Magellan basin off Tierra del Fuego. Other exploratory work is needed in Rio Negro and Neuquen provinces, Comodoro Rivadavia in Chubut state, as well as in the Salta basin.

Success in the onshore part of the Argentina share of the Austral basin (Magallanes basin) continues to elude the explorers. However, offshore (in North-Sea-type environments) success has come twice. Esso found oil in 1981 200 km off the coast of Argentina as did Shell et al. 17 km off the coast (Fig. 12–6).

The Arauca field in northeastern Colombia is tagged as the nation's best oil strike in the past 18 years. More work is planned for this area of the Llanos basin, just south of the Arauca River.

Ecopetrol is stepping up its exploration/development program in 1982. The company shot 1,500 line miles onshore and 10,000 line miles of offshore seismic surveys in 1982. In 1981, exploration resulted in drilling 61 wells, of

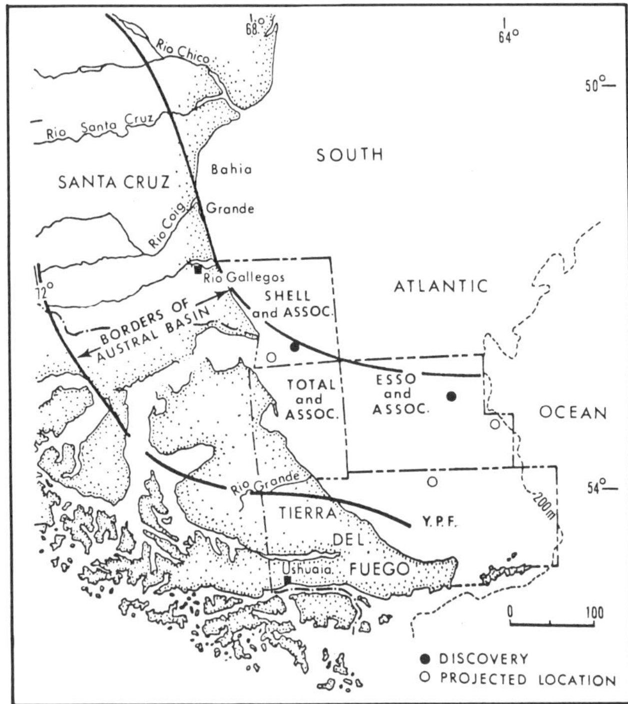

Fig. 12–6 Austral-Magallanes basin (after Arthur A. Meyerhoff and Anthony E.L. Morris)

which 16 were discoveries. Most of the wildcats were drilled by international companies working under association contracts with Ecopetrol. The company expected to shoot about 6,000 line miles in the Caribbean Sea in 1982 and 3,700 line miles in the Pacific Ocean, where exploration has been sparse. Colombia hopes recent discoveries in the Eastern Plains basin along with increased production from secondary recovery operations in the Middle Magdalena Valley basin will help reach production goals.

Petroperu plans to drill 16 wildcats and 89 field wells in 1982 on the northwestern coast of Peru. Nine wildcats and 22 field wells also are to be drilled in the northern jungle. Occidental Petroleum has found oil in two previously untested structures in the Amazon jungle. These structures are 12 miles apart. The 1 Ceci flowed 1,700 b/d of oil from two zones in the Chonta at 11,600 ft. The 1 Jibarito flowed 120 b/d of oil.

The Royal Dutch/Shell Group plans to explore a 7,772 sq mile area in Peru's southeastern jungle. Others want to work in the northern jungles. Petroperu with Belco will drill off the northwest coast on a tract of 988,400 acres, where the water depth is about 600 ft.

Venezuela's biggest challenge lies in the Orinoco Belt. About 600 delineation wells must be drilled. Ultimate aim is production of 1 million b/d of oil from the heavy oil belt. A total of $8 billion is targeted for two projects here. One will produce 100,000 b/d of oil from the Orinoco through existing systems by the mid-1980s. The other will develop by 1988 production of 170,000 b/d of oil, 8–10°, for upgrading on site to a light synthetic crude.

Ecuador's new petroleum law, designed to encourage foreign companies to resume exploration in jungle areas of the Andes Mountains, may complete its 2-year passage through the legislature. If everything is worked out, more foreign firms will be interested in acreage in the eastern jungles. Almost all of the country's production comes from the Oriente region. Oil moves across the Andes to the coast through the trans-Ecuadorian pipeline.

CEPE expects to spud the first of four wildcats in the jungle area during the 1982–83 season. Seismic surveys show a number of structures that CEPE believes are very similar to those in Peru, currently producing about 90,000 b/d of oil. This is virgin jungle, and all movements are by helicopter.

The CEPE-Texaco operation, where most of the nation's production originates, made 250,000 b/d of oil in 1981, mostly from the Shushufindi and Sacha fields, accounting for 155,000 b/d combined. New fields—Culebra, Yulebra, Yuca Sur, and Auca Sur—have been brought on stream (Fig. 12–7). The Cononaco field also will be on stream eventually.

There is renewal of exploratory interest in Jamaica, both onshore and off. The main part of the onshore effort is a four-well drilling program by the state-owned Petroleum Corporation of Jamaica (PCJ). PCJ has collected a great deal of seismic data during the past 5 years, and the results were good enough to warrant some exploration. Total 1982 spending for exploration will be $40 million.

The 1 Hartford was the first hole drilled. It was dry at 9,956 ft. The second test is the 1 Windsor. Projected depths for all of the wells is 10,000–11,000 ft.

Offshore, on the 980,000-acre Arawak block on Pedro Bank (a group of keys off the island's southwest coast), Union Texas Petroleum and Agip have plugged a dry hole. The well went to 15,053 ft near a gas seep.

PCJ intended to offer acreage by the fourth quarter of 1982 for bid off the southern coast, where a 3,100 line mile seismic survey was just completed. It also plans seismic work off the east coast of Jamaica.

Petro-Canada is providing drilling management services as a consultant to PCJ, and the Canadian government company's role may expand. Petro-Canada International Assistance (Pciac) has sent teams of petroleum specialists to Jamaica to discuss potential exploration ventures. Pciac was set up in Canada's national energy program to spearhead that nation's drive to help countries explore for and develop their energy resources in conjunction with the work of the international banks.

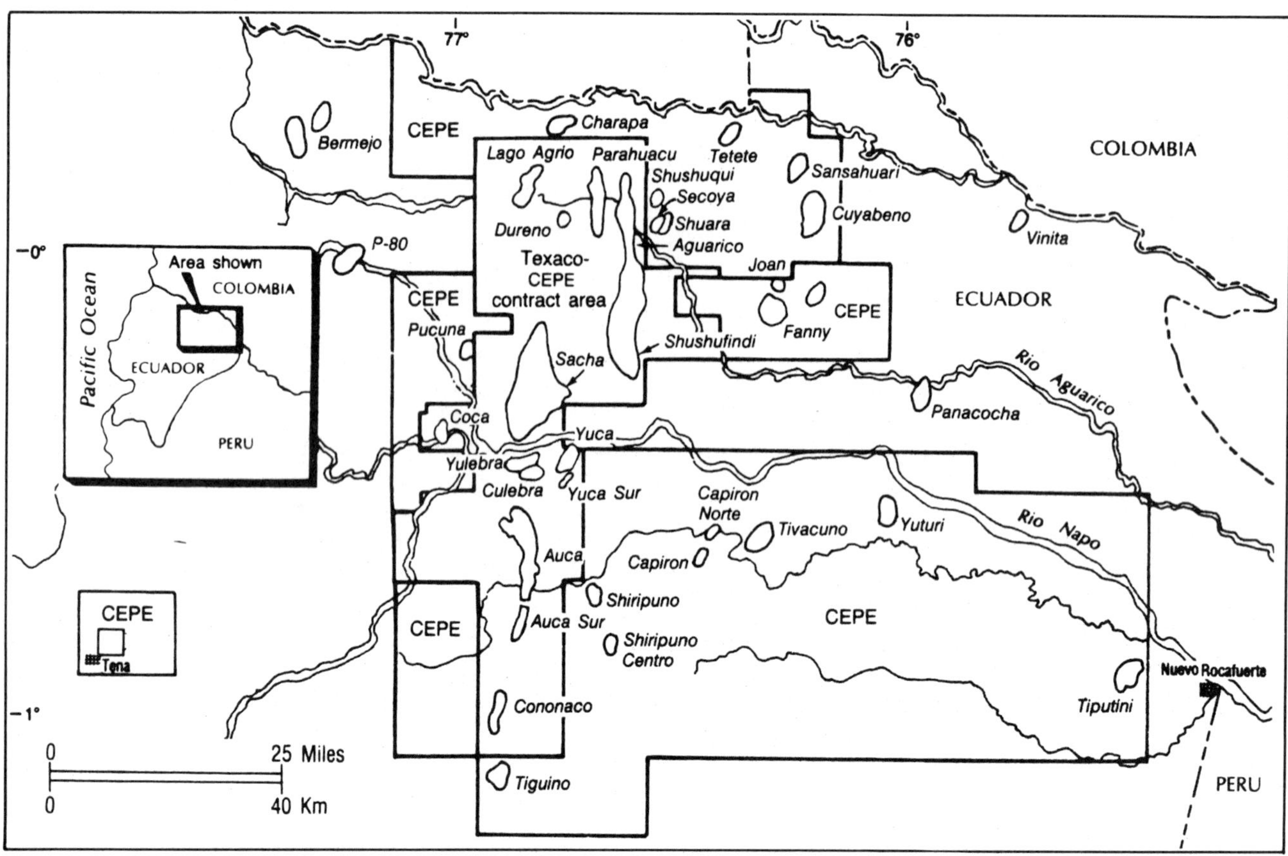

Fig. 12–7 Producing fields in the Oriente province of Ecuador (courtesy *OGJ*, June 21, 1982)

REFERENCES

1. Aceuedo and Dautt of Pemex, *Oil & Gas Journal* (July 20, 1981), p. 91.
2. A. Meyerhoff, *Oil & Gas Journal* (May 21, 1980), p. 121.
3. A. Meyerhoff and A. Morris, *Oil & Gas Journal* (October 11, 1977), p. 109.
4. *Offshore* (June 6, 1982), p. 151.
5. A. Meyerhoff, *Oil & Gas Journal* (October 20, 1980), p. 226.
6. Ibid.
7. Ibid.

13
The Eastern Hemisphere —North

Many of the world's largest oil and gas fields are in the northern part of the Eastern Hemisphere. Most are in the USSR, but some prolific oil fields from Europe's North Sea have been added to the giant and supergiant roster. Large oil and gas fields also are being reported in China as more and more information is disclosed. In other words, exploration is active across the entire length of the northern portion of the Eastern Hemisphere.

Large fields found in Europe in recent years include Italy's Malossa, discovered in 1973 and the Groningen gas field in Holland, one of the largest gas accumulations in the world as well as the field that opened the North Sea offshore theater. Other major discoveries are the huge Ekofisk (1969) in Norway's North Sea empire and other Norwegian North Sea finds such as Cod, Tor, Edda, Eldfisk, Statfjord, the huge Forties field (1970) in the North Sea waters of the United Kingdom, plus Thistle, Ninian, Beryl, Brent, and many others. As Figs. 13–1 and 13–2 show, the exploratory theaters of Europe and the USSR are many and varied.

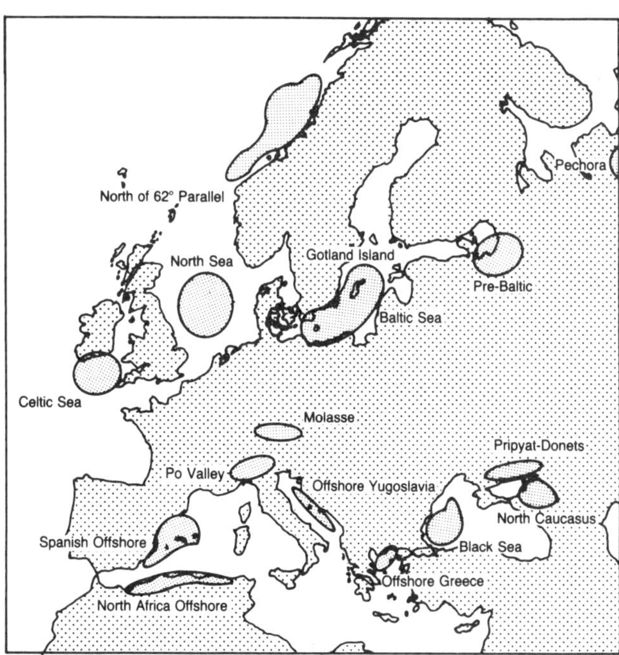

Fig. 13–1 European exploration theaters (from *IPE*, PennWell)

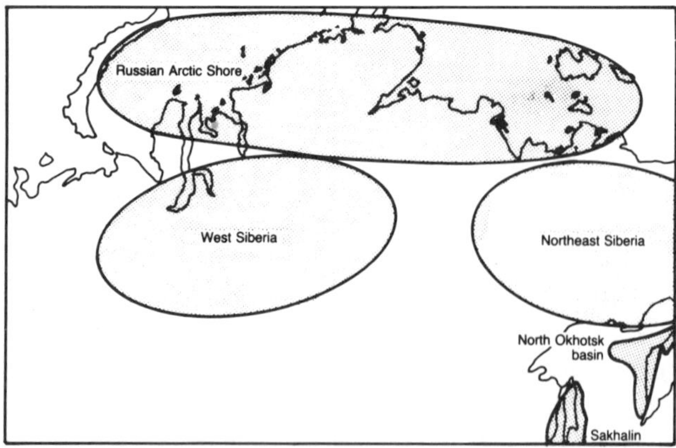

Fig. 13–2 USSR exploration theaters (from *IPE*, PennWell)

EUROPE

France

A significant wildcat is planned for the Mediterranean continental shelf off France in 1983. Cie. Francaise des Petroles and Elf Aquitaine plan to spud in the Gulf of Lyon, a deepwater basin considered to be the largest potential exploratory target remaining off France. The objective is 19,800 ft in as much as 4,620 ft of water, which will cost about $49 million.

Esso found gas in its 1 Ledieux wildcat on the Oloron Navarrenx permit in the Aquitaine basin of southwestern France. More tests are planned. Total Exploration began drilling on the government owned Pelchelbronn acreage in Alsace Province in eastern France. This test is to determine if more oil can be produced from an area that already has produced more than 44 million bbl of oil.

Greece

This nation's first domestic oil and gas production began in 1981, but policies by the new government cloud future work in Greece.

The Prinos field, 12 miles off Kavala in the Aegean Sea, has gone on stream with production at 28,000 b/d of oil. Recoverable Prinos reserves are estimated at 74 million bbl of oil and 4 million bbl of natural gas liquids from four Miocene sands. Geologists think more pays are in deeper zones that can be explored and tapped. South Kavala field, 7 miles south, also is now on production. South Kavala has recoverable gas reserves of about 30 billion cu ft from the Miocene sands.

Public Petroleum opened a large new gas field off the western coast of Greece in the western Peloponnes in the Ionian Sea. This field is 2 miles offshore and is estimated to be four or five times larger than the South Kavala field. Oil also may lie under the gas zones, so deep prospects are to be drilled between the islands of Paxos and Zakynthos.

In the summer of 1982, North Aegean Petroleum, operator for a group in the Aegean Sea, spudded the North Prinos wildcat. The location is 1 mile northeast of the Prinos oil field and ½ mile east of 4 Prino, a 1976 discovery which found light crude oil shows. The new well will be drilled down 12,000 ft. At least two more wildcats are planned on the 420,000-acre license area held by the group.

Studies prior to the Greek discoveries indicate an extension of the Paleogene Thrace basin in Turkey into the

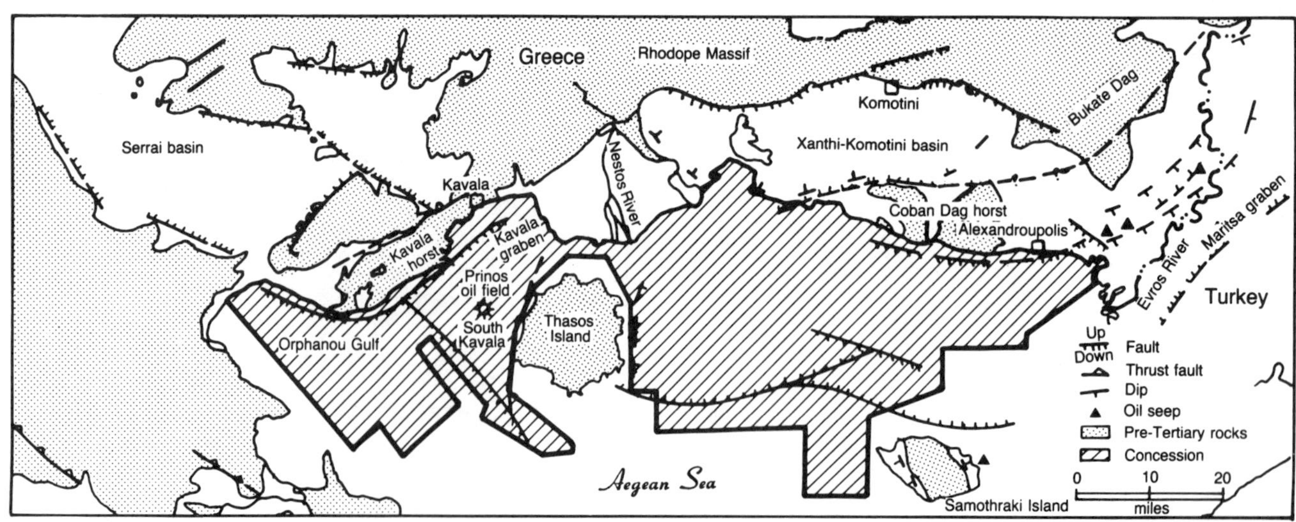

Fig. 13–3 Tectonics of the Northern Aegean Sea (after Oceanic Exploration, courtesy *OGJ*, May 14, 1973)

northern Aegean Sea as a complex graben essentially south of Thasos Island and north of Samothraki Island (Fig. 13–3). Smaller grabens were predicted northwest of Thasos and in Orphanou Gulf. Later geophysical work proved the Tertiary basins in the northern Aegean Sea were larger than first thought.

Ireland

Phillips Petroleum made a discovery in its second hole on Block 35/8 in the Porcupine basin off Ireland. Flow was 925 b/d of 40° gravity oil and 4.8 MMcfd of gas from pay at 13,110–13,220 ft. Various seismic surveys off Ireland may uncover some structures for future exploration, possibly yielding new fields like the Kinsale Head gas field off the southern coast.

Mediterranean

Montedison plans several confirmation wells near its 1 Vega discovery, 14 miles off Sicily in the Malta Channel. The discovery well, which flowed up to 4,200 b/d of 15.5° gravity oil during production tests, lies under two offshore concessions covering 73,500 acres.

Oil production began on Agip/Deutsche Shell's Nilde field southwest of Sicily in 1982. Output will be about 10,000 b/d of oil. This was Italy's first offshore oil discovery, found in 1980.

The Mediterranean nations Italy and Spain continue to lead the way in offshore exploration, development, and production. Bright spots in these nations include a new high in exploratory action off Italy and the beginning of the deepest water development on Spain's Casablanca field.

Most of Spain's exploratory work is centered in the Gulf of Valencia, where four fields are producing. The Bay of Biscay and the Gulf of Cadiz also are attracting new wildcat interest. Casablanca is Spain's biggest producer. Offshore oil production accounts for 90% of the nation's output, 68,000 tons in 1981. Current production from three wells in Casablanca is 18,000 b/d of oil. Production will reach 40,000 b/d when a three-well development plan is completed.

A group led by Occidental of Spain spudded A-1 Vizcaya wildcat in the Bay of Biscay off northern Spain in 1982. The Jurassic test will go to 15,400 ft. This is the first of two wells to be drilled by the group in which Canada Northwest Energy of Calgary is a partner. Canada Northwest is doing the Casablanca drilling, too.

The settlement in the International Court in The Hague of the dispute with Libya over the contested Gulf of Gabes should open the way for renewed exploration and development off Tunisia. Isis field, with estimated reserves of 110 million bbl of oil, was discovered in the disputed Centre Oriental permit in 1974 by Total Tunisia. Production off Tunisia is limited to Ashtart field at 45,000 b/d of oil.

Italy

Italy's Malossa gas condensate field is the deepest hydrocarbon deposit on production in Europe. The discovery of this field set off a flurry of exploratory drilling in the Po basin of northern Italy, but nothing as big as this field has been found. The field has produced 16 million bbl of condensate since discovery in 1973. Pays are in the 17,000–21,231 ft range.

The North Sea

Recent highlights of activity in the North Sea include the revival of the long dormant West German sector of the North Sea with some promising discoveries; discovery of oil in the traditional gas province of the Netherlands North Sea; first production of oil from Gorm field, the second to go on stream in Danish waters; and the first commercial hydrocarbon discovery in the Norwegian Sea north of the 62nd parallel. In addition, exploration in the U.K. waters has built up to the highest level since 1977, and the number of rigs is growing. Seventy-three rigs are active in the North European play. The U.K. dominates with 30 rigs. Four new rigs have moved into Norwegian waters. Fifteen rigs are working in Dutch waters.

Appraisal work is under way in Norway's biggest gas prospect, the Troll field, Block 31/2. Reserves for the field are estimated at 70 trillion cu ft, based on the results of six wells drilled. Troll is a shallow reservoir covering a large area that would require multiplatform development.

With the government planning to open up the Traenabank and extend the drilling season into northerly waters, the areas north of the 62nd parallel may start to command an even greater amount of Norway's drilling money. Exploration for hydrocarbons on the Norwegian continental shelf north of the 62nd parallel began in 1981 (Fig. 13–4). Two areas were opened: the Haltenbank off mid-Norway and the Troms 1 area off northern Norway.[1]

Interest in the Jurassic play of the southern Norwegian North Sea has been refired as a result of the discovery of Saga Petroleum AS in Block 2/2. Saga's well flowed 1,323 b/d of oil and 342 Mcfd of gas, 20 miles northeast of the Ekofisk complex.

In British waters, the British National Oil Corporation (BNOC) has named the discovery in the northeast corner of its Thistle field block Don. The field, with only 100 million bbl of oil reserves, is considered marginal at this point.

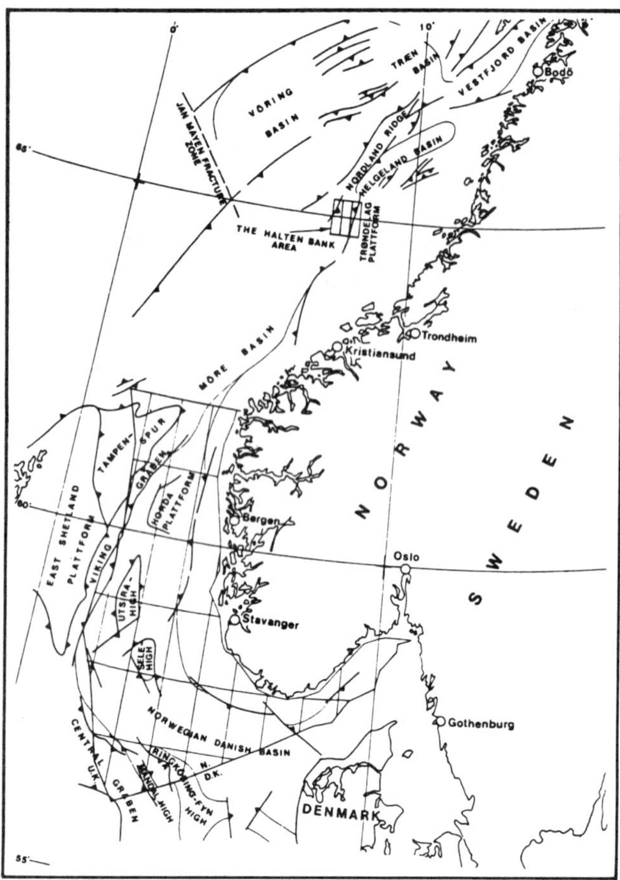

Fig. 13–4 Structural framework of the Norwegian continental shelf south of 68°N (modified after Ronnevik et al., 1977, and Jorgensen et al., 1980, courtesy *OGJ*, April 12, 1982)

An oil find by Hamilton Brothers Oil and Gas, 4 miles west of the Argyll field in the North Sea, has been confirmed as commercial and is called Duncan. This field lies 80 miles off Yorkshire.

BNOC also has a discovery on Block 16/21B, adjacent to the North Sea Sun group's Balmoral field. The new well was drilled on a structure contiguous with Balmoral, and it flowed 5,100 b/d of oil from the Paleocene.

Crude oil production from the U.K. North Sea was more than 2 million b/d in 1982. The British have 20 producing fields in this area. In the less glamorous sectors of the North Sea empire, Holland got its first regular supplies of offshore oil, and Denmark is busy installing producing gear.

Off northern Norway

The industry rates the petroleum potential as high off northern Norway. The Troms 1 area has had five wells drilled in three of the allocated blocks. Five important geological structural features have been found in the Troms 1 area:

1. The Troms/Finnmark fault zone, northeast-southwest trending zone of normal faults separating the Upper Paleozoic to Cenozoic sedimentary basins from the more stable platform area to the southeast.
2. The Hammerfest basin, an east-west trending basin with considerable deposits of Upper Paleozoic to Cenozoic sediments. The basin is subject to only minor Mesozoic tectonics, resulting in a series of small, east-west trending normal faults.
3. The transition zone between the Hammerfest and the Tromso basins, an intensely tectonized zone, dominated by extensional forces and tilted fault blocks.
4. The Tromso basin, a very deep basin containing immense deposits of Upper Paleozoic/Mesozoic/Cenozoic sediments. The basin is dominated by salt tectonics with the salt probably Permian in age. Salt diapirs have been traced by seismic from sea floor to a depth of more than 10,000 m.
5. The Senja ridge, a structural high of probably post-Jurassic age, bordering the Tromso basin toward the west.[2]

Most of the deposition in the area took place in Lower Cretaceous time. The main source rock in the area seems to be an excellent Upper Jurassic shale, which is fully mature in the transition zone, probably overmature in the Tromso basin and marginally mature in the Hammerfest basin. The main reservoir section consists of transgressive Lower to Middle Jurassic mature sandstone deposited in a high-energy coastal environment. The main targets in the area are tilted fault blocks.

Of the five wells drilled, two have found gas in good reservoirs, and one has dead oil shows and excellent reservoir. Another well has oil shows in a tight reservoir, and one was a duster.

This portion of the North Sea has strong winds, rough seas, cold weather, and depths of more than 200 m. The area is also more than 2,000 km from European markets. Any discoveries of commercial value here will have to face those hard facts.

History

The petroleum saga of the North Sea began in 1969 with the discovery of the huge Ekofisk complex in Block 2/4 in Norwegian waters. Vast reserves of oil and gas have been found during the years since 1969. Current production is averaging 1.8 million b/d of oil in the U.K. sector, enabling that nation to outstrip consumption for the first time in history.

Exploratory work is the highest since 1977 in the U.K. sector. In 1981, 48 wildcats and 161 development wells were drilled. This surge resulted in a dozen discoveries, the most since 1976.

Ekofisk is in the Central graben in the southern part of the Norwegian portion of the North Sea. The complex now contains six fields. North Sea reserves total more than 18 billion bbl, with some estimates as high as 40 billion bbl. Ekofisk is on a simple north to south oriented anticline.

North Sea outlook

The industry is convinced there are undiscovered giants in the North Sea, particularly in the untouched areas to the north and northwest. Seismic activity helped delineate the subunconformity trap in Block 34/10 in the Norwegian part of the North Sea. Future improvements in seismic should help uncover more structures to drill for hydrocarbons.

The 34/10 Delta structure lies 200 km west-northwest of Bergen and 20 km east-southeast of Statfjord field (Fig. 13–5). It was declared a commercial field in 1980, and the current reserve estimate is about 1.5 billion bbl of oil.[3]

Most fields in this part of the North Sea (such as Brent, Statfjord, Dunlin, Cormorant) are on tilted fault blocks between the Shetland platform to the west and the Viking graben to the east (Fig. 13–6). The main reservoirs are Jurassic sands. Oil sources were the Upper Jurassic shales and seals were Cretaceous shales.

The first well was drilled in 1978, confirming the presence of oil in the Middle Jurassic Brent sand near the crest of the structure. The next three wells also found oil in the Brent sand and confirmed the structural concept of internal faulting.[4]

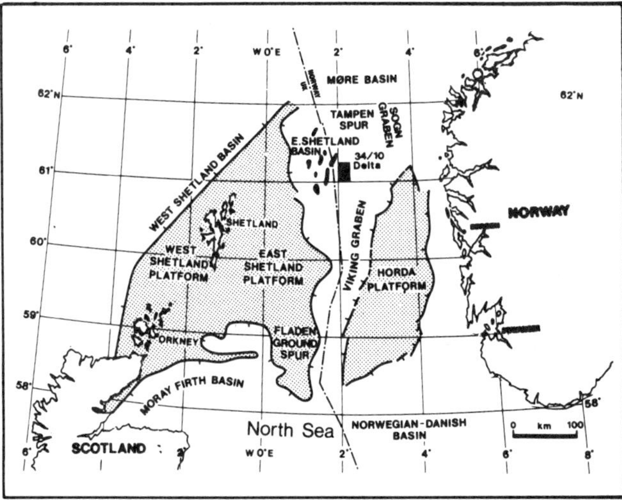

Fig. 13–5 Block 34/10—61° N, 2° E (courtesy *OGJ*, November 9, 1981)

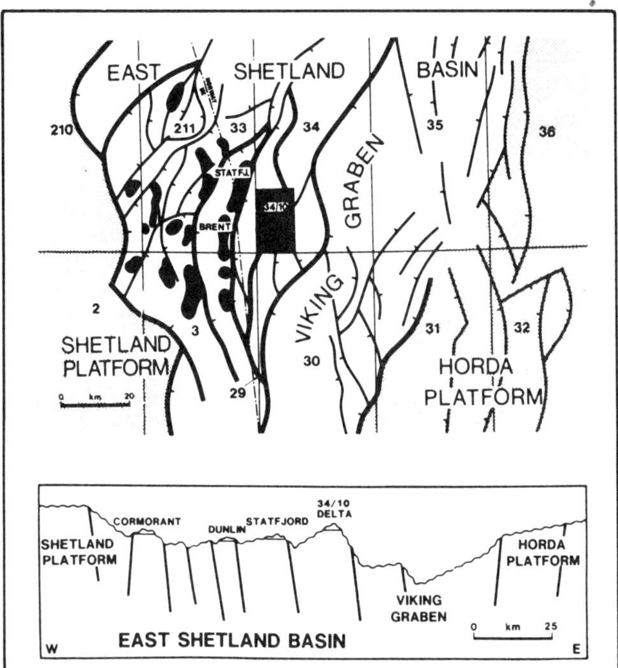

Fig. 13–6 Simplified structure map of the area around Block 34/10 with the east-west cross section from the Shetland to the Hordo platform through the block (courtesy *OGJ*, November 9, 1981)

Today's concern is that self-sufficiency in the U.K. may be shortlived. Output is to start declining toward 1990, and oil companies feel the fiscal regime should be attuned to this, particularly since discoveries are becoming smaller and more expensive to develop.

The Shell/Esso Tern field might be regarded as just the sort of field that must be exploited if Britain is to retain its spot as an oil producer.[5] This field is in 550 ft of water, northeast of Shetland. It has 140 million bbl of recoverable oil. The cost of a single steel platform development was set at $1.5 billion, so the project has been deferred.

The Eider project also has been put away for now. The BP/Phillips Andrew field, 100 million bbl of reserve, has slipped into hiding, and various projects around existing fields have been forgotten for the time being.

Meanwhile, Marathon recently completed its fifth well on North Brae near Brae field in Block 16/17a. Previous wells are being studied before plans are made to develop this field. If application were made to the Department of Energy before 1982 year-end, this could lead to a 1984 platform order and production to begin in 1987.

Most exploratory work now is aimed at the mature central North Sea area. All of the 12 discoveries made in 1981 were in this region.

Exploratory and development work in Norway's portion of the sea has proved enough reserves to maintain that nation's production at current levels for the rest of this century. This does not consider the vast potential seen in the huge area under exploration now and in the future north of the 62nd parallel. Commercial discoveries in this part of the theater are only a matter of time.

EASTERN EUROPE

Exploration in eastern Europe's Baltic Sea, after years of near dormancy, is back on track again. Oil discoveries near the Baltic Coast or just offshore in Poland and East Germany have renewed interest and inspired more exploratory drilling. The most significant of the recent eastern European finds are two Polish oil discoveries—at Karlino, completed in 1980 close to the East German border and offshore from Gdynia (Fig. 13–7).

These two discoveries bolstered the Communist bloc hopes for finding commercial production all along a 500-mile sector of the Baltic Sea's southern shelf between Latvia and West Germany. The Karlino discovery was apparently the best yet in Poland. Flow was as much as 35,000 b/d of oil. The 1 Daszewo wildcat went on production in 1981 for 7,300 b/d of oil. That was more than Poland's entire 1980 crude oil production of 6,600 b/d.

THE USSR

The Soviet Union has many basins and huge oil and gas fields. The largest oil field is Romashkino, discovered in 1948. This vast complex has produced 12 billion bbl of oil from Devonian pay at 5,791 ft. The next largest is Arlan with 2.1 billion bbl of oil produced from the Carboniferous at 4,429 ft. Other large oil fields include Kotur-Tepe, Uzen, Ust-Balky, and Samotlor. Production of oil and gas is widespread in Russia. Fig. 13–8 shows where and how much the country produces.

The critical problem for the Soviets at this stage of their oil industry's history is to find more giant oil fields in western Siberia, although discovery of another Samotlor is unlikely. Several large, unpublicized oil discoveries have been made in Tyumen Province far north of the Middle Ob district, which includes Samotlor, Fedorovo, and Mamontovo. None of the new finds is expected to rival any of Tyumen's big three.

The new northern oil fields in Tyumen, 125–150 miles north of Ob River's port of Surgut, are in the Yamalo-Nenets Autonomous District. Two of the fields are the giant Kholmogorskoye, discovered in 1973, and the Karamovskoye, which are both on production. They are producing about 85,000 b/d of oil. Development of the promising Sutorminskoye and Muravienkovskoye fields is under way.

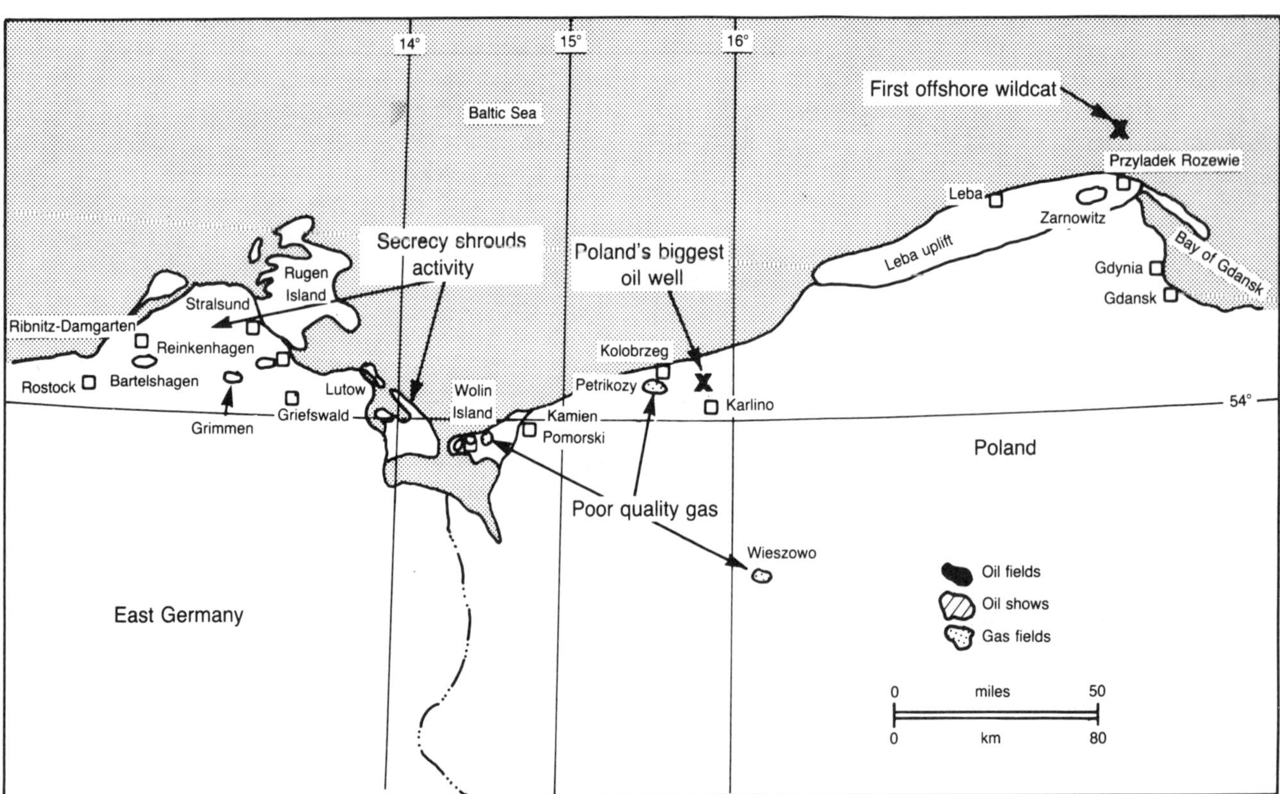

Fig. 13–7 East German and Polish sector of the Baltic Sea (from *IPE*, PennWell)

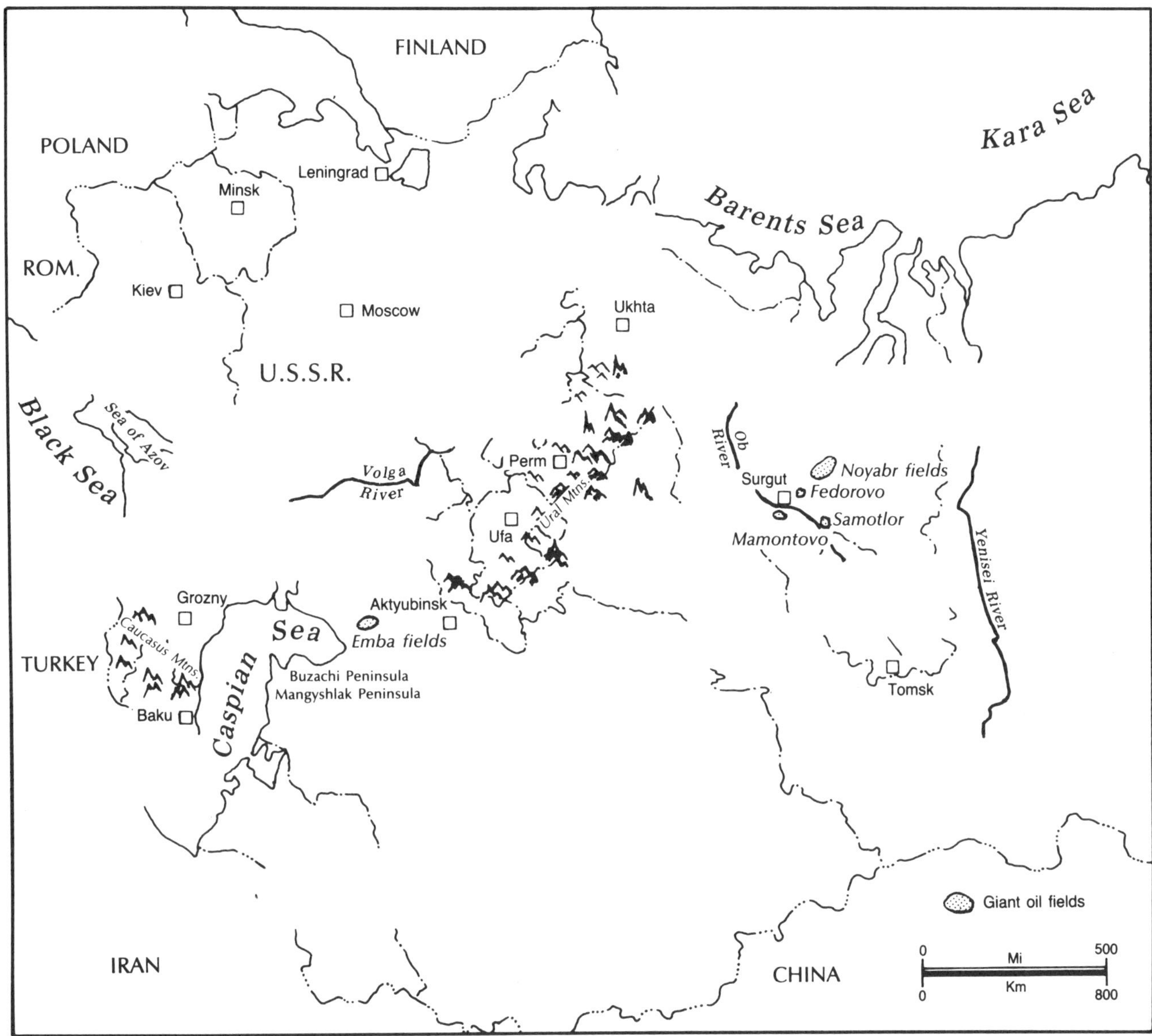

Fig. 13–8 Areas of USSR oil production (from *IPE*, PennWell)

By the end of 1985, Noyabr Association oil fields are to make 720,000–800,000 b/d of oil. Drilling in the area for 1985 is expected to be 5 million m or 16.4 million ft of hole.

Soviets repeat often than only 20% of western Siberia has been explored. They add that oil reserves no doubt will be found in other areas of the vast basin. Most of western Siberia's oil comes from Cretaceous rocks at 6,562–8,202 ft.

Most of the oil in Tomsk Province, nearby, comes from Mesozoic rocks in fields along the middle reaches of the Ob River. New fields being developed farther south along the Vasyugan River, an Ob tributary, are comparatively small. Overall, Tomsk crude production comes from Sovetskoye field on the Ob near Samotlor. The field makes 216,000 b/d of oil. Tomsk output should be up to 300,000 b/d of oil by 1985.

Apart from western Siberia, only four Soviet oil areas have been designated to achieve significant production increases during the next several years. Kazakhstan, north and northeast of the Caspian Sea, will boost to 500,000 b/d in 1985. More production will come from new fields on the Buzachi Peninsula just north of the larger Mangyshlak Peninsula; in the Aktyubinsk Province, about 400 miles to the northeast; and possibly from new discoveries in the old Emba district.

Gas reserves

The Soviet Union's explored reserves of natural gas (proved plus probable) include a small volume of what would be called possible reserves by U.S. definition. The Soviets' proved gas reserves are thought to exceed 800 trillion cu ft, compared with 500 trillion cu ft for Iran and 187 trillion cu ft for the U.S. The growth of Soviet-explored gas reserves since 1953, when the first commercial gas field was found in western Siberia, has been spectacular. Most of the new gas discoveries in recent years have been in the northern sector of Tyumen Province, near and north of Arctic Circle. Some big strikes also have been made in Soviet Central Asia.

In 1981, a big gas strike was made 200 miles northwest of Norilsk on the east bank of the Yenisei River, 300 miles north of the Arctic Circle in the Taimyr region of Krasnoyarsk Territory. Flow was 53 MMcfd from 6,713–6,999 ft. This well is in extreme northwestern eastern Siberia, but it is adjacent to and geologically akin to western Siberia's northern Tyumen Province, which holds most of the big fields in Russia. Soviet Central Asia ranks second under western Siberia in gas production and reserves.

Discovery of the giant Dauletabad field in southeastern Turkmenia near even bigger Shatlyk field apparently ensures continued gas production in Soviet Central Asia through this decade. Shatlyk, found in 1968, has estimated reserves of 1.5 trillion cu ft of gas.

A flurry of new discoveries has been reported just east of Turkmenia in Uzbekistan, where gas production temporarily peaked in the early 1970s as output from huge Gazli field began to decline.

Other Russian events

The Soviets' second deepest hole, Saatly SG-1 in Azerbaijan, has drilled below 26,246 ft. This well was spudded in 1977. The old total depth record was 49,212 ft on the Kola Peninsula, according to the Russians. Goal for the new hole is 36,089 ft by the end of 1985.

Soviet offshore exploration/production is on the upswing. By 1985, the USSR plans to boost offshore drilling activity 1½ times over the 1980 levels. The main reason for this boost is a report by geologists indicating potential oil-bearing sediments cover about 70% of the Soviet shelf area.

Most current offshore production is in the Caspian Sea, where as many as 12 oil and gas fields are being developed. Discovered in 1979, the 28th of April field east of Azerbaijan's Apsheron Peninsula has two wells on production, with more platforms and wells on the way (Fig. 13–9).

Work continues in the Bulla-More gas field, 60 miles southeast of the 28th of April field. By 1985, drilling in the Black Sea and the Sea of Azov should increase 2½ times. Interest in the Arctic and far eastern offshore areas also is increasing. Further exploration is planned in the Sea of Okhotsk near the northern tip of Sakhalin Island, where the Odoptu and Cahivo oil fields were found from 1977–80.

The USSR's far eastern Kamchatka Peninsula is expected to begin commercial gas production by the middle 1980s (Fig. 13–10). Kamchatka, nearly as large as California, will be the Soviet's second far-eastern area to achieve commercial production. Oil and gas fields lie across the Sea of Okhotsk on Sakhalin Island.

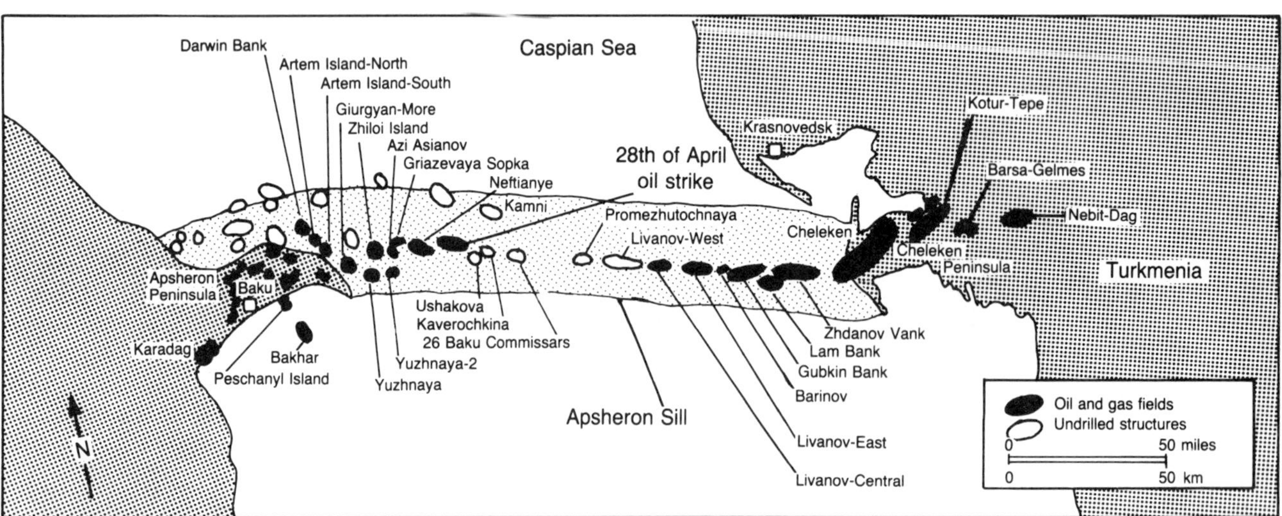

Fig. 13–9 Areas where Soviets hope to find offshore production (courtesy *OGJ*, October 8, 1979)

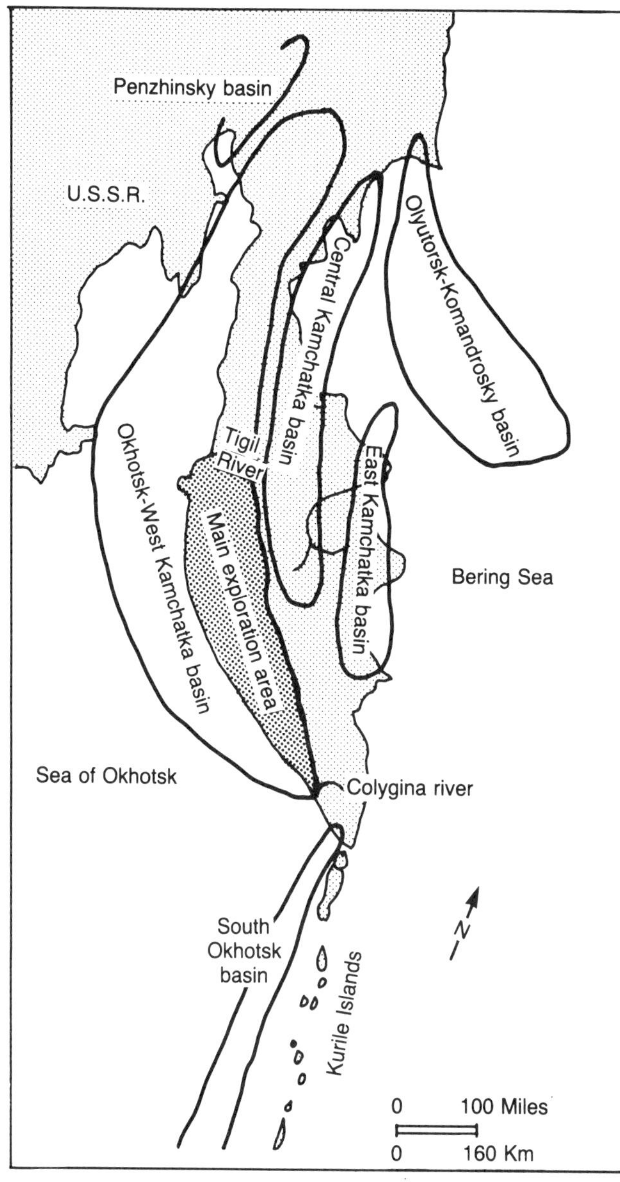

Fig. 13–10 Kamchatka area, USSR (courtesy *OGJ*, June 28, 1982)

Production will come from the Kshukskoye field on the southwest coast. This field was discovered in 1980, blowing out at 35 MMcfd of gas. The production potential is rated at 8.8–35 MMcfd. Exploration in western Kamchatka is under way from the Golygina River north to Tigil, a distance of more than 400 miles.

Hydrocarbon indications were reported on the peninsula as early as 1755. The area has possible petroliferous basins along the east and west coasts, near the northeast and northwest base, and on the southwestern tip. On the east coast, oil seeps have been found along the Kamchatka and Ozernaya rivers near the Bering Sea and farther south along the Bogachevka River near Kronotsky Bay.

The best discovery in the 1970s was near Icha on the west coast at Limanskaya. A gas well flowed from the Upper Cretaceous sand at 6,791–6,919 ft. This well indicates the possibility of finding offshore deposits of gas in the Sea of Okhotsk.

By 1980, Kamchatka had 12 potential oil or gas districts, totaling 77,000 sq mile onshore and 54,000 sq miles offshore. Eight locations had evidence of oil, and more than 20 had gas indications. Thickness of potentially petroliferous Mesozoic and Cenozoic rocks ranges from 3,000–8,000 m.

The Caspian Sea

This sea will remain Russia's prime offshore exploration target through this decade. Fig. 13–11 shows the abundance of Caspian Sea targets. Prospects for major finds in the Caspian, especially in its southern basin, are still bright. Soviet geologists believe each of several large undrilled southern basin anticlines may have gas reserves of 1 trillion cu m (35.3 trillion cu ft) or more.

Potential of deep Mesozoic reef formations in the southern Caspian Sea has been compared recently to Mexico's Reforma and Golden Lane areas. A vast majority of the inland sea's more than 200 attractive structures detected by geophysical surveys through the late 1970s have never been tested.

The South Caspian basin embraces the offshore area of the southern part of the Caspian Sea and its adjoining Kura and West Turkmenian intermontane basins, together with the coastal area of Iran. It is composed of Mesozoic and Cenozoic rocks 15–20 km thick and the Neogene-Recent stratigraphic interval reaching 8,000–10,000 m. The land sectors have widely developed, clearly marked anticlinal zones complicated by faults and mud volcanism. The basin interior is characterized by maximum downwarping in Pliocene-Quaternary times. The roof of the Middle Pliocene series has subsided 3,000–6,000 m. Quaternary deposits exceed 2,000 m in thickness. In the eastern part of the area are ridges 60 km long and 30 km wide. The South Caspian basin has a broad range of oil and gas reserves, occurring in the section from Mesozoic to Recent, and main pays are the Middle and Upper Pliocene sandstones.

Land areas contain many oil fields. The first offshore fields developed in Russia were in this basin—the Neftianye Kamni, Artem Island, Uyzhnaya, Darwin Bank, Zhiloi Island, Bulla Island, Cheleken Offshore, Zhdanov Bank, and others. The basin's depression in the middle of the offshore area is promising but has not yet been drilled.

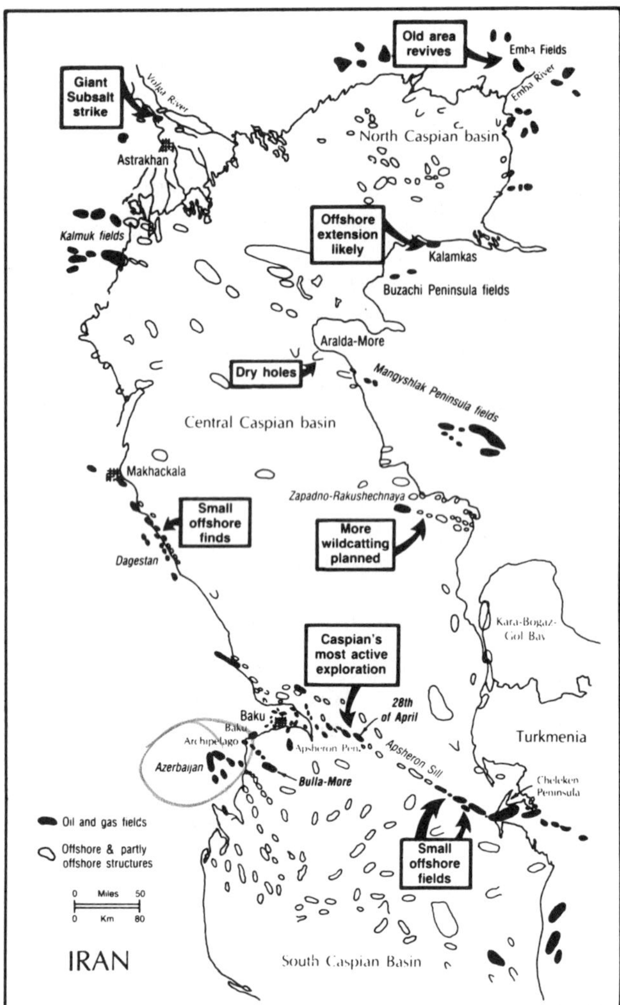

Fig. 13–11 Caspian targets (courtesy *OGJ*, October 12, 1981)

Far Northeast

The USSR recently claimed its first oil discovery on the Chukchi Peninsula, opposite Alaska. This strike is in the Beringovsky district on the southeastern part of the Chukotsk Peninsula, close to the southwestern coast of Anadyr Bay. Drilling in this area has been under way since 1963. By the late 1970s, about 25 strat tests and exploratory wells had been drilled in the province, which includes extensive offshore area in Anadyr Bay and the Bering Sea.

Many oil and gas seeps and more than 60 structures have been found in the region. This is indeed one of the nation's highly promising future oil and gas provinces, akin to the prolific prospects indicated for its neighboring Alaska.

CHINA

The annual oil production in the People's Republic of China has risen from 2,420 b/d in 1949 to 2,170,000 b/d of oil in 1981. A total of 30 prospective petroleum basins are in China. All produce or are able to produce from Triassic or younger strata, with some small amounts of production from the Carboniferous and Permian sediments. Six have sizable commercial production, seven are marginal, six have potential, and the rest are untested for the most part.[6]

Twenty-three basins are onshore with only three known to contain marine reservoirs or source materials. Production ranges in age from Carboniferous through Pliocene. One basin is partly onshore and offshore, with production from both marine and nonmarine Tertiary beds. Six basins are entirely or mostly offshore, with potential production from both marine and nonmarine Tertiary rocks.[7]

Sedimentary rocks are widespread throughout China. Besides the 4 million sq km onshore, many sedimentary basins are on the offshore shelves and islands. These offshore areas of China offer some of the most promising places to look for future oil in the world today. Eight large sedimentary basins are onshore, each with an area of more than 100,000 sq km (Fig. 13–12).

Some of the basins have been confirmed as rich oil and gas areas. Others are assumed to be prolific in nature. Most of the basins are still inadequately explored, and some have never been explored. Indeed, a virgin territory for the oil industry lies in China. One of the most prominent characteristics of these Chinese basins is the complexity of tectonics and sedimentation.

Oil and gas reservoirs in China are classified as follows:

1. Anticlinal oil and gas accumulations.
2. Buried hill oil and gas accumulations, mostly found in eastern China. These are found in the Renqiu structural zone and Xinliongtai structural zone in the Bohai Gulf basin.
3. Oil and gas accumulation zones on an upheaval of the basement with thin sedimentary cover. Large oil fields are usually in this type of trap. These are found in the Caoyanggou-Fuyu swell in the Songliao basin and the Luzhou swell in the Sichuan basin. A large reserve has been found on the Luzhou swell with pays in the Permian and Triassic. Oil reserves in the Daqing swell account for 80% of the total discovered oil in place in the Songliao basin.
4. Monoclinal oil and gas traps generally developed on the slope of a nonsymmetrical basin with rather large and widespread oil and gas fields.

Foreign help

China does not downplay the role of outsiders in recent offshore exploration. Beijing gives wide, approving pub-

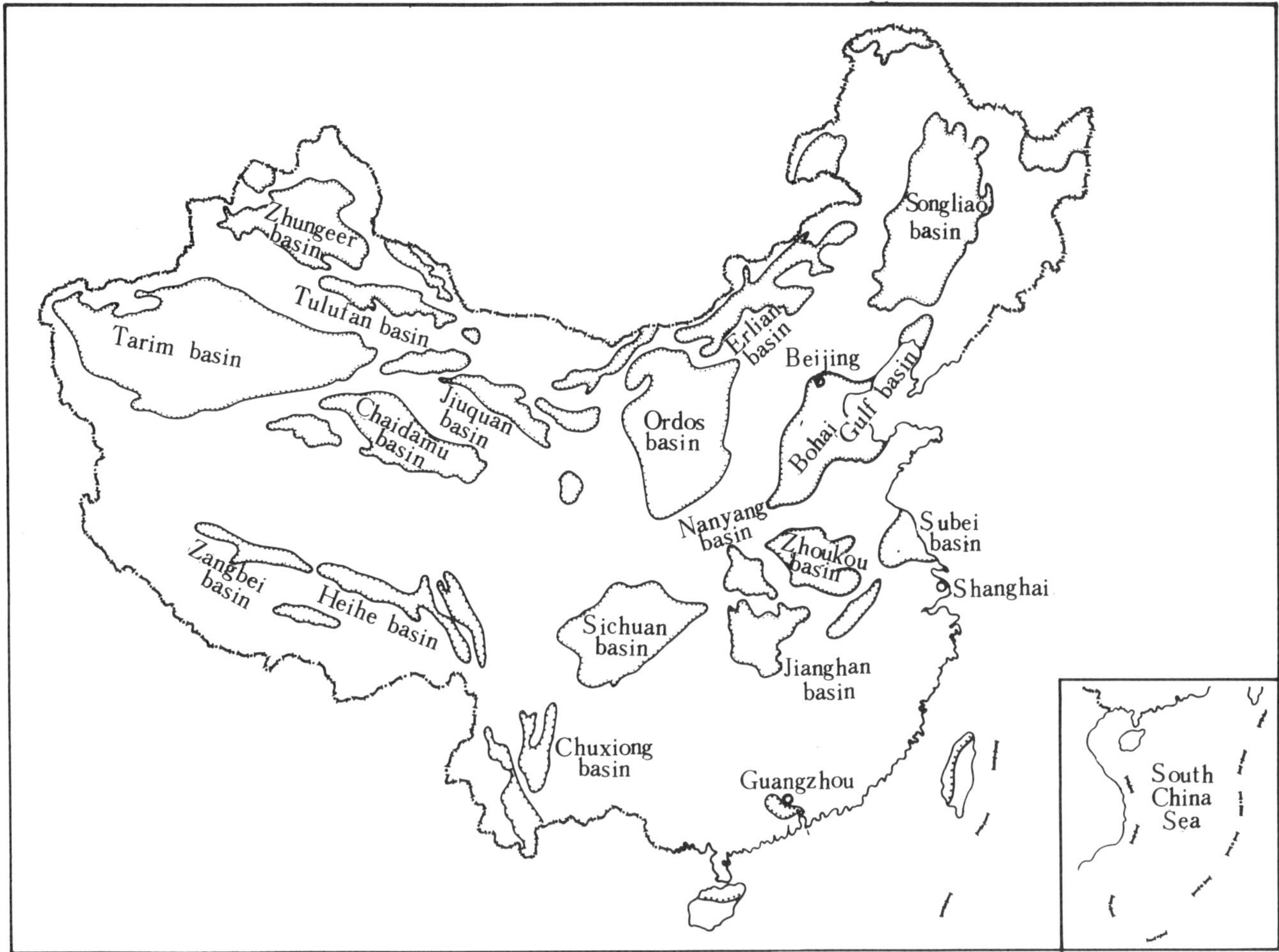

Fig. 13–12 Sedimentary basins of China

licity to the contributions by Japan, America, France, and other countries from other nations. China assures foreign interests that their participation in offshore oil development will be just as welcome as it has been in exploration.

Beijing says the first stage of offshore exploration—geophysical surveying—has been completed. The second stage (drilling on a large scale) will now proceed.

A well drilled in 1981 in the northeastern part of Beibu Gulf, a joint project of the Petroleum Company of China and Total of France, has made 4,672 b/d of oil and 45.5 MMcfd of gas. This well has two pay zones, both rich in oil and gas. In Bohai, a well drilled jointly with Japan has produced 7,300 b/d of oil from a Mesozoic sandstone. Oil is similar in grade to that of Saudi Arabia.

More than 100 wildcats have been drilled in Bohai. Oil and gas have been found in 13 structures and faults. Three offshore platforms are on production. Of 17 wildcats drilled in Beibu Gulf, 8 have produced commercial oil.

Development around the new Bohai discovery could cause production in the area to climb to 300,000 b/d of oil. This would permit China's oil exports to Japan to begin going up again before 1985. Officials say this is the best discovery yet off China, and it opens broad vistas for future successful exploration in the gulf. More than a dozen major geological structures had been identified in the Bohai by 1978. The gulf is about the size of Lake Superior, covering some 32,000 sq miles.

Other optimism about China's future reigns. China has designated a sector of the East China Sea off Zhejiang Province as its most promising future offshore oil and gas exploratory theater. A long strip of the coastal shelf here contains the largest oil- and gas-bearing structures ever found in China, according to government officials. Offshore work reportedly has uncovered six large petroliferous basins. Other promising areas are the Bohai and south Yellow Sea basins to the north and the South China Sea's Zhujiang

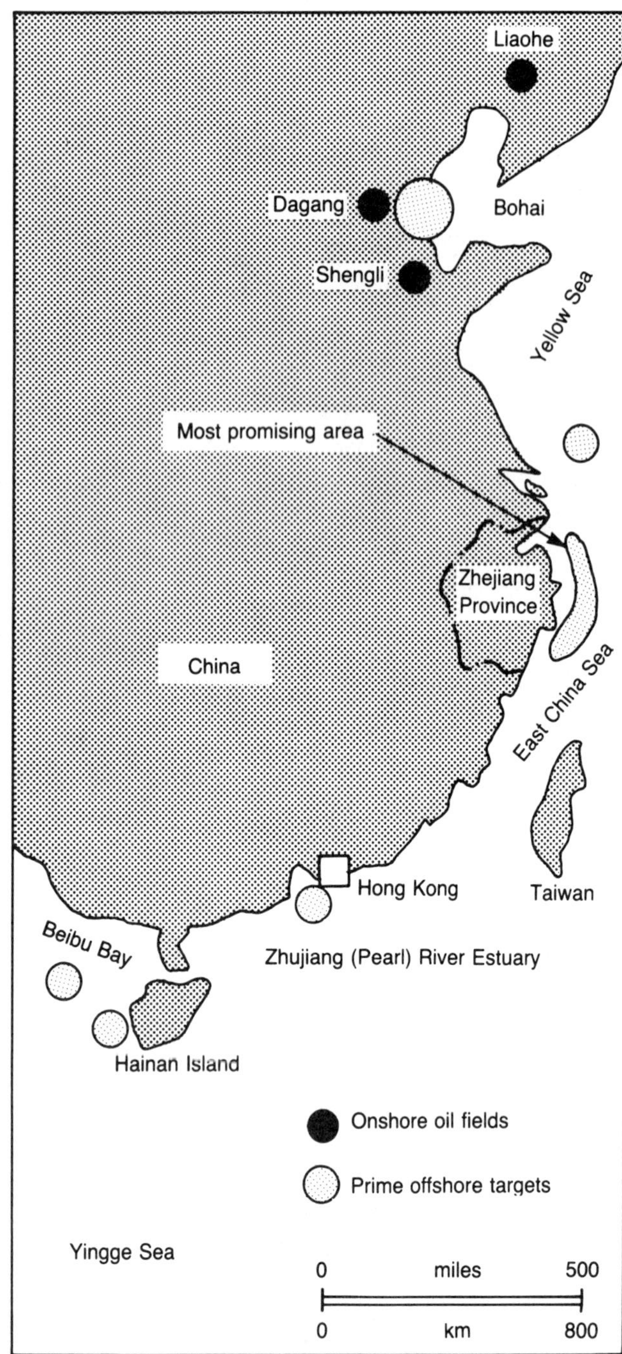

Fig. 13–13 Prospects of China (from *IPE*, PennWell)

(Pearl) River estuary, the Beibu Bay or Gulf of Tonkin, and the Yingge Sea, southwest of Hainan Island. These areas are shown in Fig. 13–13.

Reserves

A.A. Meyerhoff estimates China's total potential oil reserves at 100 billion bbl of oil, perhaps even 130 billion bbl.[6] He indicates a conservative potential oil reserve of 42 billion bbl onshore, but discoveries in the unexplored northwestern part of China could boost this figure as much as 60–70 billion bbl of oil. Meyerhoff maintains that the region with the greatest potential of all basins is the Tarim basin, followed by the Junggar and Qaidan basins.

The major gas reserves in China are in the Sichuan and Qaidam basins. The produced, proved, and probable gas reserves of the fields in the Sichuan basin were estimated at more than 18,500 billion cu ft in 1969. Ultimate gas recovery is set at more than 150,000 billion cu ft. For the onshore areas as a whole, 200,000 billion cu ft would be a very conservative figure. Offshore areas could add at least 100,000 billion cu ft of gas.

Offshore outlook

Drilling in one of the two exploratory areas off China could begin in 1983, if bids from the 35 oil companies are sorted out in time and contract negotiations concluded in 1982. Seismic surveys in an area of 401,600 sq km began in 1979, and about one-third of the area was selected for bidding. This area covers 150,200 sq km and is divided into 43 blocks.

The first 26 blocks cover the South China Sea Pearl River basin and a portion of the north Yellow Sea, while the rest cover a part of the southern Yellow Sea, Beibu Gulf, and Yinggehai basin. Of the 26 blocks, 22 in the South China Sea are in the seismic areas where Mobil, Exxon, Chevron/Texaco, and Phillips were the operators while the four blocks in the north Yellow Sea are in the areas where Elf-Aquitaine was the operator.

REFERENCES

1. N. Hollander, *Oil & Gas Journal* (April 12, 1982), p. 168.
2. T. Gloppen, *Oil & Gas Journal* (April 6, 1982), p. 114.
3. G. Saeland and G. Simpson, *Oil & Gas Journal* (November 9, 1981), p. 381.
4. Ibid., p. 385.
5. R. Steven, *Offshore* (June 20, 1982), p. 54.
6. Meyerhoff and Willums, UNESCAP 1976.
7. Ibid.

14
The Eastern Hemisphere —South

The southern portion of the Eastern Hemisphere contains most of the world's oil reserves. Production and reserves are dominated in this vast region of the globe by the lush fields of the Middle East. But giant fields also are scattered on and off the shores of Africa, India, Indonesia, and off Australia and the Philippines.

AFRICA

The leading oil producer in Africa is still Nigeria, but Libya has the largest reserves of estimated oil resources. Estimated proved reserves in Libya stand at 22.6 billion bbl of oil, 16.5 billion in Nigeria, 8 billion in Algeria, about 3 billion in Egypt, almost 2 billion in Tunisia, 1.5 billion in Angola-Cabinda, 1.3 billion bbl in the Congo Republic, and smaller volumes in other African nations. Many countries on the continent are still oil barren, but exploratory work is under way in many.

Nigeria remains the biggest producer of oil on the continent with 1.4 million b/d of oil. Production in Africa's big three—Nigeria, Libya, and Algeria—nosedived in 1981. Those producers of high-quality crudes adamantly maintained prices out of line on the high side until very late in 1982.

Important gains were made in 1982 despite losses in Libya and Nigeria. Cameroon production rose 48%. Congo and Tunisia made smaller gains.

Ivory Coast, with development of its first oil field in progress, saw its output climb 600% in 1982 to 7,000 b/d of oil. Phillips Petroleum, operator for another offshore group, completed its first Espoir field well for 6,335 b/d of oil in 1982 in deeper waters. More production rises are expected in late 1982. Oil reserves rose 1.02 billion bbl to 56.17 billion in 1982, while gas reserves climbed 3.2 trillion cu ft to 211.7 trillion cu ft.

The largest oil fields in Africa are Algeria's Hassi Messaoud S., discovered in 1956 with pays at 10,500 ft and cumulative production at 48.8 million bbl; Angola's Limba (1969), cumulative production 50.6 million bbl; North Malongo (1966), 171.9 million bbl; South Malongo (1966), 51.8 million bbl; West Malongo (1969), 186.4 million bbl; Congo's Emeraude (1969), 105.6 million bbl; Egypt's Karrem, 663.4 million bbl; Rudeis, 144.5 million; Nubian, 107.3 million bbl; Ramadan offshore (1974), 192.9 million bbl;

Belayim (1955), 324.5 million bbl; Belayim offshore (1961), 297.0 million bbl; Gharib (1938), 235.2 million bbl; Gabon's Grondin (1973), 143.4 million bbl; Gamba (1963), 136.6 million bbl; Libya's Bu Attifel (1968), 457.3 million bbl; Jebel (1962), 153.6 million bbl; Nasser (1959), 1.9 billion bbl; Raguba (1961), 544.5 million bbl; Kotla (1963), 2.7 billion bbl; Amal (1959), 685.8 million bbl; Bahi (1968), 314.7 million bbl; Dahra (1959), 422.3 million bbl; Defa (1969), 771.8 million bbl; Gialo (1961), 1.7 billion bbl; Samah (1961), 336.2 million bbl; Waha (1960), 798.9 million bbl; Intisar A (1967), 654 million bbl; Intisar D (1967), 903 million bbl; Nafoora-Augila (1966), 739 million bbl; Nigeria's M'Bede (1966), 124 million bbl; Obagi (1964), 213.2 million bbl; Delta (1965), 123.9 million bbl; Delta South (1965), 207 million bbl; Meren (1965), 371 million bbl; Okan (1964), 338 million bbl; Ekpe (1965), 152 million bbl; Ugh-Olomoro (1963), 235 million bbl; Ugh-Kokori (1960), 271 million bbl; Forc-Jones Creek (1967), 330 million bbl; Forc-Forcados/Yokri (1968), 391 million bbl; Phl-Bomu (1953), 321 million bbl; Phl-Imo River (1959), 385 million bbl; Tunisia's El Borma (1964), 325 million bbl; Ashstart offshore, 110 million bbl.

Angola

Sonangol, Angola's state oil company, has opened more offshore acreage for exploration. It has invited companies to submit proposals for production sharing agreements for offshore Block 8 south of Luanda. Other agreements have been made for six offshore blocks.

Sonangol has signed exploration/production sharing agreements with Angola Cities Service and Marathon Petroleum Angola covering Block 9, a 1.2 million acre offshore tract calling for an outlay of $26 million for the first 3 year exploration period. Angola-Cities Service will be the operator.

Texaco Latin America/West Africa recently opened its second offshore Angola field and has put it on production. The Essungo field produces 2,800 b/d of 34° gravity oil from the Eocene at 6,350–6,850 ft. Seven new wells have brought the field output to 17,000 b/d of oil. The field is on the 1 million acre Block 2 concession, about 200 miles north of Luanda. Site is in the Congo basin about 9 miles off Angola's northwestern coast in 100 ft of water. The company also operates the Cuntala field on the block and continues its exploration program in the area. Cuntala makes 2,000 b/d of oil.

Gulf Oil Exploration has operational responsibility for Cabinda Gulf Oil, which operates Angolan oil fields. A joint project by a subsidiary of Gulf and Angola's national oil company has earmarked $1.2 billion to more than double oil production off Cabinda. Cabinda is an enclave of Angola surrounded by Zaire and Congo.

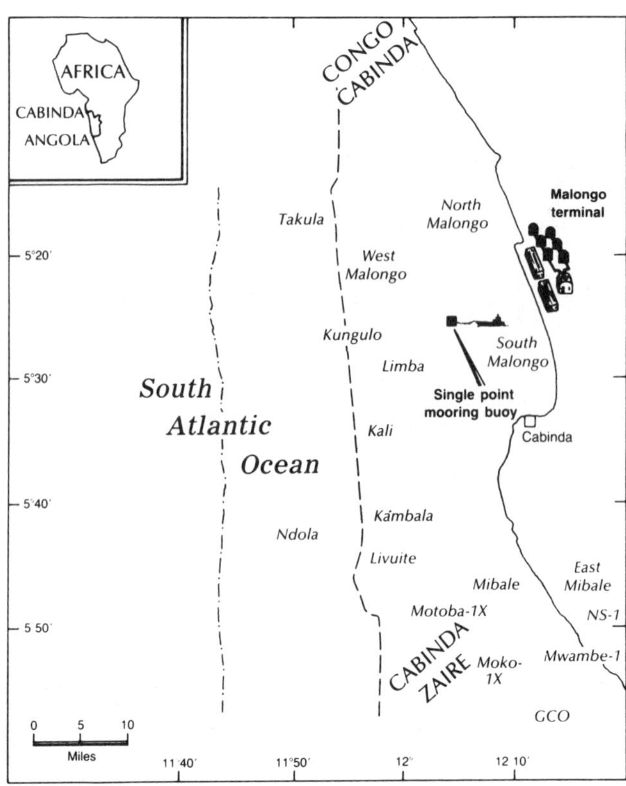

Fig. 14–1 Cabgoc/Sonangol acreage off Cabinda (from *IPE*, PennWell)

The venture by Cabinda Gulf and Sonagol plans to increase production from its offshore Cabinda fields to 200,000 b/d by 1985. This project will increase production from seven of Cabinda's main fields—North Malongo, South Malongo, West Malongo, Kungulo, Limba, Kali, and Kambala. The new Takula field was to go on stream in late 1982. Production from all Cabinda fields was 100,000–120,000 b/d of oil in early 1982. Drilling plans call for 25 new wildcats, 15 appraisal wells, and 145 field development wells.

Cabgoc notes that the Takula field, 25 miles northwest of the Malongo terminal, was discovered in 1980 on Block 57-2. Flow was more than 2,000 b/d of sweet, 35° gravity oil from the Vermelha sands at 3,500 ft. Seismic data indicate the pay found in the Takula discovery well may cover about 6,000 acres. If so, the field could contain 100 million bbl of oil by primary recovery (Fig. 14–1).

Cameroon

Cameroon became an oil producing nation in 1978, although there had been exploration in the Gulf of Guinea since the 1950s. The largest field is Kombo South Center, discovered in 1976. Cumulative production is more than 18 million bbl of oil from pays at 5,900 ft. All of the field is offshore. Oil production by Elf and Shell of Cameroon began in 1977 and increased to more than 80,000 b/d at the end of 1980. Only six of some 30 fields were on production. Mobil and Total made a large gas discovery in 1979 at Sanaga Sud in the southern offshore area. This discovery was followed by the Kribi L-1 discovery.

Congo

The Yanga-Sendji field off the Congo is being developed. Production is from Middle Cretaceous carbonate reservoirs. Congo aims at a production rise to 100,000 b/d of oil here. The Congo continues to see increased production, thanks to the efforts of the chief area operators, Elf Aquitaine and Agip. Likouala field was brought on stream in April 1980, adding another 30,000 b/d to output in the Congo.

Egypt

The ultimate potential of the Gulf of Suez is a long way from being fully realized. Probably all of the 1 billion bbl oil fields have been found, but there is still a great chance of finding medium sized fields, even on acreage which has been relinquished several times. The geology of the Gulf of Suez has been hard to unravel. Drilling over the past couple of years in the gulf has shown that the area still has some surprises (Fig. 14–2).

The cause of the trouble is a layer of evaporites in the Upper Miocene which extends across the entire Suez area.

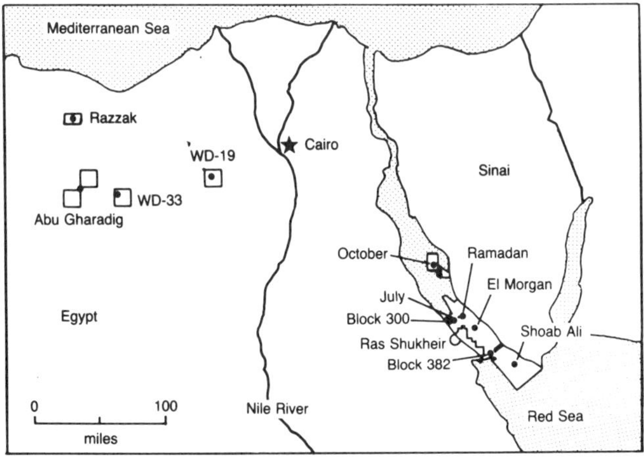

Fig. 14–2 Egyptian activity (from *IPE*, PennWell)

This has a highly absorbent effect on seismic, making it very hard to get good definition on the oil bearing structures below.

There are three main oil objectives: the Kareem and the Rudeis sands in the Miocene and the Nubian sands in the Cretaceous. The Kareem and Rudeis structures can be identified relatively easily. Under the unconformity, however, the tilted Nubian fault blocks are almost invisible on seismic profiles. There are recorded instances of major faults with throws between 2,000–4,000 ft, which have been established by drilling but do not show on seismic.

It is surprising that Nubian oil fields, which are invariably found behind leading edge faults, were only found for the first time in 1973. The Nubian potential is such that already 60% of offshore oil production comes from that formation.

This geological complexity has insured that the area is still in its first phase of exploration, enhanced by the fact that it is only now that access to the eastern gulf is possible. Amoco, who is a 50-50 partner with Gupco, is the senior established explorer in the Gulf of Suez. It found three new fields in 1980 (Gupco found four in 1981). Two of the recent Gupco finds are of interest because they found new pays, increasing hopes for more exploration and production in multipays in the future in the gulf.

Well GS 160-3 got 4,800 b/d of oil from the Nukhul sands, a stratigraphic trap at the base of the Miocene. The GS 206-1 discovery found 1,600 b/d from the Cretaceous Lower Senonian sand, generally tight in the gulf area. This discovery points the way for more exploration into that sand. The key here is to develop the controls to be able to predict those areas of the Lower Senonian with good porosity.

After a number of dry holes on its Hurghada permit at the mouth of the Gulf of Suez, Mobil tested 4,000 b/d of oil from two pays at its 1 Geisum wildcat. This discovery has increased interest in the relatively unexplored Red Sea area which has not had any discoveries.

Egypt expects oil production to reach the 1 million b/d mark in 1984–85. Crude oil reserves have been hiked to 4 billion bbl. Reaching this goal will depend largely on the outcome of operations in the Gulf of Suez. Current production is 680,000 b/d, most of which comes from a cluster of offshore fields. By 1983 this total should be 700,000 b/d.

In the 6 years since Egypt started its open-door policy for oil companies, 70 agreements have been signed with 49 companies from 18 countries. Most successful of these has been a group led by Deminex of West Germany. It has made a major discovery in Zeit Bay (Fig. 14–3) at the southern end of the gulf. The field will probably be on production in 3 years and have a peak output of 200,000 b/d of oil.

Some limited production also is under way in the Western Desert. Phillips' local affiliate, Wepco, has three

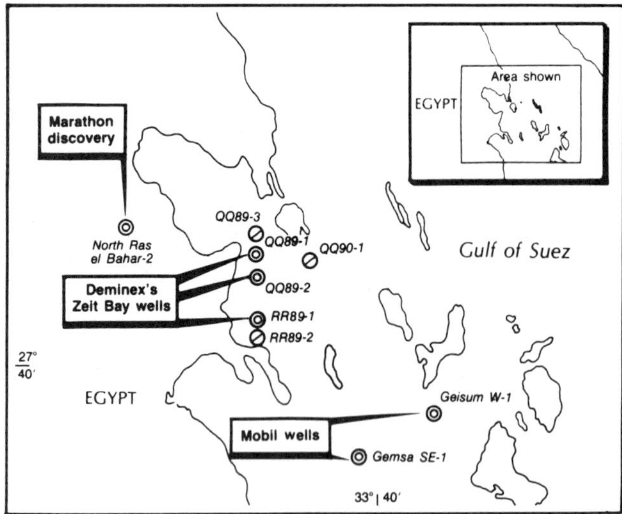

Fig. 14–3 Suez Gulf action (courtesy *OGJ*, April 12, 1982)

small fields while Amoco is producing oil and gas from Abu-Gharadig field and oil from Razzak and WD 33 fields. The Western Desert always has been a disappointment to Egyptians, who for many years hoped that the rich oil fields of Libya's Sirte basin extended into western Egypt. Shell did get a significant find in 1982 at 1 Badr al-Din. Flow was 5,900 b/d of oil from 12,000–13,000 ft. The oil was in the Lower Cretaceous Kharita zone, a pay found throughout the desert, but previously only a gas zone. The discovery should renew interest and hopes for the Western Desert.

Other Egyptian areas

Based on the experience of the Nile Delta area, opinion in some quarters expects gas is more apt to be found on the recently awarded North Sinai concessions in the Mediterranean Sea. Mobil found gas at El Temsah at 25 MMcfd and 1,440 b/d of condensate in 1981 in one pay and 42 MMcfd of gas and 203 b/d of condensate from another. Location is on the Damietta permit, adjacent to the North Sinai acreage (Fig. 14–4).

A North Sinai Agip affiliate, IEOC, won Blocks 3, 4, 5, 6, 16, 22, and 28. Total won 15 and 27, and BP obtained 21. Interest in this area is centered on large periclinal features associated with the Syrian Arc. Exposed onshore, these are of late Cretaceous age with a Jurassic core. Although only primitive seismic is available of the offshore features, they could have as much as 500 sq km of closure and could potentially yield 1 billion bbl of oil. To the west of the Nile Delta, Elf plans more work at its North Alexandria gas field discovery of 178. Gas was found in the Abu Madi formation.

Ivory Coast

A supertanker converted to a floating storage unit is off the Ivory Coast as a key element in a Phillips Petroleum plan for development of Espoir field. The tanker was converted to storage in Rotterdam. This tanker was connected to the buoy, 2 miles northeast of the Dan Duke production platform. Shuttle tankers will move oil from the floating

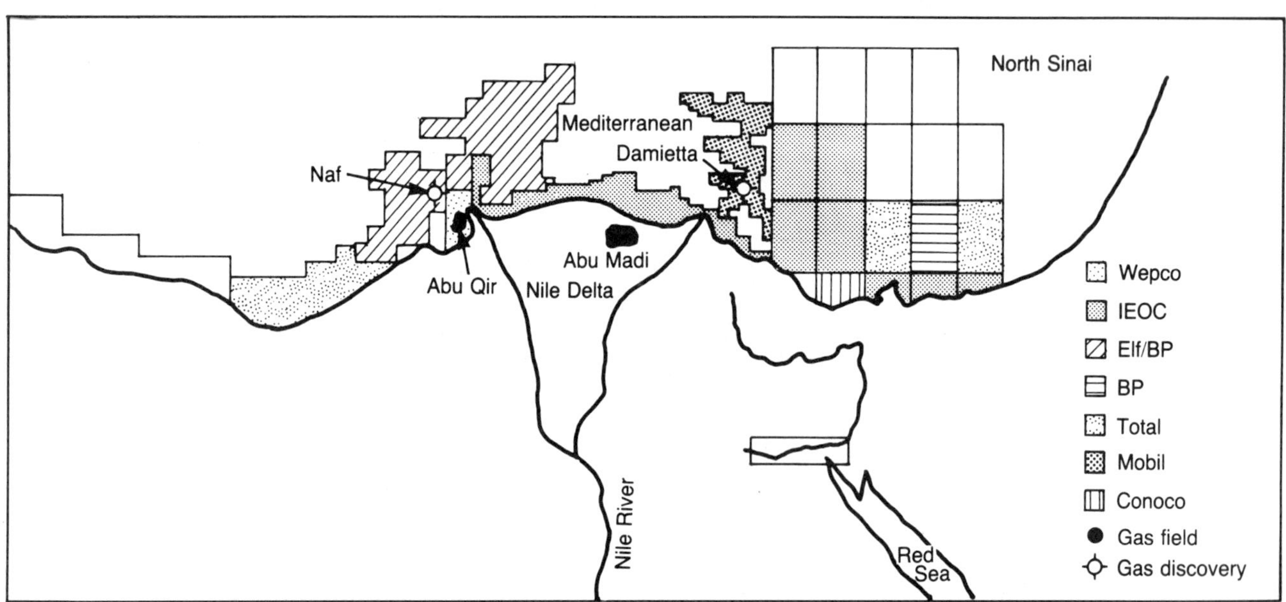

Fig. 14–4 Egyptian Mediterranean concession map (courtesy *Offshore*, July 1981)

storage vessel to market. Phillips Ivory Coast began production in late 1982. It completed the C1-5 in 21 days. The A-5 and C1-5 wells are two of the four wells drilled as part of an early production program. The third and fourth wells are on the way.

Ivory Coast is the newest member of the West Africa oil producing club. Production from Exxon group's offshore Belier field began in 1980. Then the discovery of Espoir field by Phillips pushed this tiny nation to the brink of an oil boom. Espoir is a giant field, 13 miles offshore, 35 miles southwest of Abidjan. The field was to be producing 50,000 b/d by the end of 1982. From 1985 onward, the field should average 300,000–500,000 b/d of oil. Estimates of the field's reserves vary from 2–8 billion bbl.

Libya

Despite this nation's many supergiant oil and gas fields, no further mention of Libyan oil activity will be made in this book due to political upheavals in the country and pullouts by the petroleum industry.

Nigeria

Gulf Oil Nigeria (Gocon) will quicken its exploration and production pace in Nigeria in the next 5 years. Gocon and its majority partner, Nigerian National Petroleum, plan capital outlays of about $1.3 billion during 1981–85, to expand drilling operations in the eastern and western sectors of the Nile Delta. Gocon also hopes to recover an additional 200 million bbl of crude oil from fields off Nigeria with two water injection projects.

With proven reserves of about 19 billion bbl of oil and 115 billion bbl of potential oil reserves, Nigeria controls about 34% of Africa's reserves and 3.2% of the entire world proved reserves. It is also thought that Nigeria's gas potential is enormous.

Exploration for oil in Nigeria began in 1908 when shallow drilling took place in heavy oil outcrops of the Cretaceous sands in the western coastal part of the nation. The first commercial oil production came in 1955 at Oloibiri. An intensive drilling program followed. Success ratios rose to 80% in the 1970s. Exploratory drilling was highly successful. There were more than 120 oil fields in Nigeria in 1980. Production is now 1.4 million b/d of oil. Nigerian crude oils have low sulfur and a high but variable paraffin content.

The total sedimentary thickness of the Niger Delta has been estimated at 12,200 m of section at the depocenter of the delta. Commercial oil production in Nigeria is from the deltaic sandstones of the Agbada formation where entrapment is mostly in rollover anticlinal structures formed by growth faults. However, the main source rock is thought to be the shales of the underlying Akata formation.

It would seem that the era of windfall discoveries is over in Nigeria. Exploratory efforts must now be focused on the hard to find stratigraphic traps, possibly in the transition zone from the delta to the delta front environment.[1]

Sudan

Development plans for Sudan's first commercial oil production, complete with downstream facilities, have solidified. Chevron Oil of Sudan and White Nile Petroleum have solicited lump sum turnkey bids from selected groups of contractors to build a 341-mile pipeline and a 25,000-b/d refinery. Output is expected to reach 25,000 b/d by 1985–86.

Chevron continues its exploratory program in the Sudan (Fig. 14–5). The company plans to develop the Unity trend, covering Unity and Talih fields in its 76,600 sq mile concession in Sudan's central plains. Chevron has identified 10 reservoirs in the Unity structure underlying the two fields. In addition to these discoveries, Chevron has drilled nine delineation wells in the area, with two more under way. Development drilling should begin in 1983. Production should begin in 1985–86. Plans call for about 30 producing wells and a waterflood program.

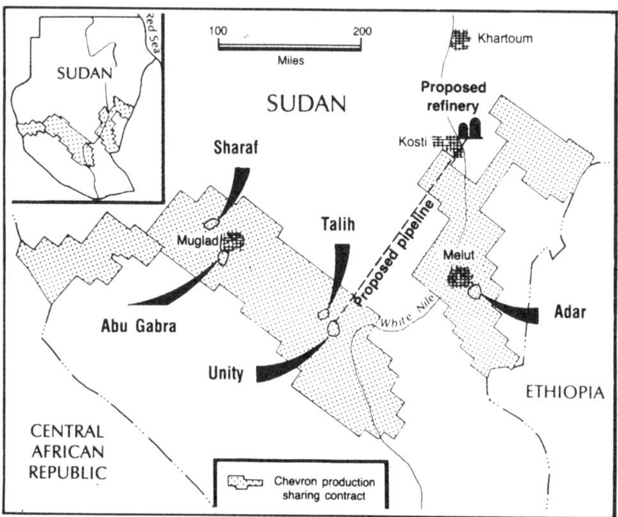

Fig. 14–5 Chevron's best strikes in Sudan (courtesy *OGJ*, May 17, 1982)

Other Africa

Tanzania has large sedimentary areas suitable for oil and gas generation and accumulation. Geological and geophysical work in the past few years has been rather widespread both offshore and onshore. In 1982, the discovery of commercial gas was confirmed, a total of six exploration licenses covering 73,544 sq km was granted, and more licensing is on the way.

The Tanzania coastal and offshore sedimentary basins cover an onshore belt stretching from Kenya to Mozambique. They extend eastward beyond the islands and join the great sedimentary basin found under the Indian Ocean. Only nine wildcats have been drilled in these basins to date.

An important discovery was made at Songo Songo south of Mafia Island. The reservoir appears to be a major deltaic sand body capped by shales of Upper Albian age. Operator is Tanzania Petroleum Development.

Morocco has stepped up its offshore exploratory effort. Onarep let contract to a group that involves Atlantic Resources, Seismic Profilers AS, and Merlin Geophysical to conduct a seismic survey of all Atlantic offshore areas not covered by existing leases in waters as deep as 3,280 ft. This program began in August 1982 and will be completed in mid-1983. The Atlantic continental shelf off Morocco includes several sedimentary basins that have had little exploration. Only 10 wildcats have been drilled in these basins in the past 15 years. About 1,000 miles of Morocco's coastline is not covered by existing leases.

THE MIDDLE EAST

Tremendous reserves of oil make the Middle East the pivotal point around which the oil industry spins. Saudi Arabia, the world's largest coffer of oil, has 164.6 billion bbl of oil reserves and 114 trillion cu ft of gas reserves. The country produced 9.6 million b/d of oil in 1981. The next big one in this vast region of deserts, mountains, and little population is Kuwait with 64.4 billion bbl of oil and 30 trillion cu ft of gas. Then comes Iran with 57 billion bbl of oil and 484 trillion cu ft of gas. Abu Dhabi is fourth with 30.6 billion bbl of oil and 19.5 trillion cu ft of gas. Iraq is next with 30 billion bbl of oil and 27 trillion cu ft of gas. The rest of the giant oil reserves are scattered among the Neutral Zone, the Emirates, and Syria.

An earlier chapter discusses the size, production, and reserves of all the giants in the Middle East as well as in other parts of the world. The overall picture in the Middle East continues to be one of production cuts in the face of slackening world markets and set against a background of continuing investment in exploratory and development projects.

Offshore work is scattered throughout the region. Off Saudi Arabia, Aramco continues an active exploration and development program using 12 jack up rigs. The nation has expansion projects under way on the offshore Zuluf and Marjan fields, while contracts also have been let to gather associated gas from Berri and Safaniya fields. Aramco had five new oil discoveries in 1980 in Saudi Arabia. Two of these were offshore, Dawl and Salsal.

Oil and gas reserves in Saudi Arabia showed another increase in 1981. Aramco estimates Saudi proved oil reserves at 116.747 billion bbl, a gain of more than 3.3 billion bbl from year-end 1980. Probable reserves, including proved, slipped to 177.229 billion bbl. Gas reserves, including dissolved, associated, and nonassociated gas, increased to 70.81 trillion cu ft. Probable, including proved, amounted to 111.852 trillion cu ft. The government said the figure was 114 trillion cu ft.

Exploration

Aramco is evaluating a number of 1981 discoveries (Fig. 14–6). A wildcat on the Amad prospect about 46 miles southeast of Jawb field, had oil shows, and the potential discovery area was being tested in 1982. Three shallower oil

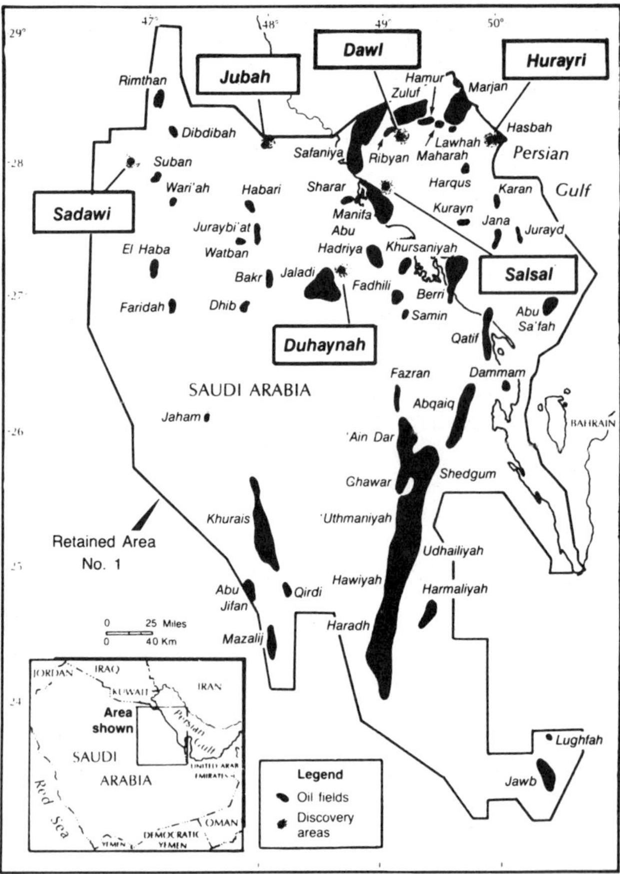

Fig. 14–6 Potential strikes being evaluated by Aramco (from Arabian American Oil Company, courtesy *OGJ*, June 14, 1982)

reservoirs were found in Jurayd and Zuluf fields in the Persian Gulf, and a deeper oil discovery was made in Abqaiq field onshore.

Saudi Arabia's Eastern Province has grown up around development of the country's massive oil reserves. To counterbalance this, the government is well into a megadollar program to bring oil-related development to the west coast and to build a crude export terminal on the Red Sea. At the heart of this plan is the building of parallel crude oil and gas liquids lines across the Arabian Peninsula. This has caused a new industrial city to spring up south of the old town at Jeddah.

Abu Dhabi

This nation is the largest of the United Arab Emirates. It is also a hot spot for exploration and development of new production. Discovery of nonassociated gas in the deep Permian Khuff zone has given a major boost to exploration in the lush offshore waters and on the extenisve onshore concessions (Fig. 14–7). The $5 billion project in the Upper Zakum field is moving to completion.

The Upper Zakum oil field off Abu Dhabi is on stream. Initial capacity was 300,000 b/d, reaching 500,000 b/d by 1984. The field is 80 km northwest of Abu Dhabi City. The Upper Zakum reservoir is contained in a giant anticlinal structure 30 km long and 20 km wide. The main reservoir is the Lower Cretaceous Thamama limestone at 7,000–8,000 ft. The field is made up of three zones containing an estimated 40 billion bbl of oil in place. Gravity is 34–37°. There are several smaller fields under development as well as an extension of the gas gathering network.

A major discovery has been made at Umm al-Lulu, a 17,770 ft wildcat. This discovery flowed 24,000 b/d of oil from seven pays, one of the most prolific tests in years.

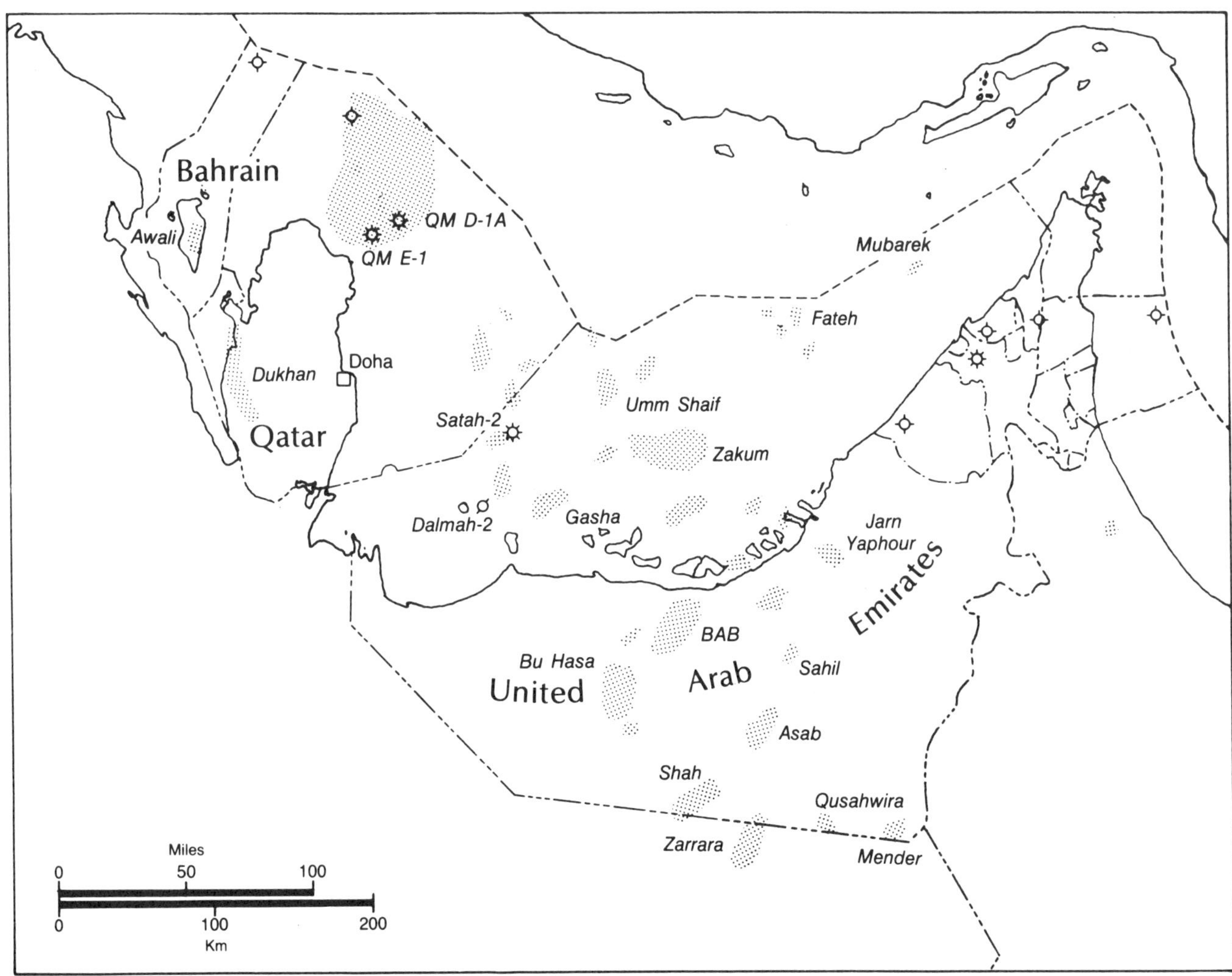

Fig. 14–7 Bahrain, Qatar, and United Arab Emirates producing areas (from *IPE*, PennWell)

Three more wells are planned for the Khuff gas zones. Explorers are onshore probing the Khuff. Wells at Bu Hasa and in the Jarn Yaphour structures have reached the Khuff at more than 20,000 ft. Results are not yet conclusive.

Adnoc has wet gas reservoirs in the Thamama C and Thamama F at the Bas field. The Japanese firm Mubarraz Oil tested 5,000 b/d of oil and 24.7 MMcfd of gas at a wildcat 112 miles off Abu Dhabi in 65 ft of water. Commercial development may begin by 1985.

Sharjah

Production has begun by Amoco Sharjah Oil at the Saaja gas condensate field in this United Arab Emirate area (Fig. 14–8). Exports will begin soon. Most of the gathering and storage facilities are complete. When all 20 field wells are completed, the output would be 800 MMcfd of gas and 80,000 b/d of liquid hydrocarbons. Amoco is reported to have 10 trillion cu ft of gas reserves in Sharjah—7 trillion in Saaja field and 3 trillion in Mawaid field.

Bahrain

Bahrain National Oil will drill six deep gas development wells as part of its effort to boost nonassociated gas production to 950 MMcfd by the end of this decade. The six holes will be drilled below the producing Khuff formation and will test formations as deep as 20,000 ft. The nation now produces about 340 MMcfd from 11 wells in the Khuff formation. Industry demands more gas.

Dubai

Production from the four offshore fields in Dubai averaged an estimated 350,000 b/d of oil in 1980. These fields are operated by Dubai Petroleum. Dubai Petroleum completed field drilling on the Southwest Fateh 2 platform. This platform has produced 28,000 b/d of the estimated 200,000 b/d field total.

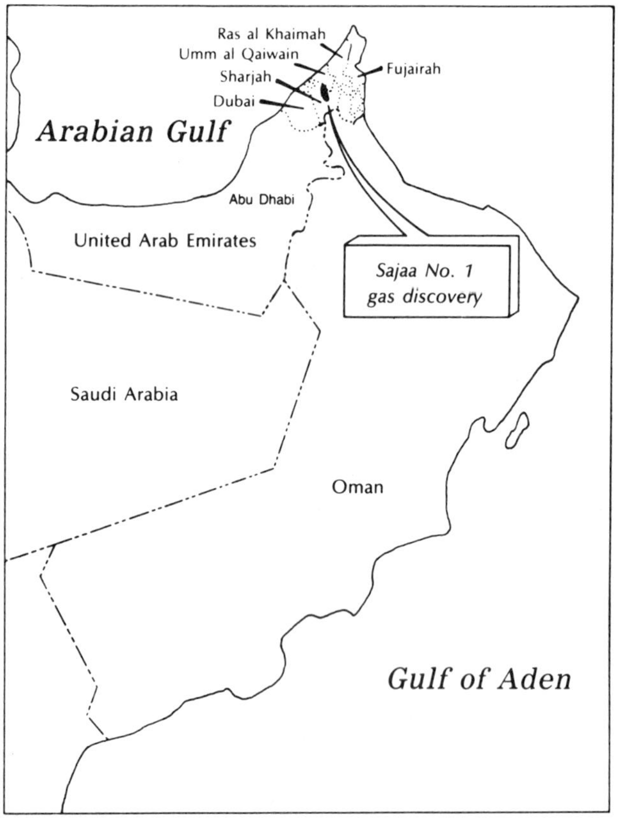

Fig. 14–8 Sajaa gas area (from *IPE*, PennWell)

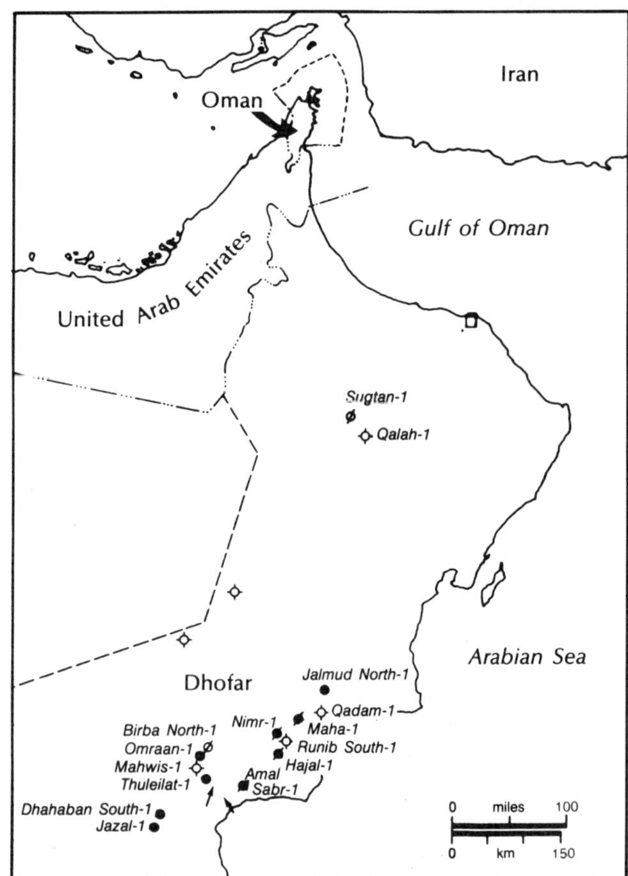

Fig. 14–9 Oman action (from *IPE*, PennWell)

Oman

Exploration here continues at a brisk pace (Fig. 14–9). The petroleum ministry signed a production sharing agreement with Amoco for a 14.7 million acre concession along the Oman Mountains thrust belt. The company will spend $70 million on geophysical exploration over 7 years and drill a wildcat in 1983. Amoco also is participating in exploration of a 5.2 million acre concession along the southeast coast of Oman. Elf has become operator of its third concession, a 6.7 million acre tract in eastern Oman.

Qatar

Qatar is proceeding with plans to develop its huge offshore North Dome gas reserves (Fig. 14–10). Reserves in this vast field are estimated at 300 trillion cu ft of gas.

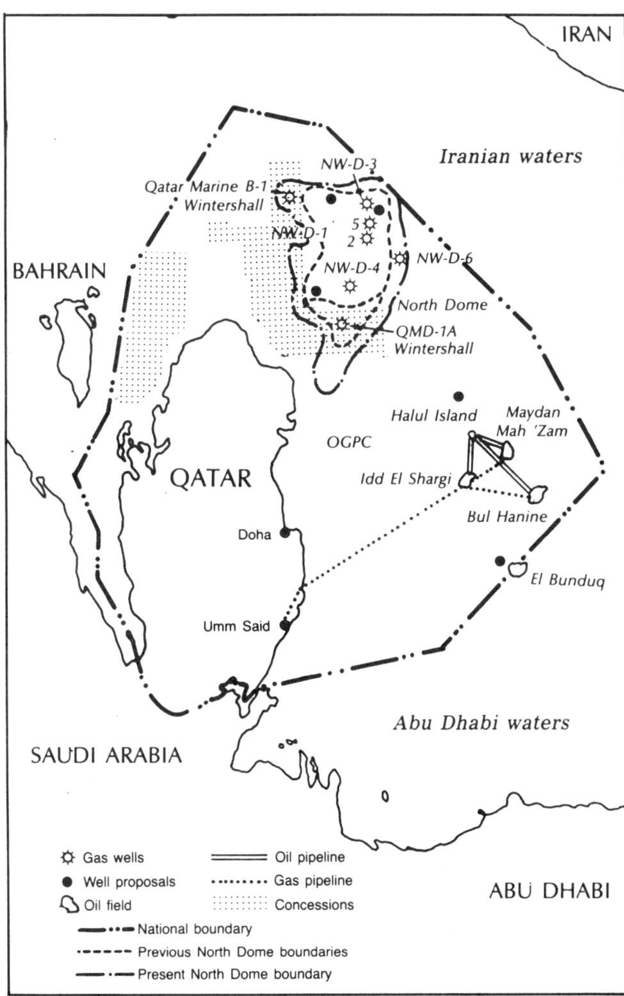

Fig. 14–10 Qatar offshore development (from *IPE*, PennWell)

Iraq and Iran

These nations, though big oil producers and possessing vast reserves, will not be discussed in this book due to the war conditions and virtual shutdown of the industry in each nation.

ASIA-PACIFIC

Plans and projects ranging from exploration/production through the other sectors of the industry abound in the Asia-Pacific region. One hot spot is India where the government awarded Chevron a 7,200-sq-mile concession off the west coast. This acreage is about 100 miles north and on trend with the Bombay High oil field. Chevron will begin work in 1982. Drilling includes three wildcats during the initial 2-year term of the permit.

Oil India has resumed exploratory drilling in the Mahanadi area off India. Two drillships will drill six tests 30 miles off Paradeep.

In the Philippines, oil production will increase with the opening of the Cities Service Matinloc complex late in 1982. Development of the Matinloc-Pandan-Libro oil discoveries is under way in the Northwest Palawan service contract area, in which Cities holds a 39.6% interest. Production began in 1982.

New production has gone on stream from Cities' 45% interest in a concession in Pakistan. The Khaskeli wildcat flowed 1,005 b/d of oil. Two appraisal wells were then drilled. Union Texas Petroleum found oil in the Badin area of Southeast Pakistan in 1981.

Pakistan's government signed an agreement with a group that includes Oxy for a 250 sq mile exploration concession in the North Potwar area. Oxy will run seismic surveys and drill two exploratory tests under terms of the agreement.

Union Oil's big deal in 1981 was the start-up of production from Erawan field in the Gulf of Thailand. Natural gas began moving to the Bangkok area in 1981. In 1982 Union was delivering 120 MMcfd and 4,400 b/d of condensate ashore.

The company is continuing development of Erawan, where it has installed 11 platforms. Union has found five more gas condensate fields in the Gulf of Thailand since 1981 and now has major interests in 11 fields there (Fig. 14–11).

In Turkey, the government in 1982 awarded eight exploration licenses in Hakari, Mardin, Siirt, and Urfa provinces to Union Texas Turkey. Plans call for seismic work.

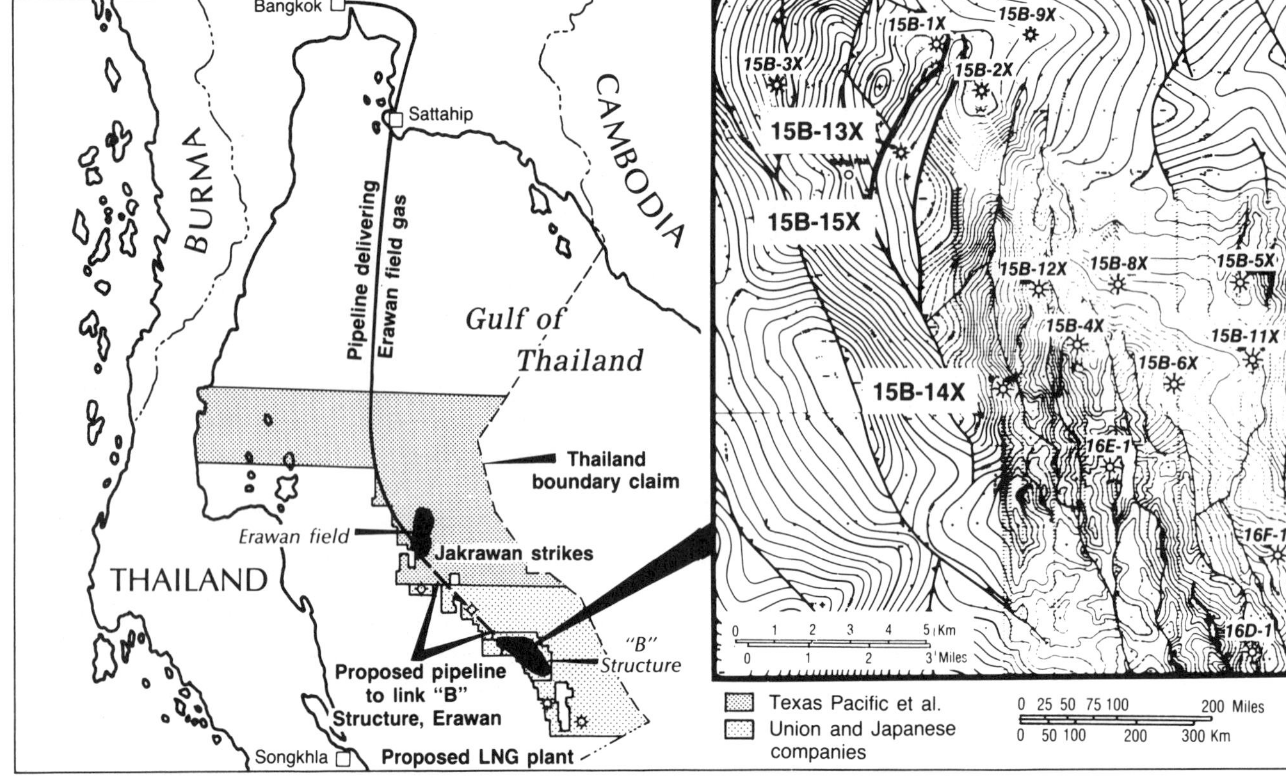

Fig. 14–11 Development in the Gulf of Thailand (from *IPE*, PennWell)

India

Exploration off India is growing. At least 17 rigs will be operating offshore in 1983. Production at the Bombay High field is 160,000 b/d. Thirty platforms have been installed in this field (Fig. 14–12). Of these, 3 are process platforms, 16 have four to five wells each, and the others are single-well structures. Each well produces 3,000–4,000 b/d of oil.

Two new oil strikes were made in 1981 by India's Oil and Natural Gas Commission (ONGC). One is on the B-57 structure, 35 km east of Bombay High field. The other is in the northern Palk Strait in the offshore Cauvery basin. More wells are being drilled to evaluate these finds.

India will drill 566 onshore wells and 325 offshore wells under its current 5 year oil exploration/production budget, which has been hiked to $6 billion. ONGC is drilling in the Godavari basin in the Bay of Bengal. Surveys are being run off the east coast in the Mahanadi basin, onshore and offshore. Deep wildcats will be drilled in West Bengal and Tripura states by ONGC with Russian money.

Three fault bounded pericontinental basins—the Kutch, Cambay, and Narmada—in western India assume importance for future explorers in view of their location in the major oil province of the Indian subcontinent (Fig. 14–13).

Cambay is one of the major onshore oil and gas regions of India.[2] Most fields are in the mature stage of development. Main pays are the Eocene and Oligocene sands of varying origin. Traps are mostly combination type generated by complex interplay of paleohighs, distributary channels, lagoonal delta, offshore bars, and delta lobes, folding and faulting.

The Narmada basin at this point is not considered to be a prospective oil and gas area. But its offshore extension where it crosses the offshore extension of Cambay basin appears to be prospective. Of all the marginal rift basins, the Kutch basin has a long history of more than 12,000 ft of sediments. Prospective wedgeouts could be expected on the western flank of the basin high. In the western part of the basin and in the offshore areas, stratigraphic prospects do exist. Mid-Jurassic carbonate (reefal) and Lower Cretaceous deltaic sediments are the main exploratory objectives. Tertiary prospects are mostly in the offshore shelf. Mid-Miocene to Paleocene carbonates are the targets for exploration.

Pakistan

No offshore oil or gas fields have been discovered yet off Pakistan. Some wells have been drilled in the Indus

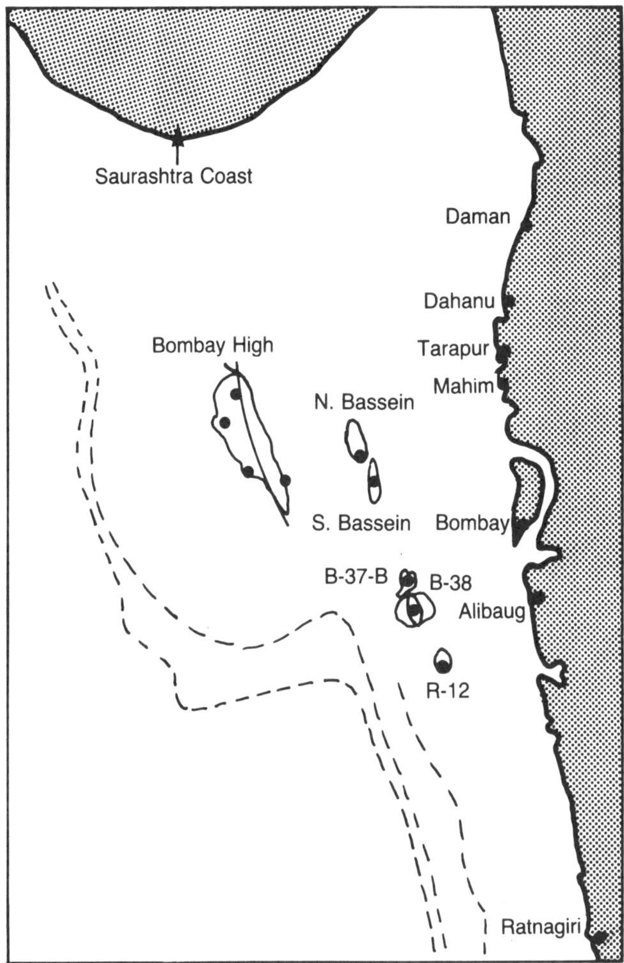

Fig. 14–12 Bombay offshore basin (courtesy *Offshore*)

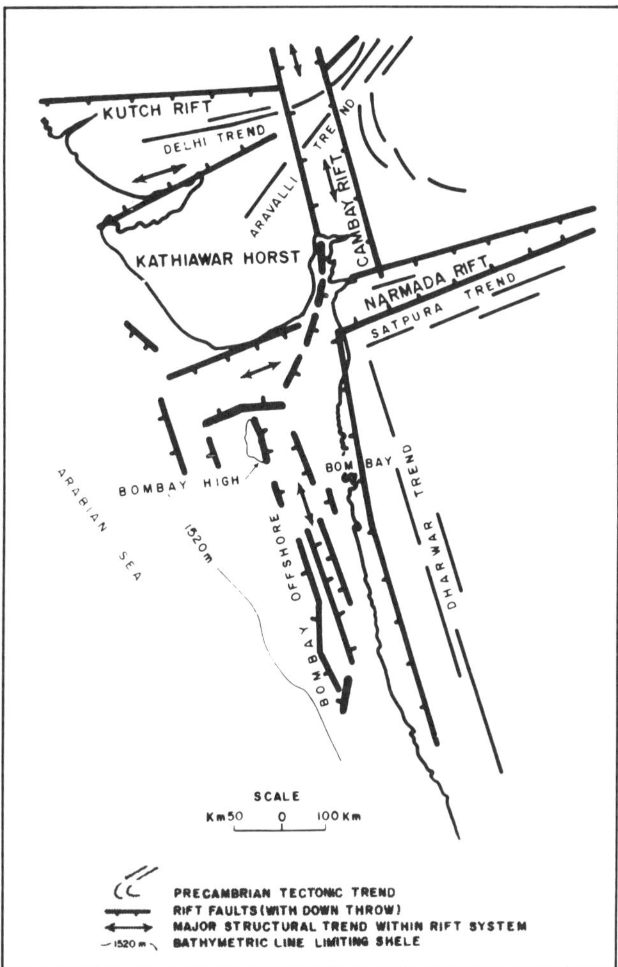

Fig. 14–13 Tectonic map of the western margin of India showing the three rift basins and major Precambrian Tectonic trends (courtesy *OGJ*, April 19, 1982)

offshore basin without reporting any potential reservoirs. But this is a large offshore basin and should not be written off until more wildcatting has been done. The hingeline zone and the possible presence of reefal buildups on the outer margin of the carbonate platforms could be potential areas in which to seek out hydrocarbons. The area definitely needs more drilling.

Indonesia

Indonesia is pushing exploration and production projects. Action in both areas is almost booming (Fig. 14–14). At the end of 1981, Pertamina had 71 exploration/production areas under contract to foreign firms, covering some 820,000 sq km or 202 million acres. Eighteen of the contracts have commercial production.

New discoveries have sparked a renewed interest in the prospects of the offshore areas of Southeast Asia. Most of Indonesia's action is centered around Southeast Sumatra where Natomas Iiapco heads operations. The company expects to add about 50 million bbl of oil to its reserves. The recent discovery at Sundari field brings the number of oil fields in Natomas Southeast Sumatra territory to 12. The well flowed 973 and 327 b/d of oil from 4,328–6,108 ft. The Natomas territory now extends 40 miles and includes 33 platforms and 175 wells producing 100,000 b/d of oil.

Iiapco started up the Zelda D platform at 2,700 b/d of oil from one of nine platform wells planned for this area of Sumatra. Titi field was another big find for Natomas. This is the most promising discovery since Krisna, now producing 11,000 b/d from four of its nine planned wells. The first Titi well flowed 3,909 b/d of oil from 7,066–7,130 ft.

Eleven excellent producing fields are off Java. Exploratory work also is under way in the Natuna Sea, an area expected to join the family of oil areas in Indonesia.

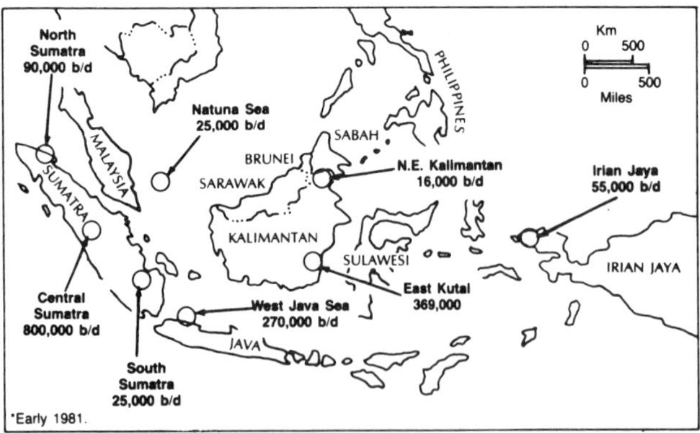

Fig. 14–14 Indonesian oil production areas as of early 1981 (from *IPE*, PennWell)

Indonesia is pushing for partners in its oil, gas, and petrochemical development. The country planned to drill 800 wells in 1982 and spend $4 billion.

The last of four oil fields in the Beruk-Zamrud project in central Sumatra has been put on stream by Caltex Pacific Indonesia. Zamrud is the latest field to go on production. The other three—Beruk, Northeast Beruk, and Bungsu—were placed on stream in 1980. They produce an average of 25,337 b/d of oil from 16 wells. Of the 11 wells in Zamrud, eight are making 21,300 b/d of oil.

Drillers have stormed into the South China Sea around Natuna Island. A large part of the offshore area north of this island is claimed both by Indonesia and Viet Nam. Conoco had the first hit at Udang Island. Production began 2 years ago at 40,000 b/d of oil. North of Udang Island, Marathon has oil at its Kakap block.

In the Malacca Strait off Sumatra Hudbay has confirmed its oil discovery of 1980. On Sumatra itself, Mobil, Caltex, and Stanvac continue to make discoveries. Conoco has an important find in Irian Jaya at Wirriagar on the Vogelkop, 130 miles from nearest production and in an entirely different basin. In the Java Sea, Iiapco extended the Krishna and Yvonne fields and proved the Farida field to be commercial. This makes 10 fields in the southeast Sumatra area with another one about to become a producer, Titi. On the adjoining block, ARCO made a major find at the P field in 1980.

Indonesia awarded nine production sharing contracts during the first half of 1982. These awards covered more than 90,780 sq miles. Companies awarded the contracts plan to spend $524.1 million to explore the areas. The outlook is for the high level of action in Indonesia to continue for most operators during the next few years if the government does not cut incentives in its petroleum laws.

Activity among a few major international oil companies may plateau into 1983 and early 1984 as those firms tighten the belts on their world exploration/development budgets. But after that, Indonesian exploration/development by major companies will increase. The hottest areas are the Java Sea, South China Sea, and off Kalimantan. Onshore work centers on Sumatra and Kalimantan. Drilling in 1982 should be a record at 284 wells completed (Table 14–1).

A stable government, reasonable contract terms, and good relations between operators and Pertamina add up to an attractive investment picture. Other enticements are excellent recent discoveries, good prospects in many underexplored sectors, onshore and offshore, and favorable drilling conditions that include shallow water and pleasant weather.

The history of petroleum exploration in Indonesia began in 1880. This was the pioneering phase which lasted until the 1930s. Early work was confined to the onshore areas, mostly on the larger islands.

The second phase, the modern onshore phase, covered the years 1930–65. Whereas the first phase of exploration was done mostly with surface mapping, shallow coring, and the drilling of oil seeps, the second phase was characterized by the introduction of geophysical techniques, especially seismic surveys. Early fields were shallow, but in five of the 11 producing Tertiary basins, the largest field was found before the seismic phase began.

The third phase of exploration began in 1966 with the signing of the first offshore production sharing contract. Throughout the middle 1970s, exploration activity was low due to a general reorganization of the Indonesian state oil company, Pertamina. The lastest phase began in 1978 with government assurances of fiscal stability and conciliatory incentives. By mid-1980, Indonesia was enjoying an exploration boom.

Geologically, Indonesia is divided into four large regions termed tectonic plates (Fig. 14–15). The Eurasian cratonic plate contains most of the island system—Sumatra, Java, Kalimantan, and a part of Timor and Sulawesi (the Calebs).[3] The Indian-Australian plate contains the rest of the island system—Timor and Irian Jaya—as well as the bounding Java trench. The Philippine Sea and the Caroline plates bound the Indonesian land areas to the northeast.

Superimposed upon these huge plates are some 28 Tertiary basins. These basins can be classified into four categories: outer arc, foreland, cratonic, and inner arc basins. Outer arc basins are generally considered as unfavorable habitats for hydrocarbons. Foreland basins contain prolific deposits of oil and good potential remains for future discoveries. Cratonic basins are present on the Sunda shield (Kalimantan area) and the Australian craton (including Irian Jaya). These basins may contain excellent reservoir rocks and source rocks deposited in lacustrine or shallow marine environments, but the requisite structure and geothermal maturation of the organic matter is uncertain. These basins may have excellent oil prospects, but considerable risk also is present.[4] Most of the inner arc basins are in deep waters and have not yet been drilled.

SUMMARY OF PETROLEUM INFORMATION FOR INDONESIA

Table 14–1

Basin Date of first discovery Plays	Annual production, bo/d (MMcfgpd)	Cumulative production, 10^6/bbl/oil (10^9 cfg)	Estimates reserves,* 10^6/bbl/oil (10^{12} cfg)	Largest field % Annual production % Cumulative production % Reserves	Prospects Short-term production projection
North Sumatra 1885 Regressive clastic rocks Carbonate platforms	19,400 (600)	381	134 (13.7)	Rantau (1929) 48 57 51	Oil production decline; gas production increase; small new fields expected on and offshore; moderate prospects
Central Sumatra 1939 Transgressive clastic rocks	779,840	4,567	4,437	Minas (1944) 43 53 36	Oil production decline; relatively small new fields expected; moderate prospects
South Sumatra 1896 Transgressive clastic rocks Regressive clastic rocks Carbonate platforms	32,000	1,494	190	Talang Arkar-Pendopo (1922) 2 24 4	Oil production decline; small new fields expected; moderate prospects
Northwest Java 1969 Transgressive clastic rocks Regressive clastic rocks Carbonate platforms Pinnacle reefs Fractured igneous rocks	225,506 (58.7)	498 (21.4)	1,432	Jatibarang (1969) 8 11 14	Oil production increase; good prospects, especially offshore
East Java 1888 Transgressive clastic rocks Regressive clastic rocks Carbonate platforms	561	165	4	Kawengan (1894) 51 of cumulative production	Oil production decline; most discoveries expected offshore; moderate prospects
Kutei 1897 Regressive clastic rocks Deltas	359,653 (530)	920 (442)	1,100 (7)	Handil (1974) 46 22 54	Oil production stable, gas production increasing; good prospects, especially offshore
Barito 1937 Transgressive clastic rocks Regressive clastic rocks Carbonate platforms Fracture igneous rocks	5,121	112	30	Tandjung (1937) 100 94 100	Oil production decline; poor prospects
Tarakan 1900 Regressive clastic rocks Deltas	14,000	260	82	Pamusian (1905) 28 72 ? reserves	Oil production decline; prospects moderate on and offshore
Salawati 1936 Carbonate platforms Pinnacle reefs	73,765	187	497	Walio (1973) 60 38 67	Oil production stable; prospects moderate on and offshore
Waropen Little exploration thus far Transgressive clastic rocks Carbonate platforms					Prospects moderate
Arafura Little exploration thus far Carbonate platforms					Prospects moderate
North Ceram (Bula) 1897 Regressive clastic rocks Carbonate platforms	927	13	5	Bula (1897) 100 99 100	Oil production decline; prospects poor
Bone 1975 3 shut-in gas discoveries					Prospects poor to moderate
Natuna 1974 Deltas(?)	26,000	6	200	Udang (1974) 100 100	Oil production increase; prospects good
Remaining basins					Oil prospects poor to moderate

* Estimates by J. Riva Jr. (courtesy *OGJ,* March 8, 1982)

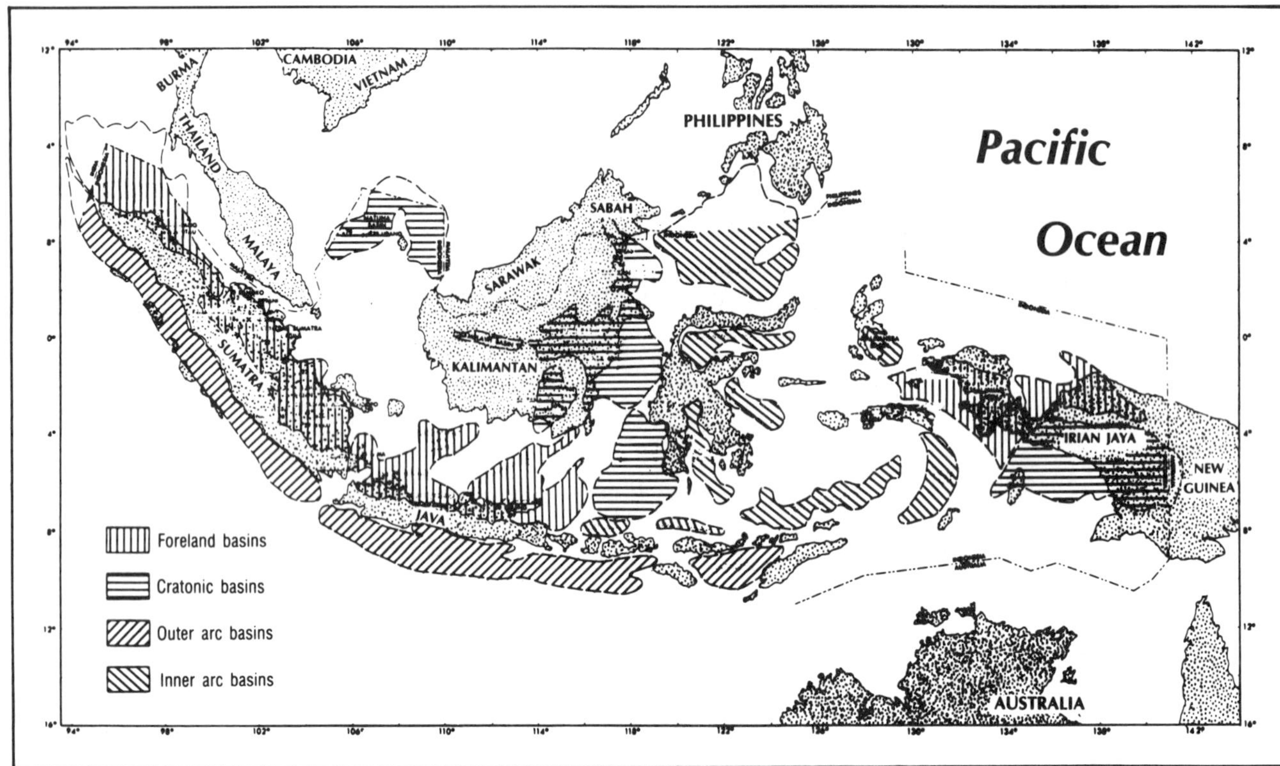

Fig. 14–15 Sedimentary basins in the Pacific (courtesy *OGJ*, March 8, 1982)

Some in the industry feel that, if domestic demand continues to increase at its current rate in Indonesia and other forms of energy are not substituted for oil, it is possible that by the 1990s the nation will need all of its oil at home.

Malaysia

After Shell moved off Sabah in 1958 to search for oil, it went through a 13 year drought without discoveries. The company finally hit in 1971 with the Erb West discovery. Oil finds then followed at a rapid pace with the Samarang, South Furious, Erb South, St. Joseph, Barton, and Ketam fields. Other small oil finds were made at Lokan, Southwest Emerald, and Southeast Collins in the course of exploration from 1979 to 1981. Shell put Samarang on production in 1974. It was then tied into the Labuan terminal in 1977.

The company then took the next major step in the development of Sabah production when it put the offshore North Sabah trunk line in operation in 1980. This 24-in. line connects South Furious to the terminal. Combined production from Samarang and South Furious, called the Labuan blend, reached 54,000 b/d in 1980.

Only the primary development phase has been completed at South Furious. Erb West, St. Joseph, and Barton are just entering that phase, which means years of work. Shell says these fields are highly complex and faulted. Shell has several rigs working off Sarawak and Sabah.

In the South China Sea off Sarawak, Shell is pushing to get gas ashore for the huge Bintulu LNG plant. The key offshore field is E-11. Shell began drilling there from a 20 slot platform early in 1981. E-11 also will be the riser station for future gas development.

When E-11 is drilled up, Shell will move to F-23 by the end of 1983. The third field, F-6, will be drilled in 1985–86. The Borkor field was to be added to the system in 1982. Drilling began in mid-1982. Also, a sixth jacket was placed in the Tukau field and drilled.

New Zealand

Exploration in New Zealand will gain speed as a result of the nation's round of licensing in 1981. The government says at least 30 wells will be drilled during the next 5 years in 16 new petroleum concession areas off the west coast. Most of the licenses are for 5 years.

The new concession areas cover a region of Late Cretaceous and Tertiary sediments lying unconformably on a pre-Cretaceous basement. Source material is thought to be mainly derived from Late Cretaceous and Early Tertiary coal measures and marine shales deposited during a period of widespread subsidence and marine movement. Sands throughout the sedimentary column provide potential reservoirs capped by shale formations.

Onshore lies the Kapuni gas condensate field, found by Royal Dutch/Shell-BP and Todd in 1959. Reserves are estimated at 670.89 billion cu ft of gas and 22 million bbl of condensate. Offshore lies the Maui field, a 1969 gas condensate discovery with 5.3 trillion cu ft of gas in reserve and 69.2 million bbl of condensate.

In mid-1982, New Zealand awarded four exploration licenses on North Island and one off South Island in the Great South basin (Fig. 14–16). Eighteen offshore and eight onshore licenses were awarded in a year's time, 1981–82. More acreage was to be let in late 1982. The Great South basin is considered to be highly prospective. Three of the six wells drilled there in the 1970s had shows.

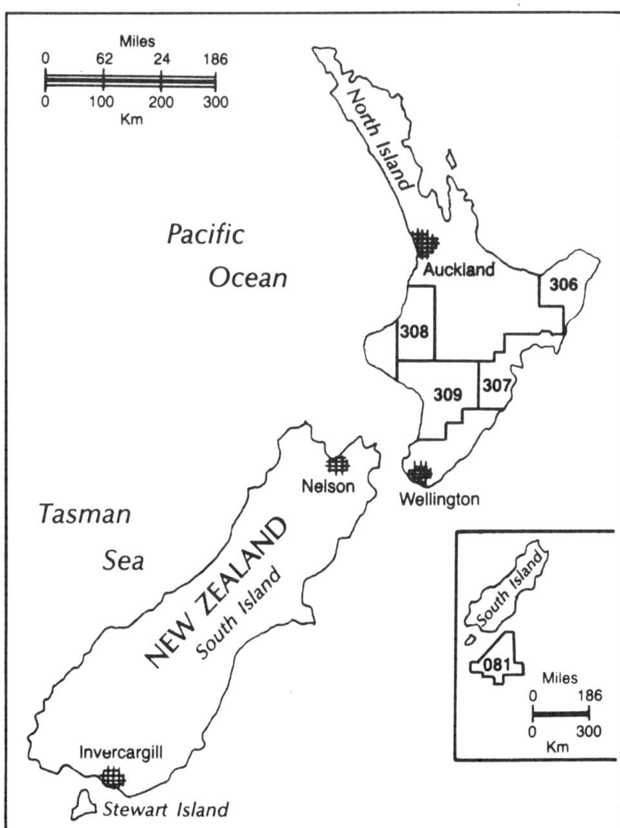

Fig. 14–16 New license areas in New Zealand (from New Zealand Ministry of Energy, courtesy *OGJ*, July 5, 1982)

Philippines

The first sandstone reservoir tapped off the islands was an oil discovery for a group led by Cities Service. All prior production had been from reefal limestone reservoirs. The 1 Galoc, the Philippines' northernmost oil discovery, flowed 1,850 b/d of 35° gravity oil from 7,338–7,364 ft and 1,800 b/d of oil from 7,255–7,285 ft. This well lies off Palawan Island in 1,100 ft of water (Fig. 14–17). Late in 1981 the group tested gas from two zones at 1-A South Galoc, 3 miles south of the discovery well.

Reserves at Matinloc field are set at 10 million bbl of oil. Oil will be recovered in 4 years with initial production rates at 15,000 b/d of oil. Cadlao field was put on stream in 1981 at 9,000 b/d of oil. This field is 27 miles west of the Palawan coast and 246 miles southwest of Manila. It produces from a Lower Miocene reef at 4,920 ft.

The one-time 40,000-b/d Nido field barely produces 3,000 b/d of oil now. It may become noncommercial by

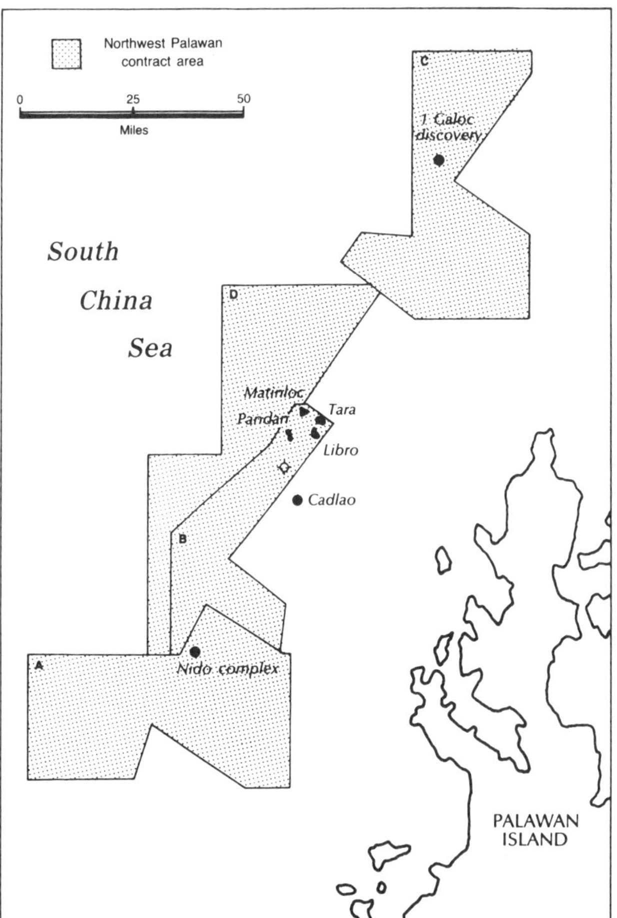

Fig. 14–17 Offshore action in the Philippines (from *IPE*, PennWell)

1983. However, the Cadlao project has added new life to Palawan hopes for increased production from this area.

Thailand

Gas discoveries and development work still propel Thailand's offshore action. Thailand stepped into a new energy era when the first power ever produced from natural gas in the kingdom flowed into Bangkok's electrical grid in 1981. The recent finds north of Erawan field in the Gulf of Thailand, where reserves are figured at 8 trillion cu ft of gas, fired the new interest. The latest discovery was made in 1982 at Union Oil Co. 1 Trat in Block 11. This discovery flowed 12 MMcfd of gas and 432 b/d of condensate. This is Union's 10th discovery in the Erawan field area.

Bangladesh

There is no doubt Bangladesh is a gas prone area, but the possibility of oil deposits cannot be ruled out. Oil-bearing strata in Assam, the Barail formation, occur in Bangladesh under a thick Neogene sequence. Geologic transverses and satellite photos suggest such strata may come reasonably close to the surface in the eastern part of the frontal-fold belt of the Chittagong Hill tracts and toward the rim of the Surma basin.

There are many untested structures in the nation. A more precise, comprehensive knowledge of the petroleum geology of the country is needed, as well as more deep drilling because the tertiary geology here is very complex.

Fiji

A group of U.S. and Canadian independents launched a wildcat program on a 2-million acre concession, lying mostly off the Fiji Islands in the South Pacific in 1982. The first test was on Viti Levu, the largest island. It was to test seismic prospects in Lower Miocene reefs at 4,500–6,500 ft. The block is about 20 miles east of Suva, Fiji's capital city. Geologists note Block 7 looks like a cabbage patch with reefs everywhere. Gas/condensate and oil seeps are known off Fiji close to seismically defined reefs.

Chevron and Mapco drilled two offshore wildcats in 1979, the Bligh Water and the Great Sea Ranch to 8,997 and 9,315 ft, respectively. Both were dry. These wells were 60 miles apart and 160 miles north of Viti Levu in a different geological area.

Although the Chevron holes did not test the Miocene reef, the Bennett group, leaders in the new program, believes the results of the first two wells show that Fiji shares the same structural and stratigraphic trends with all other producing areas of the western Pacific, including Indonesia, Java, New Guinea, Malaysia, and the Philippines. Production in these areas is mostly reefal with some oil in sandstones. The group further believes that the Chevron tests indicate that Fiji has more than 25,000 ft of Plio-Miocene sediments offshore that wedge updip into near shore reefal sections.

The group, which includes several companies from Canada and the U.S., found that rocks exposed on Fiji consist of cavernously weathered Eocene and Oligocene shelf carbonates as well as Lower Miocene with rocks. These range from deep-water, organically rich shales to pinnacle reefs, fringing reefs, barrier reefs, patch reefs, and lagunal carbonates. The same system has been identified offshore.

Afghanistan

The most prospective areas of central and south Afghanistan for the discovery of hydrocarbons are near Ghomel in the Katawaz fault block where a syndepositional high seems to exist in marine Mesozoic formations.[5] Attractive prospects for oil exploration in this area are Oligocene anticlines up to 30 miles long. Reflection seismic work is needed here to uncover favorable structures.

Factors favorable for the accumulation of hydrocarbons are deposition of thick marine Late Paleozoic, Mesozoic, and Paleogene sediments containing source rocks; the probability of varying thickness within the Mesozoic formations by which a primary migration of hydrocarbons to preexisting highs could have been initiated; deposition of clastic porous and permeable rocks in bituminous Mesozoic-Paleogene formations; formation of almost undisturbed Oligocene anticlines in the areas of the Ghomel high.

Australia

A combination of sliding oil prices and soaring interest rates threatened what promised to be a record year for the search for oil and gas in Australia in 1982. However, recent oil discoveries in the Northwest Shelf, Bonaparte Gulf, and the Bass Strait will continue to attract offshore attention as will Victoria, the Northern Territory, South Australia, and Tasmania.

At least $921 million will be spent on offshore development in 1983, most of it on the Northwest Shelf. The 6 Goodwyn on the Northwest Shelf hit oil in 1982 flowing 3,500 b/d. This is the best flow yet in the area. Also, Esso-BHP's 1 Tarwhine, 17 km southwest of Barracouta oil and gas field in the Bass Strait, flowed 2,600 b/d of oil and 2.9 MMcfd of gas.

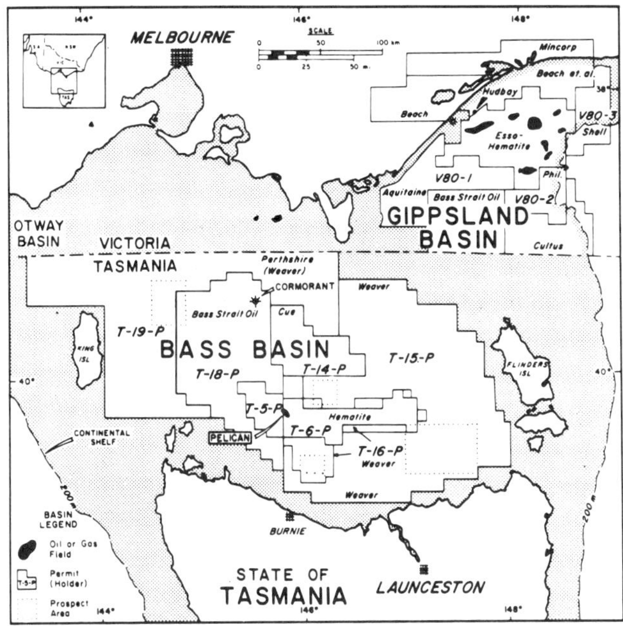

Fig. 14–18 The Bass Strait (courtesy *OGJ*, January 4, 1982)

In the Bass Strait basin of Gippsland, four new fields—West Kingfish, Cobia, Fortescue, and Flounder—are being developed. This will increase known commercial oil reserves in the basin by some 600 million bbl.

A group led by Woodside Petroleum has another discovery off Western Australia at 1 North Scott Reef. This well flowed 57 MMcfd of gas during production tests at about 14,000 ft. The flow was one of the largest yet in Australia. The Scott Reef could add considerably to the Northwest Shelf group's gas reserves. Early development, however, is not expected due to plentiful supplies. Location is 185 miles off Western Australia and 620 miles north of the Northwest Shelf project.

More oil and gas have been found in the Cooper basin of Queensland and South Australia. The 1 Jackson wildcat, a Delhi Petroleum well, flowed 2,500 bbl of 39° gravity oil on a drill-stem test of the Hutton sandstone at 4,687–4,773 ft. This is being called the most significant discovery in Australia since the 1 Blina oil find in 1981, which was apparently the first commercial find in Northwest Australia's Canning basin.

In the Cooper basin, a group led by Delhi found gas at 1 Wanara. It flowed 7.5 MMcfd of gas from the Permian Toolachee. Pancontinental Petroleum found gas at 1 Dingo in the Amadeus basin of Central Australia. At least 80 wildcats were drilling in Western Australia in 1982. About half of these were offshore.

Triton-Delhi had a discovery at 1 Lenara in the Permian Toolachee formation. The well flowed 7.5 MMcfd of gas on two tests. Location is in the Nappacoongee-Murteree area of South Australia and 26 miles southeast of Moomba gas processing plant.

Australia's big fields are in the Gippsland basin off the southeastern coast. The Halibut field, discovered in 1967, produces at 7,700 ft from 17 flowing wells. Cumulative production is 496 million bbl of oil. Another large one is Kingfish with 21 wells and a cumulative production of 761 million bbl of oil. This field was discovered in 1967. Then there is Mackerel, a 1969 find, with 17 wells. Cumulative output is 100 million bbl of oil. All of these fields are in Bass Strait. The big field on the western side of Australia is at Barrow Island, opened in 1964. There are 294 wells and the cumulative production is 176 million bbl of oil.

The Bass Strait, separating Tasmania from the Australian mainland, is the source of more than 400,000 b/d of oil from its eastern portion in the Gippsland basin. The initial recoverable reserves in this basin alone exceeded 3 billion bbl of oil and 8 trillion cu ft of gas.

Geologists now feel the Bass basin has good prospects of becoming a producing province for the strait. This basin contains essentially the same reservoir section as the lush Gippsland basin. Potential reservoirs range from 3,500 ft to 10,000 ft and deeper. Water depths are in most cases less than 220 ft. The first discovery of note in the basin was at Pelican in 1970. However, this basin with a prospective area of over 15 million acres now has had only 18 wildcat and confirmation wells drilled.[6] Three of these wells are within the Pelican field limits while oil was found in the Cormorant well in the northern part of the Bass basin.

REFERENCES

1. E. Egbogah and D. Lamberta-Aikhionbare, *Oil & Gas Journal* (April 14, 1980), p. 184.

2. S.K. Biswas, *Oil & Gas Journal* (April 19, 1982), p. 228.

3. J. Riva Jr., *Oil & Gas Journal* (March 8, 1982), p. 306.

4. Ibid., p. 309.

5. A. Schreiber, et al., *Oil & Gas Journal* (September 13, 1971), p. 110.

6. Weaver, et al., *Oil & Gas Journal* (January 1, 1982), p. 154.

15
Offshore

The global offshore petroleum industry is big, and it continues to grow.

Since the world's first commercial well was drilled out of sight of land in the Gulf of Mexico in 1947, offshore operations have grown into a sizable segment of the petroleum industry.

More than one-fifth of the world's total hydrocarbon reserves are in offshore fields. With continued depletion of small onshore fields in consuming nations, giant reserves (over 500 million bbl of BTU equivalent gas per field) now appear to represent nearly 80% of the world's reserves. Twenty-two percent of the world's giant-field reserves and production have been found offshore, and more than 80% of the reserves and production correspond to giant fields.

The growing importance of offshore oil and gas has been the subject of much debate concerning offshore international boundaries, mineral rights, and environmental concern, but this has not yet resulted in development of a set of worldwide standards.

About 90% of the offshore reserves are in inland seas or interior shelf areas. Slightly more than half of the offshore reserves are in onshore fields which extend offshore. These represent the early discovery and development phase of offshore activity, and they have continued with substantial additions to the present (Fig. 15–1). The other half of the offshore fields are totally offshore and began important development in the 1950s. Fields that are totally offshore appear to be of smaller size, on average, than those partially offshore.[1]

Fig. 15–2 is a comparison of cumulative reserves and average field size of worldwide giants, offshore giants, and offshore fields of 100–500 $\times$ 10^6 bbl of BTU equivalent. This comparison is shown including and excluding the unique Middle East. Cumulative reserves and relative field sizes are analyzed on the basis of basin and trap type, reservoir age, reservoir lithology, depth, and relative abundance of supergiants (fields over 4 $\times$ 10^6 bbl BTU).

In the 1980s, often referred to as the deepwater decade, the petroleum industry is no longer viewing the ultradeep environment as an energy alternative but as an energy necessity. Great strides have been made in new equipment designs and technical capabilities to meet these awaiting targets. New and renewed subsea completion and production systems are moving from drawing boards to ocean floors as quickly as assignments arise. Also responsible

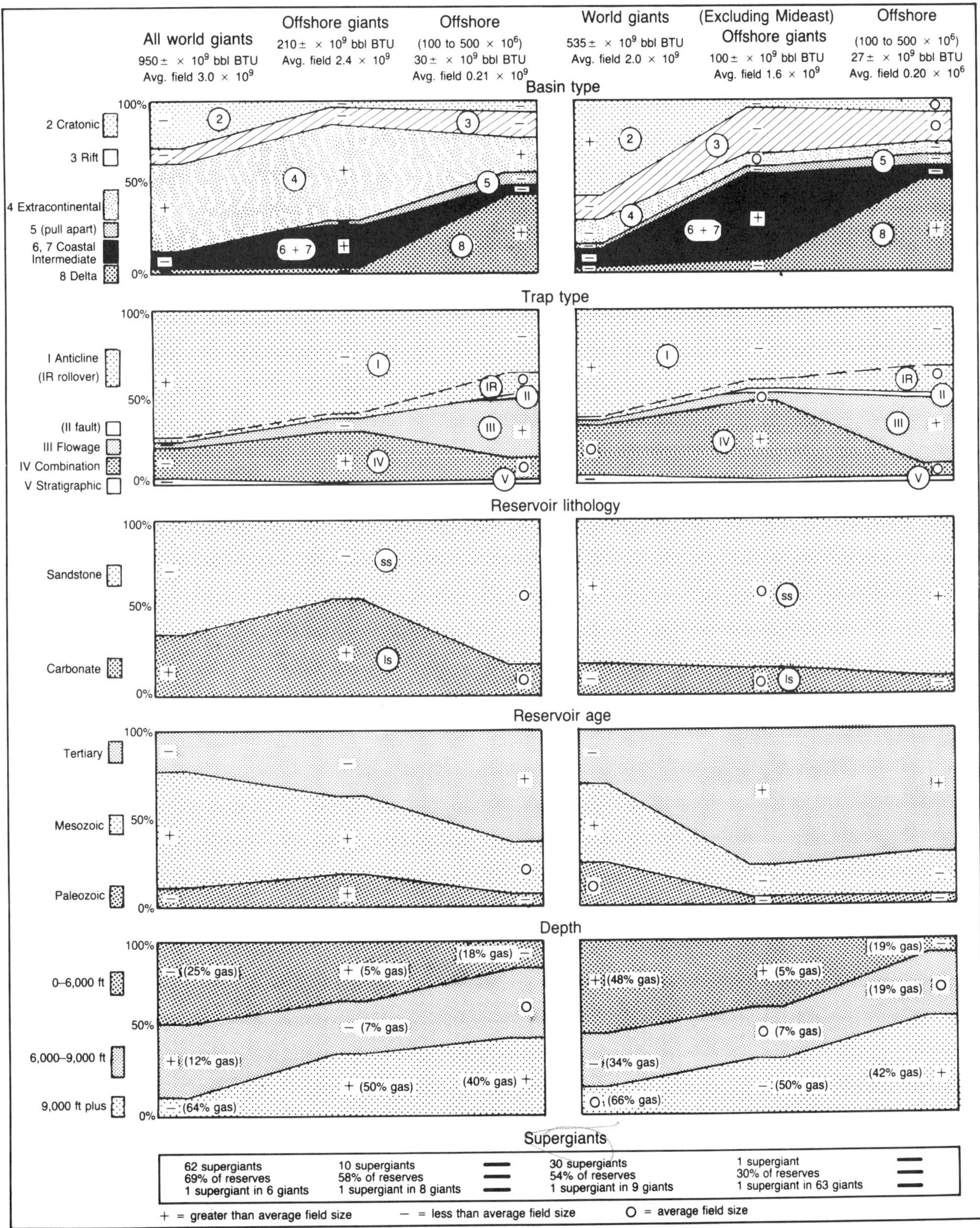

Fig. 15–1 Oil and gas discoveries—cumulative produced and proven areas (courtesy *OGJ*)

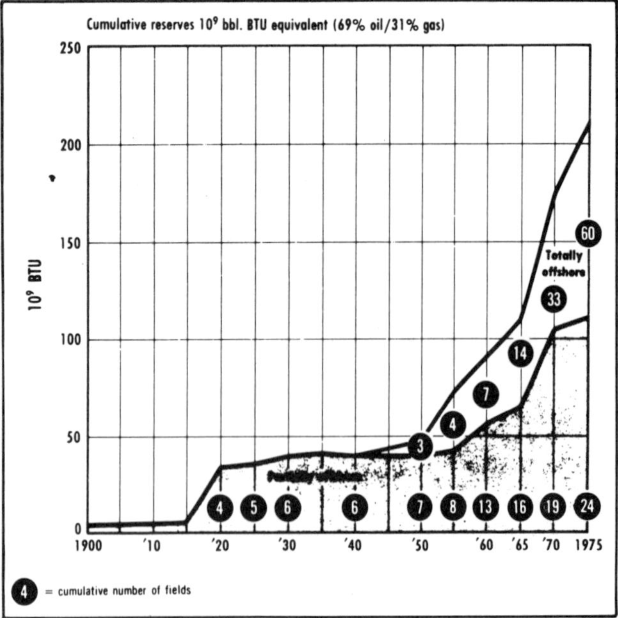

Fig. 15–2 Offshore giants (courtesy *OGJ*)

are the giant discoveries made off Alaska and Canada which have instigated great advancements and have sparked new operating concepts in ice technology—seen by many in the industry as the up-and-coming offshore experience.

Contracts abound in all facets of the supply side of the field as the industry voices greater demands for effective tools, rigs, platforms, and coastal facilities.

Oil prices and supply needs will continue to dictate the next offshore moves. Some nations around the globe continue to work toward their own goals of self-sufficiency. Others are striving to supply a hungry market, and some are trying to take their first wet steps.

The outlook for offshore exploration/production in the non-Communist world is definitely positive through the 1980s (Fig. 15–3).

But offshore drilling contractors must weather a slump in rig demand that likely will persist into 1983.[2] Petroconsultants of Geneva predict a steady climb in the number of offshore wildcats drilled to nearly 1,500 wells in 1990 from less than 1,000 in 1981.

Offshore production will increase to 17.9 million b/d by 1990 and will account for 35% of total non-Communist output. By comparison, offshore production amounted to 10.3 million b/d in 1980, or 23%.

Rowan Companies of Houston indicates there will be more up than down years in the drilling cycle, and through the end of the century offshore drilling will be a growth business. The likelihood, however, of a significant rig surplus in the short term will discourage potential entrants into the business and squeeze out some existing operations.

Petroconsultants predicts most offshore exploration, drilling, and production will take place in waters of 300 ft and less. Activity will remain strong on continental shelves. Deepwater activity will continue to a lesser extent, but no more than 10% of total outlay for offshore drilling will be spent in waters 600 ft and deeper.

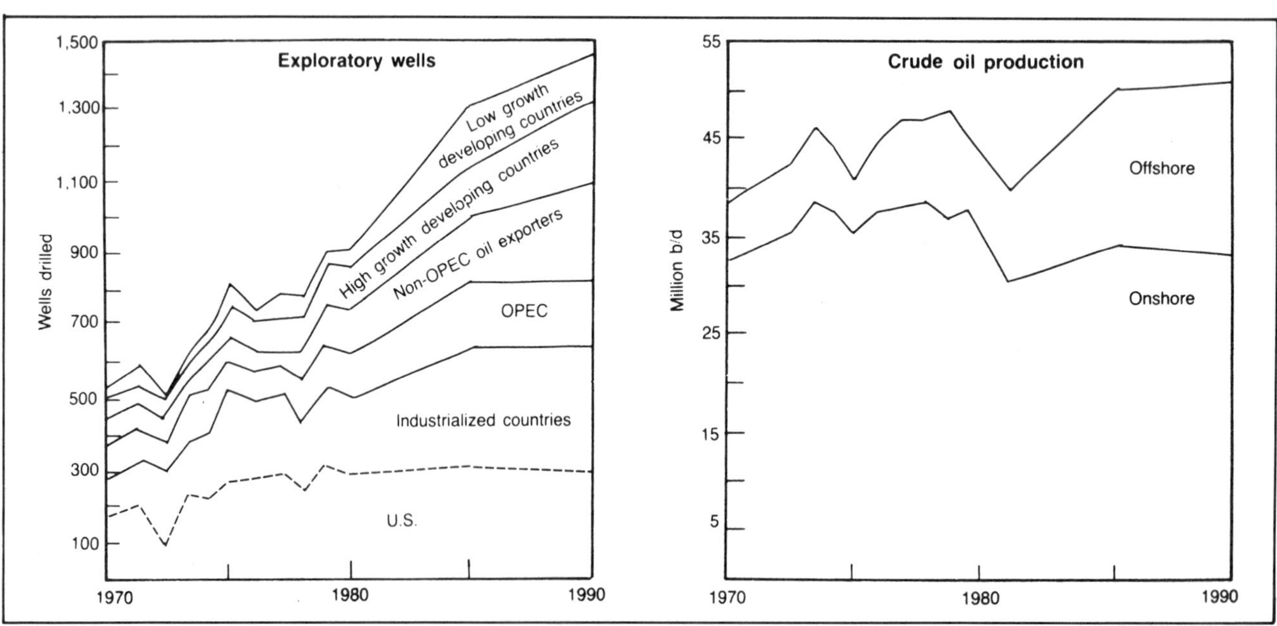

Fig. 15–3 Outlook for non-Communist offshore areas (from Petroconsultants Inc., courtesy *OGJ*)

Production off industrialized nations is expected to increase to 5.8 million b/d by 1990 from 3.8 million b/d in 1981. The largest increase is expected from Norway—2.2 million b/d from 509,000 b/d in 1981.

U.S. offshore production will remain essentially flat throughout the 1980s. Decreases in Gulf of Mexico production will be offset by increases off southern California.

U.K. production will increase in the mid-1980s to 2 million b/d but will fall by 1990 to 1.8 million b/d. The predicted drop is due to continued heavy taxes, which restrict drilling.

High-growth developing countries—mainly Argentina, Brazil, Chile, and Colombia—have the largest predicted increase in offshore wildcats. Their share of total non-Communist wildcat drilling is expected to increase to 16% from 12% in 1980.

Offshore and onshore production for the category will be almost 2 million b/d by 1990, compared with 890,000 b/d in 1981.

Of the 47 countries in the low-growth developing country category, those in West Africa have a potential for significant crude oil production. Petroconsultants predicts that by 1990, Benin, Cameroon, Equatorial Guinea, Ghana, and the Ivory Coast will be producing at rates of 15,000 b/d (Equatorial Guinea) to 360,000 b/d (Ivory Coast).

By 1990, the top five producers in the low growth category will be India, Ivory Coast, Cameroon, Turkey, and Sudan. They will be producing more than 80% of the group's crude oil. Offshore production is expected to increase to 25 million b/d in 1990, with Saudi Arabia the main contributor. Non-OPEC oil exporting countries will increase production substantially.

Such is the state of the industry. However, to appreciate the strong and ever brighter days ahead, here's a breakdown of each nation's potential.

NORTH AMERICA

Oil and gas output from the Gulf of Mexico—the number one U.S. offshore producer—continued its slow, downward trend in 1981. Although the Texas coast increased its output in 1981, Louisiana production could not reverse the general decline. It also was not a good year for the U.S. East Coast in terms of successful wildcatting activity. However, operators are still supporting the East Coast future and are

Table 15–1

MEAN ESTIMATES OF U.S. RESERVES BASIN BY BASIN

Basin	Oil (billion bbl)	Gas (tcf)
Georges Bank	1.4	5.7
Mid-Atlantic	3.1	14.2
South Atlantic	0.9	3.6
Eastern Gulf of Mexico	0.2	0.5
Western Gulf of Mexico	2.4	26.1
Santa Barbara	1.3	2.5
California inner basins	0.6	0.5
California outer basins	0.5	0.9
Santa Maria	0.7	1.7
Outer Santa Cruz	0.2	0.2
Point Arena	0.1	Negl.
Eel River	Negl.	Negl.
Oregon-Washington	0.3	1.5
Gulf of Alaska	0.4	2.2
Kodiak	0.4	2.0
Lower Cook Inlet	0.4	2.2
Bristol Bay	0.2	1.0
St. George Bank	0.4	2.3
Navarin	0.9	5.6
Norton	0.2	1.2
Hope	Negl.	0.3
Chukchi	1.4	6.4
Beaufort shelf	7.8	39.3

Courtesy *OGJ*

looking toward the new acreage off Massachusetts for another chance.

The year 1981 was extremely successful for activities along the U.S. West Coast, where production start-up on three fields is just the beginning of their aggressive offshore program. High hopes also have been expressed for the potential basins off Alaska, expected to hold some giant possibilities (Table 15–1).

Gulf of Mexico

Despite an impressive increase in annual crude and condensate production off Texas, the sliding output rates off Louisiana continued to push total Gulf of Mexico production downward for the second year in a row in 1981.

The U.S. Geological Survey reported enhanced recovery techniques are being used in two-thirds of the 435 fields still producing in the gulf to bolster declining oil and gas production. Crude production is continuing to fall from a peak year in 1972, while gas production leveled off in 1981 after a long climb (Table 15–2).

The USGS reports conventional waterflooding is used to maintain pressure in about one-third of the wells producing, and a combination of water, steam, or gas is used in another third of the wells. Most of the injection well projects are on a small scale due to the economics of gulf production and substantial primary flow in most of the wells.

Crude production from gulf wells in 1981 was expected to drop to about 250 million bbl and gas production to 4.5 trillion cu ft, slightly higher or nearly the same level as 1980 production. At the end of 1980 a total of 435 fields were listed as producing, not including 16 of which have been depleted. Another 31 fields were in the process of development.

A total of 482 fields have been discovered in the Gulf of Mexico. Of the 451 fields for which the USGS has calculated reserve potentials, 355 are primarily gas producing

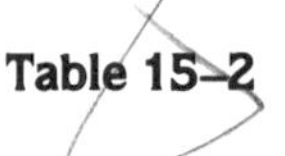

Table 15–2

MEASURED-INDICATED RESERVE ESTIMATES FOR GULF OF MEXICO OCS FIELDS

	No. of fields (1)	Original recoverable reserves		Cumulative Production		Remaining recoverable reserves	
		Oil (MM bbl)	Gas (MMMcf)	Oil (MM bbl)	Gas (MMMcf)	Oil (MM bbl)	Gas (MMMcf)
Mustang Island-Padre Island	9	0	380	0	0	0	380
Matagorda Island	5	0	161	0	1	0	160
Brazos	8	6	670	2	320	4	350
Galveston	7	27	830	17	640	10	190
East Breaks-Garden Banks	3	18	340	0	0	18	340
High Island	59	137	7,700	17	1,900	120	580
West Cameron-Sabine Pass	57	200	14,700	70	7,400	130	7,300
East Cameron	38	160	6,700	80	4,300	80	2,400
Vermilion	50	350	10,500	180	6,700	170	3,800
South Marsh Island	33	550	9,600	260	5,200	290	4,400
Eugene Island	51	1,140	11,300	700	7,200	440	4,100
Ship Shoal	33	910	8,500	610	5,200	300	3,300
South Timbalier-Bay Marchand	18	1,060	4,500	840	2,600	220	1,900
South Pelto	4	90	350	50	150	40	200
Grand Isle	10	810	3,600	640	2,200	170	1,400
West Delta	15	1,140	3,600	790	2,500	350	1,100
South Pass	8	610	1,500	360	800	250	700
Main Pass-Breton Sound-Viosca Knoll	20	640	2,700	360	1,600	280	1,100
Mississippi Canyon-Ewing Bank	7	190	1,250	10	10	180	1,240
Total	**435**	**8,038**	**88,881**	**4,986**	**44,220**	**3,052**	**40,160**

(1) Represents 419 of the 466 active (December, 1980) fields and 16 formerly productive fields abandoned before December 31, 1980. Source: USGS Open File Report No. 81-604 (courtesy *OGJ*)

and 95 are primarily crude producing. The remaining recoverable reserves from oil and gas fields are 3.1 billion bbl and 40.2 trillion cu ft. Original reserves, current plus cumulative production, are 8 billion bbl of crude and 88.9 trillion cu ft of gas.

The USGS says the deeper reaches of the Gulf of Mexico—a territory where jurisdiction is not yet established—may contain 9.11 billion bbl of oil and 18.77 trillion cu ft of gas as mean estimates of undiscovered resources in place (Fig. 15–4).

The range of estimates for these deepwater areas is 2.24–21.99 billion bbl of oil and 5.48–44.4 trillion cu ft of natural gas, with the low estimates corresponding to 95% probabilities and the highs to 5% probabilities. USGS says only a small portion of the study area can be exploited with current technology. However, it expects the industry will operate in water depths of more than 10,000 ft by 2000.

Water depths of the study area range from 98 ft on the continental shelf off the Rio Grande to 12,270 ft in the deep abyssal plain of the west central gulf. More than 75% of the study area lies in water depths exceeding 10,000 ft. The study area covers 58,940 sq miles.

USGS suggests that overall recoveries from the region from all recovery methods may average 28% for oil and 80% for gas. But the agency points out that these factors cannot be applied to gross estimated reserves.

The USGS gulf report estimates that the study area contains 188,140 cu miles of sediments. There are six assessment areas in the total:

1. Rio Grande margin—1,130 sq miles with 3,040 cu miles of sediments and mean resource estimates of 340 million bbl of oil and 1.09 trillion cu ft of gas.
2. Sigsbee escarpment—760 sq miles with 2,770 cu miles of sediments and mean resource estimates of 370 million bbl of oil and 850 billion cu ft of gas.
3. Perdido fold belt—760 sq miles with 2,700 cu miles of sediments and mean resource estimates of 1.74 billion bbl of oil and 3.22 trillion cu ft of gas.
4. Sigsbee knolls—230 sq miles with 380 cu miles of sediments and mean resource estimates of 510 million bbl of oil and 1.07 trillion cu ft of gas.
5. Campeche escarpment area—2,140 sq miles with 3,970 cu miles of sediments and mean resource estimates of 930 million bbl of oil and 1.84 trillion cu ft of gas.
6. Abyssal gulf basin—53,920 sq miles with 175,280 cu miles of sediments and mean resource estimates of 5.22 billion bbl of oil and 10.7 trillion cu ft of gas.

USGS says a geologic analysis of the six areas results in speculation "that oil and gas may be contained in discontinuous or isolated sandstone reservoirs surrounded by and interbedded with low permeability shales, in other clastic reservoirs, and in shelf carbonate reservoirs." However, secondary and tertiary recovery methods are expected to be very expensive and difficult to apply in deepwater areas.

U.S. East Coast

The waning spirit for exploratory projects off the U.S. East Coast was given a much needed jolt in mid-1981 with the initiation of wildcatting in the controversial Georges Bank and the noted industry interest for deepwater tracts in the Baltimore Canyon during the last lease sale. Offshore tracts in the deepwater Jurassic trend that stretches along the edge of the Baltimore Canyon trough (Fig. 15–5) captured the interest—and most of the high bids—of the 21

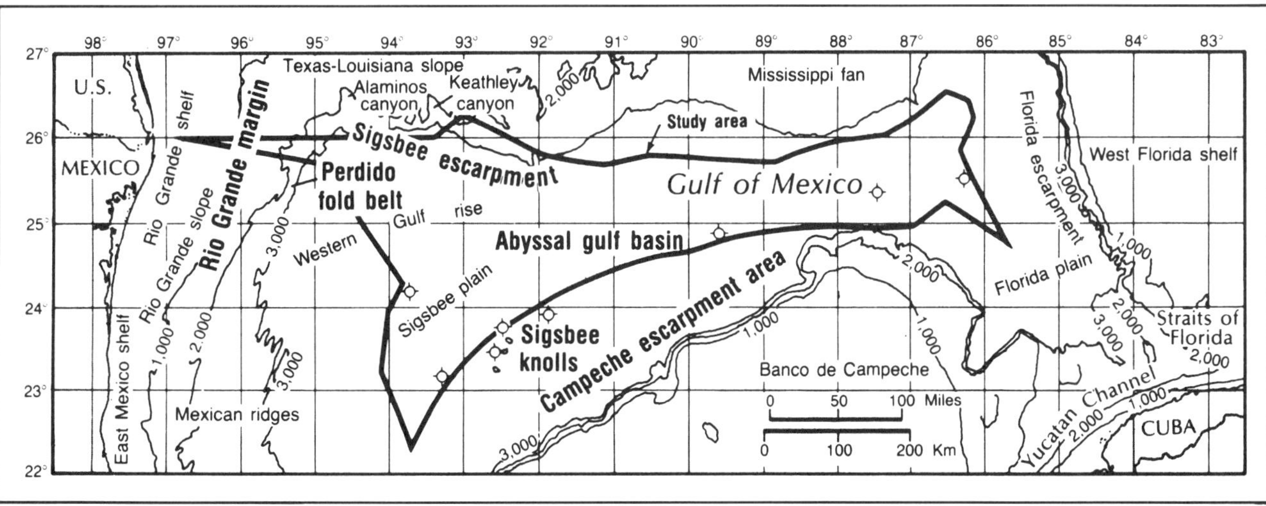

Fig. 15–4 Undiscovered resources in the Gulf of Mexico as assessed by USGS (courtesy *OGJ*)

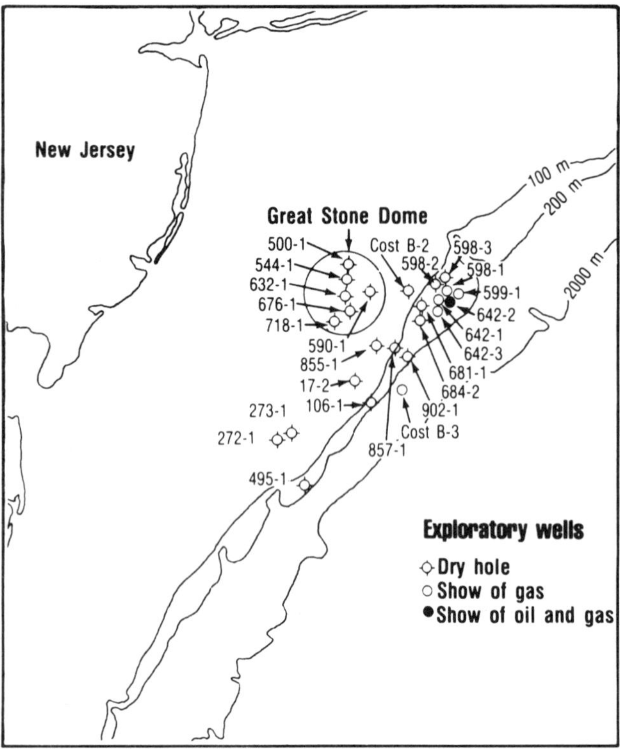

Fig. 15–5 Exploratory targets in Baltimore Canyon (courtesy *Offshore*)

petroleum companies participating in OCS Sale 59 held in December 1981. The auction closed with a total of $424.9 million in high bids on 98 of the 253 tracts offered. Shell and Shell-led groups showing an overwhelming enthusiasm for the basin as they placed 75% of the high bids. Seeing the deep tracts as a means of testing and developing their deepwater technology, Shell already plans to contract the *Discoverer Seven Seas* vessel in mid-1983 to drill 10,000–14,000 ft wells in the canyon.

The deepwater reef grabbing the industry's attention is thought to have formed around 100–140 million years ago. The formation is from the Jurassic age and is the same geological area that is producing great quantities of oil and gas in Mexico. Experts are still not certain, however, if similar conditions exist to the north.

The industry turnout at the sale was especially revealing since the canyon still maintains a dismal track record in wildcatting with no commercial successes and no current exploration activity on tap. Probably a more potent instigator for the response was the revised USGS estimates, which reported there may be 962 million bbl of oil and 7.6 trillion cu ft of natural gas in this new territory.

Prior to the sale in the fall of 1981, Tenneco confirmed its interest in the producing future of the Baltimore Canyon by requesting an extension to its 5-year lease on Block 642, due to expire in May 1981. Although no drilling has been conducted on the block since October 1980, Tenneco is hoping to convince Texaco and Exxon to join a four-block area that would include three tracts the USGS has described as producible.

There have been encouraging shows in the proposed four-block unit. Three wildcats have been drilled on Block 642, all testing gas and two testing condensate. One wildcat flowed gas on Exxon's Block 599 more than 2 years ago; Texaco's Block 598 is remembered as the site of the first East Coast noncommercial discovery back in 1978.

Ironically, Tenneco's request came days after Mobil announced it would abandon its second canyon duster which brought the total number of mid-Atlantic dry holes to 23. Mobil was the last operator to exit the basin after drilling a wildcat on Block 544 in 218 ft of water. The company has no immediate plans to return to the canyon and has since turned its attention, along with Exxon and Shell, to the adjacent East Coast basin—Georges Bank.

Slow start

After 7 years of legal battles and court injunctions, offshore exploratory drilling finally got under way in the Georges Bank trough off Massachusetts in July 1981. Exxon and Shell were the first to arrive and have since each reported a disappointing first try.

Exxon announced the first bank well a duster in November 1981 after drilling to 14,118 ft on Block 133. The company since has moved the Alaskan Star drilling rig 20 miles to the northeast on Block 975 to drill a second wildcat to 15,000 ft. Plagued with downhole problems since their arrival, Shell surrendered its first bank effort after an anchor line broke in December 1981 during a winter storm. The company decided to plug the Lydonia Canyon Block 410 well temporarily when heavy seas snapped the cables. The hole had reached 13,000 ft on its way to a targeted 17,000 ft.

Mobil was the third company represented in the basin, moving the Rowan Midland rig in to drill its first Georges Bank well on Block 312 about 160 miles east of Nantucket. The wildcat is in 250 ft of water and is targeted to 18,500 ft.

California

The U.S. West Coast is on the threshold of a new offshore chapter that offers the prospect of the discovery and development of giant oil fields. The two offshore fields that came on stream in 1981 are partly responsible for this bright outlook.

All current signs point to development of significant oil reserves in the offshore area extending northward from

Point Conception through the offshore portion of the Santa Maria basin. The 1981 lease sale in this region broke all California records, including biggest bonus for a single tract. Operators expect to find several giant fields in primary targets that include the Monterey sector. This area is the main pay in three fields discovered by Exxon in the Santa Ynez Unit, including the Hondo field about 25 miles to the southeast.

On the production scene, the two offshore fields that came on stream in 1981 are playing a key role in pushing California's production to a record high. Projected production rates for California in 1981 indicate an output for the year of about 383.1 million bbl—7.7 million more than the previous record of 375.4 million bbl set in 1968. A significant share of the credit for the strong production showing belongs to two giant offshore fields: Hondo in the Santa Barbara Channel and Beta in San Pedro Bay. Together they are producing nearly 40,000 b/d.

At Hondo, Exxon is producing 32,200 b/d from 17 wells. The oil represents the first installment of what will ultimately be 500 million to 1 billion bbl of oil produced from the Santa Ynez Unit. The company plans to install another platform some 3 miles west of Hondo in more than 1,000 ft of water, possibly surpassing Shell's record at the Cognac platform.

By the end of the decade, two other fields in the Santa Ynez Unit—Pescado and Sacate—may be producing from deepwater platforms. The Pescado platform will be installed in about 1,000 ft of water, the Sacate in about 700 ft of water. It is anticipated production from the three fields in the unit could amount to 60,000–80,000 b/d.

A USGS study for deepwater areas off southern California (Fig. 15–6) estimates in-place oil at 0–1.78 billion bbl (mean: 410 million bbl) and gas at 0–2.86 trillion cu ft

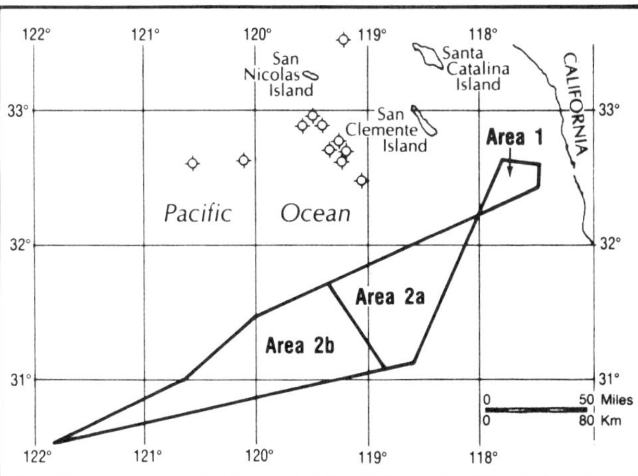

Fig. 15–6 USGS study area off California (courtesy *OGJ*)

(mean: 680 billion cu ft). This study area covers 7,500 sq miles in water depths ranging from 295–12,000 ft. The study area of 7,500 sq miles off California contains 2,347 cu miles of sediments. It is divided into the following areas:

1. 500 sq miles with 120 cu miles of sediments and resource estimates of 20 million bbl of oil and 50 billion cu ft of gas.

2a. 2,800 sq miles with 1,707 cu miles of sediments and a mean resource estimate of 390 million bbl of oil and 630 billion cu ft of gas.

2b. 4,200 sq miles with 520 cu miles of sediments and negligible resources.

Beta news

Shell began moving oil ashore from the Beta field in January 1981 and is producing 7,400 b/d. The oil is pumped ashore through a 16 in. pipeline that links the offshore field with Long Beach some 15 miles away. Shell is developing the field from a two platform complex, including the 80-slot Platform Ellen, a drilling platform in 265 ft of water, and Platform Elly on an adjoining production platform. Also on tap is another drilling platform, designated Eureka, which will accommodate 60 wells.

The Beta field is the only commercial discovery resulting from the 1975 sale at which the industry spent $417 million for 56 tracts. Shell estimates recoverable reserves beneath its two tracts at 150 million bbl and expects to be producing some 15,000 b/d by 1984 with production peaking at about 22,000 b/d by 1990.

Chevron has not announced figures for estimated reserves under its parcel, but it does plan production from Platform Edith in early 1983. Estimated peak production from the 72-slot platform in 8,000 b/d. Oil will be piped to Shell's Platform Elly and then to shore through Shell's existing line.

Additional production is coming on sream from two Union Oil of California platforms in the Santa Barbara Channel, and plans are to move oil ashore from Platforms Gilda and Gina. Gilda, a 96-slot platform in 210 ft of water in the Santa Clara field, holds the record for more slots than any existing platform. On the other hand, Gina, a 15-slot platform in 95 ft of water in Hueneme field, has the smallest number of slots of any platform on the West Coast. The company expects to produce about 20,000 b/d of oil from Gilda and 4,000 b/d from Gina.

Chevron is producing 2,089 b/d from 10 wells in the Santa Clara field and anticipates peak production of 13,000 b/d and 16 MMcf of gas in 1984. Texaco is producing from its Platform Habitat in the Pitas Point field. The 24-slot platform in 288 ft of water is the first off California to be designed for gas production.

Alaska

The campaign to find Alaska's next giant field centers on the icy waters of the Beaufort Sea. With eight important wildcat wells slated for the new drilling season, operators hope to accelerate the exploratory projects designed for the specific tracts in the area that drew highest bids of more than $1 billion at a joint federal-state sale in December 1979.

Several companies already have boarded the Beaufort bandwagon, and several more are expected to move in, as reserve figures and previous wells prove this is one of the hottest new spots off the coldest U.S. coastline.

Some of the companies expected to begin new Beaufort projects include Shell Oil, with plans to build gravel islands to test a pair of geological structures. One of the islands will be built on Tract 42 east of the Prudhoe Bay to test a structure lying beneath Tracts 40, 41, 42, 43, 98, and 99. The island will be built in 21 ft of water. The company also plans to test the area northwest of Prudhoe Bay, drilling on some of the westernmost acreage. In addition, the company will build a gravel island on Tract 47 but close to 4 other tracts. If oil is found, Shell may expand the island to convert it to a production base.

Timing for the Beaufort Sea depends on the speed with which an environmental impact study can be conducted and permits obtained. Injections of processed sea water are planned for 1984. A reservoir study indicated waterflooding of the field will result in additional oil recovery of 5–9% and also will allow gas sales with no significant loss of oil recovery. The report projects a field life of 26.6 years with gas sales and water injection and estimates ultimate recovery of the original oil in place at 45.4%, representing 9.37 billion bbl.

While Prudhoe Bay exploration holds the most interest, exploratory activity elsewhere on the North Slope has confirmed Conoco has a commercial field in the Milne Point area northwest of the Bay. The company also turned up encouraging shows at Gwydyr Bay State 2A, 15 miles east of the Milne Point activity.

West of Prudhoe, ARCO is coming on stream with the first production from the Kuparuk River field. The flow of oil last December represented the first oil from a North Slope reservoir outside the Prudhoe Bay field. The Kuparuk River is expected to attain an initial production rate of 80,000 b/d. Plans call for the field to produce as much as 250,000 b/d by 1986.

Canada

Canada's offshore exploration and development plans are suffering from the government's National Energy Policy. Established in October 1980, the policy curtailed exploration off the East Coast. Only half of the rigs drilling the summer of 1981 were active in 1982.

This comes on top of Mobil Canada's extremely successful year, in which the Hibernia oil find of 1979 was further delineated to a point where the field's reserves have been estimated at 1.8 billion bbl. Also in the area, the finds at the Hebron well indicate another massive field off Canada's east coast.

Exploration, however, was disappointing for Dome in the Beaufort during the 1980–81 season. The exciting finds of 1979 could not be production tested further because of a very early ice season. But that did not stop the company from continuing with future production projects, which include its first artificial island system based close to the Tarsiut 25 oil discovery of 1979. Dome predicts oil will flow from the Arctic by 1985.

Farther to the north and east of the Beaufort, in the Arctic Islands, Panarctic had another productive winter in 1981, which included a gas find showing an oil flow from the surface. The company's transportation system—the $25 billion Arctic Pilot Project—also nears construction start-up. Regulatory approvals are expected which would allow the ice-breaker LNG tanker system to be operational by 1986.

While Dome has been attracting all the glamour of ice taming, it is Canada's east coast offshore area which has shown the most promise. In the northern Grand Banks area, drilling took place year-round in 1981 for the first time, ending with several encouraging spots. While federal officials believe the banks area contains about 20 billion bbl of oil, Chevron contends a more realistic estimate of total reserves probably exceeds 10 billion bbl. Chevron says test results on the first step-out wells of the Hibernia field showed the producing zones vary on the potentially prolific, oil bearing structure.

Mobil, however, reported in March 1982 it could have a second Hibernia size field on the Grand Banks, which could mean previous reserve estimates are low. Based on electric logs, Mobil said the upper zone between 1,859 and 1,926 m has an estimated net potential pay zone of 46 m, while three zones between 2,530 and 2,753 m have a potential net pay of 30 m.

This good news came just days after Mobil's third Hibernia step-out well, 8 km west of the original discovery, was reported to be a dry hole. This was hardly seen as a disappointment, however, since it gave Mobil and others the valuable evidence needed to realize the massive Grand Banks field does not extend quite as far to the west as some had hoped. Mobil gambled in going just a little too far to the flanks and lost when it did not come up with any oil. But with this new information the company now has a better idea of just how far the oil field does extend, for this third well was the farthest from the original discovery.

The next step-out should prove particularly interesting. It will be 5 km northwest between the two successful delineation wells at O-35 and B-08. With a projected total depth of 5,000 m, the operation should probe both the Avalon and Hibernia formations.

SOUTH AMERICA

Exploratory work in Latin America extends from Mexico to the tip of the continent (Fig. 15–7).

In the hopes of maintaining its working relationship with disappointed risk contract companies, Brazil's state-owned oil firm, Petrobras, may soon begin to loosen its exclusive reins on the country's top producing field, Campos.

Big plans also are in store for Venezuela, where methodical exploration of its sedimentary basins are reportedly under way. Chile and Ecuador are making news with record production figures and upcoming projects. The most unexpected news from Latin America comes from Mexico where offshore exploration is on the backburner indefinitely.

Fig. 15–7 Prime Latin American exploratory targets (courtesy *Offshore*)

Brazil

Although state oil company Petrobras continued to pursue an active offshore exploration and development program in 1981, other companies operating off Brazil were not so enthusiastic.

To counter the increasing disenchantment among risk contract companies who have yet to score off Brazil, Petrobras improved contract terms and is considering leasing portions of its exclusive Campos offshoe area. Along with this impressive announcement, Petrobras also declared an aggressive 5-year exploration and development plan with the bulk of its budget and effort to go toward development of the promising Campos area. Of the $4.5 billion allocated for Petrobras' 1981–85 exploration plan, 75% will be spent offshore.

According to the plan, Petrobras will drill 806 offshore wells by 1990, or about twice the number drilled on the Brazilian continental shelf so far. Three hundred of these wells will be concentrated in the Campos basin. The average water depth will be 100 m and the average drilling depth 4,000 m.

Petrobras drilled 466 wells on and off Brazil in 1980. Risk contract companies drilled another 11. Of the Petrobras wells, 158 were wildcat, 100 were offshore. Since Brazil opened its waters to risk contract exploration, a total of 92 contracts have been signed, 40 wells have been drilled, and almost $500 million invested by private companies.

Off Brazil the trend in risk contracts appears to be near the Santos basin. BP is drilling the 1-SCS-9 wildcat on the outer fringes of one of its Santos blocks in 535 m of water—a record depth for Brazil. Several other foreign companies signed for shallow water blocks in the northeast section of the basin, like Chevron with six Santos blocks all on the land side of previous contract areas.

The estimated costs for developing the Campos permanent production system has passed $5 billion. Petrobras expects good returns on the investment, since they figure an output close to 700 million bbl from the Campos offshore fields by the turn of the century.

About half of the more than $5 billion will cover the cost of drilling exploratory and development wells for the basin's production system. The other half will cover building and installing eight production platforms which will include the Garoupa and Enchova platforms and six satellite towers. Also included is the cost of the Garoupa provisional seafloor system. This system is already producing some 35,000 b/d and will later be incorporated into the permanent production system.

Venezuela

A traditional oil producer, Venezuela now claims to be undertaking its first plan for methodical exploration of its sedimentary basins. State oil company officials are systematizing the study of geological formations in the country, and the immediate conclusions indicate the need to develop the heavy oil Orinoco Belt and to explore offshore in Caribbean waters.

Venezuela's proven and recoverable reserves now stand at 18 billion bbl, and the belt is estimated to hold reserves of about 77 billion bbl. In the last 3 years, Venezuela's offshore efforts have resulted in only seven gas shows. Gas reserves in the whole area are now estimated at 20 trillion cu ft. Production from these offshore wells has been projected to increase from the present 400 MMcfd to 11 billion cu ft by 1995.

Chile

The offshore wells in the Chilean sector of the Straits of Magellan—the area responsible for 75% of the Chilean production—produced a record 35,000 b/d in the early part of 1981. Earlier, the Chilean government decided to end the state monopoly on oil exploration and production and opened both onshore and offshore areas to risk contract bidding by private companies.

At the present time, offshore drilling is under way by Phillips Petroleum in the Magellan Straits and by a consortium led by Atlantic Richfield in the Chacao zone. With expected consumption of 6–7 million bbl in Chile this year, the state oil company ENAP hopes the straits will supply 40% of that total. Imported oil is expected to cost the country an estimated $900 million.

Ecuador

Even as the Ecuadorian government continues making plans to develop the offshore gas fields in the Bay of Guayaquil, the state oil company and Texaco have found sizable heavy oil reserves onshore.

The new discoveries in the Pungarayacu region have raised the country's reserves to an estimated 1.5 billion bbl, or enough for this small country to continue exporting oil. By early last year, Ecuador was exporting 5.4 million bbl of crude.

Mexico

The most unexpected news from Latin America came in September 1981 when Pemex decided to pull out of the offshore exploratory and development game indefinitely, bringing the future of the Mexican mecca to a startling halt. The announcement came shortly after its confirmation as the world's fourth richest nation in oil and gas reserves.

Prior to the move, production from Mexico had been a little more than 1 million b/d. The Campeche Sound output alone came to 223 million bbl in 1980 and 555 million bbl in 1981—from a basin where less than 2 years earlier there was no production. Although Mexico now plans to focus on productivity instead of production targets, rumor is that exploration in Campeche will not completely close down, since contracts for two Rowan jackups were renewed for basin work shortly after the Pemex pullout.

Total proven reserves as of the end of 1980 stood at 60.1 billion bbl, 31% better than the previous year. The increase was due primarily to the discovery of the Ich, Ha, and Ek fields in the Bay of Campeche. Mexico's proven reserves as of March 1981 stood at 67.8 billion bbl.

During 1980 Pemex spent a total of $237 million on platforms for the Campeche fields. A total of 47 exploratory, development, and water injection wells were drilled in the bay at this time with 36 of these efforts considered developmental in nature.

EUROPE

The top story from the U.K. sector of the North Sea in 1981 was the marking of million barrel days for the first time. It also marked the most active year along the Norwegian sector where exploration efforts are uncovering more and more hydrocarbon prospects.

As evident by the number of exploratory projects on tap or in progress throughout the Mediterranean Sea, several of the surrounding nations boosted the spirit of the search with vigorous development work. Spain, Italy, and Greece are developing numerous finds in the Mediterranean, and France is preparing to drill some wells to record depths.

North Sea

During 1981 three new fields came on stream in the U.K. sector, bringing the number of producing fields to 17 and edging average daily production above the 1.8 million bbl mark for the first time.

Ten fields are under development, with two to three due to begin production before the mid-1980s. However, a number of government measures have put the brake on new development projects, and it remains to be seen how long this will last. The major roadblock comes in the form of another U.K. oil tax hike (a 20% petroleum duty). The tax is destined to have the most effect on development plans, causing a number of projects to be dropped and others reexamined as commerciality comes into question.

While the tax changes are not having an immediate impact on exploration, it is likely exploration budgets will be smaller, and plans for deeper wells in more hostile waters such as north and west of Shetland will be dropped.

Recent exploration results, however, have been more encouraging. Of the 14 wildcats completed in 1981, five were declared discoveries, one a dry hole, and the rest suspended as tight holes. Hamilton Brothers made the first discovery of the year in its Argyll field, followed by Amoco's find in the Clyde field. Shell/Esso's discovery in the Forties field flowed 1,200 b/d through a ½-in. choke, and British National Oil Corp. (BNOC) tested a find close to the Buchan field which has been described as having a fair chance of proving commercial. Phillips also reported its promising Joanne field find in mid-1981 near the Fulmar field and plans to appraise the area further.

Development and initial output from the Buchan field have been plagued with problems. Finally coming in 18 months behind schedule, production from its seven subsea completions should plateau at 50,000 b/d for 5 years.

Three more fields will come on stream—Shell/Esso's Fulmar and North Cormorant and Amoco's Northwest Hutton field. Production from Shell/Esso's 400 million bbl reserve should peak at 180,000 b/d. Amoco's Northwest Hutton output is expected to be around 100,000 b/d of oil and 35 MMcfd of gas.

Marathon already is working on outline plans for the development of the North Brae reservoir, unofficially estimated to contain up to 150 million bbl of recoverable reserves. Production based on an integrated steel platform is unlikely before 1986.

Like South Brae, BP's Magnus field in Block 211/12a also is due to begin production in 1983. Once 15 platform wells have been drilled, the $2.6 billion development will produce at a peak rate of 120,000 b/d for 6 years before going into decline.

Norway

Norway's most active drilling season ever in exploration and appraisal was recorded in 1980–81 with 33 wells completed, nine more than the previous record set in 1976. The increase in activity, with an average of 11 rigs operating, brough eight significant discoveries in 2 years and saw the start-up of the giant Statfjord field.

Crude production increased to an average of 528,000 b/d, but the statistics hide a drop in Ekofisk output compensated by the continuing buildup of production from Statfjord A. Work began in summer 1981 north of the 62nd parallel, and while all three wells failed to find commercial oil, results were encouraging for the future.

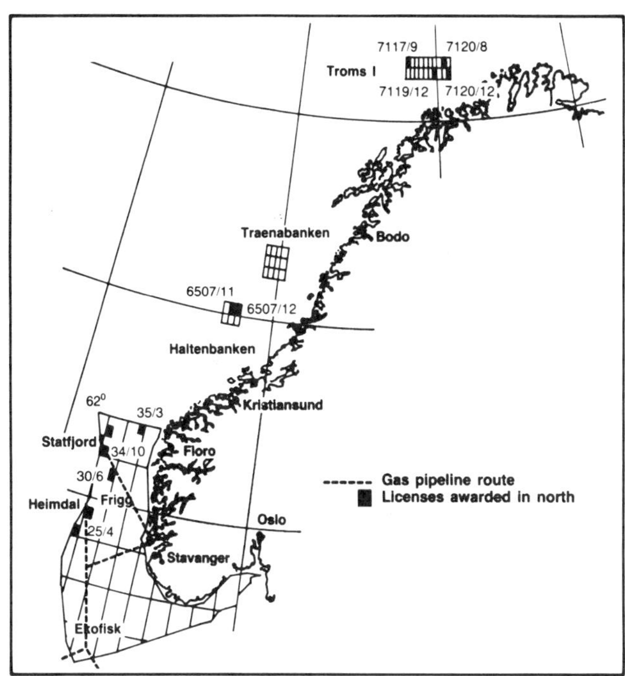

Fig. 15–8 Norwegian exploration areas (courtesy *Offshore*)

Norwegian fourth-round exploration licensing acreage has been particularly prolific in terms of exploration success since it was awarded in 1979. Continuing appraisal has established that Shell's deep-water Block 31/2 gas find has reserves of up to 50 trillion cu ft.

Exploration surprises were provided by Gulf and Saga Petroleum. Saga encountered gas on Block 35/3 north of the Norwegian Trench. Depending on future drilling, reserves could go as high as 8 trillion cu ft. Farther south in about 1,000 ft of water, Gulf tested 1,439 b/d of condensate and 32.4 MMcfd of gas from the Middle Jurassic. BP made two discoveries in Norwegian waters. One on Block 2/1 flowed up to 6,000 b/d, and on Block 30/4 tests flowed 29 MMcfd of gas and 1,100 b/d of condensate.

Offshore activity also started in 1981 on Elf's North East Frigg satellite development. Gas was to flow to the Frigg platform in late 1983 or early 1984 (Fig. 15–8).

Spain

With oil production established at four fields in the Gulf of Valencia, Spain is maintaining a lively exploration program. The most promising new areas seem to be the Mar Cantabrico off Spain's north coast and the Gulf of Cadiz. As for development, new projects are under way in Campsa's Gulf of Cadiz gas field and Chevron Spain's Casablanca field where the tallest producing installation is being used in the Mediterranean.

Ambitious development work at a cost of $113 million is taking place on state-controlled Cadiz field where another gas discovery was announced late in 1981. First production is scheduled for the end of 1983 with an estimated peak of 550 million cu m/year to be brought ashore to a terminal at Huelva.

Promising new exploration concessions south of Cadiz will be drilled by state owned Eniepsa. The firm plans two or three wells in Eolo, Neptuno, and Tartesos blocks. Campsa foresees drilling at least one well on Cadiz Permit 6.

The Gulf of Valencia, Spain's established oil province containing all of the country's producing fields, still draws industry enthusiasm. Dorada field was expanded in 1981. It produces more than 10,000 b/d from the Tarragona well. With the recent installation of permanent production facilities, the field is expected to peak at 15,000 b/d by 1983. Shell and Elf also drilled a numer of wells in the area in 1980–81 and planned to continue their ambitious drilling.

Brisk exploration continued in 1981 along Spain's north coast at Mar Cantabrico. Shows of oil and gas seem to be intriguing, but commercial-size finds have yet to be tapped.

Italy

Activity off Italy is dominated by state oil company Agip, producing 7.15 billion cu m of gas from 17 fields, which was expected to grow to 21 fields by the end of 1982. The company drilled a total of 34 wells in 1981 and made two oil discoveries in the Adriatic Sea and three gas finds in the Adriatic and Ionian seas.

Offshore Sicily is moving into the spotlight, too, after important oil discoveries. Agip's Nilde field off Marsala promises to be Italy's first commercial oil, and Perla field is due on stream.

Also off Sicily a group led by Montedison SpA discovered oil while drilling the 1 Vega wildcat 14 miles off Sicily in 413 ft of water. The well lies 5 miles from the group's developing Mila field.

Active exploration and development programs also are moving ahead in the Adriatic. Agip is developing the Barbara, Emma, and Porto Corsini gas fields while their Squalo field is producing 550 million cu m, and the David field output stands at 160 million cu m.

Also in the Adriatic, Elf Aquitaine is developing the Rospo Mare and Sarago Mare fields. Startup of the Sarago Mare is expected to start at 100 cu m and grow to 400–500 cu m by the end of 1982. Total was busy, too, with tests on their 1 Bonaccia wildcat where gas was encountered at four levels.

Greece

Greece officially became an oil-producing country in May 1981 when a consortium led by the North Aegean Petroleum Company brought in the Prinos field. Prinos is estimated to have 9 million tons of recoverable oil, and peak production will be 30,000 b/d. Later in the year, NAPC brought in the South Kavalla gas field, which had been developed in conjunction with Prinos.

The fields lie in the Thasos concession in the North Aegean Sea. Development was delayed both by technical problems and by the collapse of concession terms following the fall of the military dictatorship in 1974. Prinos and South Kavalla were developed together at a cost of $500 million. Two fixed platforms each with a 12-well capacity have been installed over Prinos; a four-leg drilling platform has been installed over Kavalla.

Exploration of promising Aegean waters is hampered by a boundary dispute with Turkey. However, determination to heat up the hydrocarbon search was obvious in 1981 when the government invited foreign companies to apply for exploration concessions in four regions of the Ionian Sea, off southern Crete, and in the area of the River Nestos. All of these are well within Greek waters.

France

In late 1982 or early 1983, Cie. Francaise des Petroles (CFP) and Elf Aquitaine were to spud the first of two exploration wells using the *Discoverer Seven Seas* drillship in the French Mediterranean in the record breaking water depth of 6,000 ft. This depth is virtually the limit of existing equipment for hydrocarbons drilling offshore. However, if the planned wells yield encouraging data on deep Mediterranean potential, the French will go ahead with draft plans to develop a system capable of drilling in 10,000 ft of water.

Apart from construction of a new drillship, this project also would require development in riser technology and the capability to drill and test in high formation temperature and pressure. For not only do the French eventually hope to drill in 10,000 ft of water, they also aim to reach targets 23,000 ft below sea level.

WEST AFRICA

Most of the African nations managed to come out ahead in the offshore exploration and production race in 1981 since politics played a less dominant role than in recent years. Nigerian production finally took an upward turn, and the high potential areas off the Ivory Coast, Benin, and

Cameroon have sparked new-found enthusiasm and a flurry of exploration projects in these once silent regions. Rig activity is bustling, and more new faces are noticeably making the West African scene.

Nigeria

Texaco and Elf hosted most of the activity off the coast of Nigeria—Africa's number 1 producing nation. Although plagued with governmental delays, offshore Nigeria should see more prosperous times.

Texaco's three new fields (Funiwa, Sengana, and Okubie) should raise the company's production standing dramatically. An estimated 100,000 b/d is expected by the time the fields come on stream.

Elf was also busy signing contracts for more offshore acreage and developing its own Otou field. A production rate of 20,000 b/d is expected from the field at its peak. Wellhead platforms are on tap for Mobil's Kobo and Unam fields, and Shell has a production platform and two nine-slot platforms planned for the KC field.

The Congo

Production is definitely on the upswing for the growing offshore region along the People's Republic of the Congo. The two leaders in the area, Agip and Elf Aquitaine, continue development of the two most significant finds last year—the Sendji and Yango fields, where startup is planned soon.

The two largest fields off the Congo, Emeraude and Loango, continue to produce at a rate close to 27,000 b/d. The third field discovery, the Likoula, came on stream in 1981 adding another 30,000 b/d to production figures.

Zaire

In Zaire, Gulf continues to hold the main operator position with its two shallow water fields, GCO and Mibale. Production is holding steady between 20,840 and 25,000 b/d.

Benin

Benin's first major oil field, Seme, is in full development. The field, in water depths ranging from 30–50 m, is considered marginal with heavy crude of about 22° gravity.

The first phase of the Seme field development is expected to cost about $142 million. Norway is guaranteeing 90% of the costs for the field, which is estimated to have 22 million bbl of recoverable reserves. The oil will be piped to shore via a 6-in. diameter pipeline to a land-based process/storage plant. From the storage plant, the oil will again be transported through a 20-in. pipeline to an offshore tanker loading facility.

The Seme field is to be in production by the end of 1982. The project is of significant importance to Benin. Unlike many other African countries dependent on oil imports, Benin will become a new exporter of oil.

Cameroon

Big news from the small Cameroon nation in 1981 was a significant gas find by Mobil/Total on their H-17 tract. Additional drilling will assess the commercial potential of the nation's only find in 1981.

Fifteen wildcats were spudded in during this somewhat dry drilling season, the same number as were completed the previous year. But, although the outcome was disappointing, operators like Mobil, Total, and Elf are still enthusiastic about the area.

Ivory Coast

Production from Esso's Belier field and Phillips' touted Espoir field have spurred a keen interest in the potential future off the Ivory Coast. Production from Espoir, already dubbed a supergiant field, was expected to begin in 1982. Although some water depth difficulties are anticipated, results from Phillips recent production tests seem to indicate it will be worth the risk.

The company dismissed earlier reports that the field contained reserves estimated at 3–8 million bbl of crude. Phillips has contracted five additional rigs to work in the region and has ordered three subsea wellheads for extended production.

Gabon

Production figures from Gabon for 1981 were down slightly. A daily production rate of 178,000 bbl was recorded in 1981, down 10% from 1980. Since proven reserves are estimated at 500 million bbl, seismic work continues to grow in this area.

Elf is the operating mainstay of the territory where its 19 fields produce over 150,000 b/d. Production started early in 1982 from its Baliste field at 1,500 b/d and is expected to peak at 7,000–8,000 b/d. The field has estimated recoverable reserves of 15 million bbl.

Angola, Cabinda

The national oil company Sonangol holds the key to Angola and Cabinda exploration, where only a chosen few continue to explore. Gulf is honing development plans for the Cabinda area, while Texaco designs drilling projects for its Cuntala, Essungo, and Etele fields. Cuntala come on stream in 1981 at 1,800 b/d; Essungo and Etele start-up is expected.

Tunisia

Tunisia's only offshore oil activity is from Elf's Ashtart field, which is moving about 49,000 b/d, and the delay-plagued development of two other finds—the Isis and Miskar fields. Shell also is moving along with its ambitious development project on the Tazerka field, which marks a new step in the evolution of floating production storage and offloading systems.

Shell discovered the Tazerka field in 1979 from two promising finds on the Gulf of Hammamet Birsa. Although the company has been tight lipped concerning production figures, it is rumored that Tunisia's second oil field, Tazerka, will be producing about 20,000 b/d. Future exploration of Tunisian waters will probably concentrate on the Gulf of Gabes and Gulf of Hammamet until the boundary dispute is resolved.

USSR

Though reluctant to admit it, Russian operators are acting more optimistically about their offshore future. Caspian Sea production has increased, according to recent reports, and the Soviet Union plans to keep it that way.

The year 1981 saw the end of a decade-long decline in offshore crude and condensate production in the USSR, which encouraged the Soviets to maintain their five-year-plan (through 1985). Reports claimed offshore crude production rose by 10,200 b/d after bottoming out at 190,000 b/d in 1979–80. Subsea gas flow hit 12.7 billion cu m, and rumors predict a further hike to 15 billion will be produced by 1985.

Official statements have asserted overall Caspian hydrocarbon production increased 5.7% in the first quarter of 1981. Overall 1985 output goals of 12.4–12.9 million b/d for crude and 21.18–22.59 trillion cu ft for gas have been announced by Soviet planners making no mention of offshore targets. Later statements, however, reveal an increase in subsea oil and gas reserves is a top Soviet economic aim during the next few years. Russia produced a world record 12.03 million b/d of crude and 15.36 trillion cu ft of gas in 1981. That put the Soviet Union in second place behind the U.S. in gas output.

Their offshore crude is about 1.7% of its total output compared with more than 3.6% in 1970. Subsea gas flow of 2.9% is only a slight change from over a decade ago.

Russia still produces all offshore oil and gas from the Caspian Sea just as it has since the early 1960s. Changes are on the horizon, however. Before the decade is over, the USSR hopes to be producing 50,000–100,000 b/d of crude and condensate from the Sea of Okhotsk off Sakhalin Island in the Far East. Substantial quantities of gas from the same area, together with modest gas output from the Black Sea, are also likely.

The only other non-Caspian subsea gas now comes from the Sea of Azov where output also is expected to climb in the coming years.

The most important factors in stemming the fall in Caspian crude production were the 1979 discovery of the 28th of April oil field and the superdeep drilling at Bulla-More, one of the Baku Archipelago fields.

At Bulla-More recent drilling from fixed platforms has extended the field dramatically. One well there in 1981 tested over 1,460 b/d of condensate and 17.65 MMcfd of gas from an interval over 20,200 ft. It is expected still deeper pay will be tapped in the area. Russia hopes to find gas/condensate deposits in the southern Caspian as deep as 25,000 ft.

EGYPT

Egypt has never been busier. Although the level of oil production dipped in a number of Mideast countries in 1981, development and exploration activities were moving at a significant rate. Probably the busiest was Egypt where the offshore business is booming. Seven fields are in the development or expansion process.

Access to the eastern Gulf of Suez has enabled Petrobel to identify a large extension to the Belayim Marine field off Abu Rudeis, and five aggressive step-out wells have tripled previous reserve estimates to 1 billion bbl of recoverable crude.

Gupco is the largest producer in the Gulf of Suez where 90% of Egypt's domestic output originates. El Morgan is producing 144,000 b/d, and the July and Ramadan fields 110,000 b/d each. The latest edition to the Gupco tally is the newly named October field comprising discoveries in Blocks GS 195, 185, and 173 and producing 100,000 b/d.

Development drilling is under way, and a fourth wellhead platform has been installed. Peak output is expected to be around 140,000 b/d.

Further north in the Gulf of Suez, Suco is developing the Ras Budran field in Block EE-85. Plans include two 9-slot platforms, one 4-slot, and one production. Production will begin at around 20,000 b/d, rising to 40,000 b/d.

Egypt's only production outside the Gulf of Suez is the Abu Qir gas field in the Mediterranean where output is 100 MMcfd. Egyptian exploration in the Mediterranean Sea, however, was given an added boost in 1981 when its 2 El Temsah flowed 25 MMcfd of gas and 2,030 b/d of condensate.

To the east of Mobil's find is the highly potential North Sinai acreage awarded to British Petroleum, Total, and IEOC. Although little is known about this area, seismic tests seem to indicate some large structures.

Mobil also has stimulated interest in the southern Gulf of Suez/Red Sea area with its 1 Geisum wildcat which flowed 4,000 b/d. Only seven wells have been drilled in the Red Sea to date by Esso, Union, and Phillips.

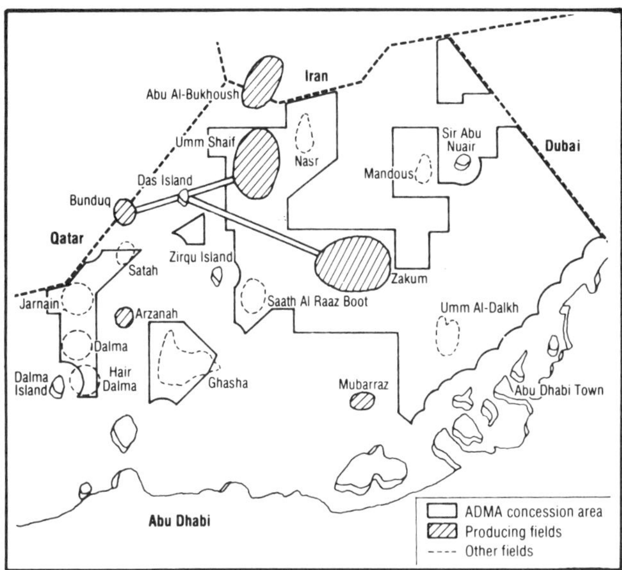

Fig. 15–9 Offshore areas of Abu Dhabi (courtesy *Offshore*)

MIDDLE EAST

Development and exploration activity in most of the countries making up the Middle East has increased dramatically amidst overshadowing influences like war and regulations.

Abu Dhabi

In January 1981, Abu Dhabi production was cut back to 1.19 million b/d from 1.27 million. Offshore fields were not affected by the cutbacks in this round, and output seems to be sitting around 600,000 b/d (Fig. 15–9).

However, development projects have the potential to more than double offshore production over the next few years. The major development on tap is the $5 billion Upper Zakum development, which will have a production potential of 500,000 b/d. The mammoth project comprises a central supercomplex incorporating water injection facilities, four separation platforms, 74 wellhead towers, and onshore processing and storage at Zirku Island. It is being completed in a single phase.

The Upper Zakum reservoir is producing 30,000 b/d. Lower Zakum is capable of producing around 200,000 b/d. Once 200,000-b/d water injection facilities have been installed at Zakum Central, production capacity will increase to 320,000 b/d.

In other areas, production from the Umm Al-Dalkh field is expected by the end of 1982, and output should reach between 20,000–30,000 b/d. The project involves a production platform and four wellhead towers. Two of them have been installed. Satah, Jarnain, and Dalma fields are in the early stages of development at a cost of around $750 million. First production should begin in 1984 and should peak to 40,000 b/d.

Other fields producing off Abu Dhabi are the Mubarraz at 20,000 b/d, Abu al-Bukhoush at 70,000 b/d, and the Arzanah at 30,000 b/d.

Dubai

Production during 1980–81 from the four fields off Dubai operated by Dubai Petroleum averaged 349,274 b/d.

The most prolific producer was Southwest Fateh at 188,631 b/d, followed by Fateh at 149,000 b/d. The Faleh field, which began production in 1978, averaged 9,163 b/d; the newest field, Rashid, produced 2,433 b/d. Current proven recoverable reserves are estimated at 1.4 billion bbl.

Other emirates

Sharjah's oil production is solely from the offshore field of Mubarak. It is shared with Iran and produces at 12,000 b/d—a big drop from its peak at 40,000 b/d. Most eyes in the area, however, are on Amoco's onshore 1 Sajaa gas and condensate discovery.

Offshore, Forman Exploration and Lasmo flowed an aggregate 14.4 MMcfd of gas in a find named al-Hamriyyah.

A second well is drilling, and Sharjah officials claim the field will be put into production as soon as possible.

Saudi Arabia

All offshore operations in Saudi Arabia have been carried out by Petromin/Aramco on a 60/40% basis since 1974, and in 1976 a full takeover of Aramco by the Saudis was agreed. However, it has only been recently that the Saudi government paid the Aramco partners compensation believed to amount to $1.5 billion for their remaining 40% stake.

The partners (Standard Oil of California, Texaco, Exxon, and Mobil) will be retained in a service capacity to explore and develop fields and to provide technical assistance. The companies have been receiving payment for their services in the form of a 25-cent discount for each barrel of oil produced, but details of the new service agreement have yet to be disclosed.

Since early 1979, production on and offshore has been running at 9.5 million b/d, 1 million bbl above the official ceiling. Offshore production alone is estimated at around 3 million b/d, mostly from the Safaniya and Berri fields.

Iran

Information is cloudy for the Iranian offshore. Prior to the 1978 revolution, crude output from the four participation areas was about 660,000 b/d. In 1982, it was estimated at 400,000 b/d.

Some restricted drilling has been reported from Ipac and Iminoco fields, and Pars gas field is still under development.

ASIA/PACIFIC

Offshore activity around the nations of Southeast Asia in 1981 showed a marked increase in enthusiasm. Indonesia increased its exploration budget. Malaysia hopes to double its output by the end of the decade. Philippines activity is mixed. Elsewhere, India will open its doors to foreign operators, and Thailand is focusing on its gas potential.

Indonesia

To demonstrate its confidence in meeting its goal of 1.82 million b/d by 1983–84, Indonesia boosted its oil and gas budget to $3.7 billion in 1982, $1.8 billion destined for offshore use. The increased outlays show the chief producing nation of Southeast Asia is enjoying its top spot and plans to hold on to it.

In 1980 Indonesia spent $2.2 billion on exploration and production expenditures. However, oil and gas interests have continued to flourish, especially in the offshore areas of the Natuna Sea. Conoco is producing Udang field, and Marathon's promising oil discoveries on Kakup A block have sweetened the success stories.

Another major reason for this country's largely improved outlook is the $30-plus/bbl that gives these traditionally small fields a more appealing commercial attraction. The fact that the country is experiencing a relatively smooth economic and political stability also enhances the outlook.

The Natuna Sea is expected to join the list of other now proven Indonesian offshore areas like Northwest Java, Southeast Sumatra, and the seas off East Kalimantan. Several majors are drawing up exploration plans. Some of the operators looking forward to future Natuna Sea work include Gulf Oil on its production sharing contract off Natuna A and Esso with its Natuna D Alpha acreage. Mobil also holds a production sharing contract for D-1 and D-2.

In Jakarta, Hudbay (Malacca Strait) Ltd. hit an oil discovery in 66 ft of water. Also, Atlantic Richfield's confirmation drilling established three wildcats in the ETA, UV, and UX areas of the Java Sea. Oil was detected in commercial quantities in separate areas adjacent to the Ardjuna producing complex.

Philippine Islands

Oil companies have developed a somewhat boomerang relationship with the Philippines since the big drive in exploration off the islands in 1973. Though a few have stayed to develop their finds, many more operators have moved in and out of the area several times during the last 5 years. It is a puzzling relationship, especially considering the future of the Philippines as a major energy producer. The half dozen discoveries under development off the Palawan coast have strengthened the wildcatting cycle, but evidence to date has determined any future finds in the area will be moderate at best.

So what makes the majors keep moving to and fro? One main reason is the economic benefits of the region where drilling terms are attractive and the Philippine government is more than supportive of any foreign company projects. Another prime component is the recent developments off Palawan, which have caused the government to revise crude production estimates, enhancing the offshore outlook.

The Philippines revised crude production estimates from three of the four fields off Palawan now range from

40,000–60,000 b/d by 1983. The lower figure was based on the Cadlao, Matinloc, and Libro-Tara fields; the higher figure was based on the potential commerciality of the new discovery off Galoc in the northeast.

When the Matinloc and Cadlao fields join the producing Nido field, the Energy Development Bureau estimates total production will reach 49,000 b/d. Cadlao field off East Palawan is expected to start up at an initial rate of 9,000 b/d and is expected to reach 15,000–20,000 b/d if everything stays on schedule.

Matinloc, under development by a Cities Service consortium off Palawan, should start producing at a rate of 15,000–20,000 b/d from two discovery wells. Reserves are believed to be at least twice those of Nido field.

Production from the waning Nido stabilized at 4,000 b/d from its 40,000 b/d output before the A-1 well was shut in due to water problems in 1981, which affected production in five other wells. Nido, the first offshore Philippine field, has produced slightly more than 12 million bbl of crude since its startup in 1979. But operators predict the field will be rated noncommercial, unless production problems are solved soon.

Malaysia

Malaysia is enjoying a prosperous time as the country continues its hopes of doubling oil production to 400,000 b/d by the end of the 1980s. This follows an 18% increase in production between 1977–78 as new fields came on stream east of the peninsula.

India

India is hoping for a new era in offshore and onshore exploration because 24 international oil companies have shown interest in exploring 17 offshore and 15 onshore blocks offered by the government. With a 10% increase in oil consumption each year and rising prices of imported crude, India had no alternative but to invite outside companies to join the oil search on a production-sharing basis. Among the 24 interested companies, 10 are from the U.S.; two each from France and Singapore; one each from Britain, West Germany, Italy, Mexico, Brazil, and Rumania.

Bombay High continues to be India's biggest oil field and may become even bigger. It produces about 160,000 b/d and 2 million cu m/day of associated gas from 50 wells on 13 platforms. The wells average 2,000–5,000 b/d each. India hopes to produce 240,000 b/d from Bombay High and the R-12 field off the Konkan coast by midyear, believing production could reach 400,000 b/d. To do this would require another set of platforms in smaller fields like North Bassein, B-37, and B-7, which are closer to land than Bombay High.

During the next 4 years the ONGC proposes to pursue exploration and development programs in areas north of the North Bassein oil field and northwest to the head of the Gulf of Cambay, in the Ratnagiri offshore area off the Konkan coast, in the deep continental shelf area flanking Bombay High to the west, and in the Lakshadweep offshore basin—all on India's west coast.

Off the east coast, exploration would be in the Godavari basin and the Andaman Sea. During 1980–85, India's state oil company plans to drill 95 exploratory wells and about 230 development wells which will require some 50 production platforms. Their ultimate goal is to produce at least 264,000 b/d by 1984–85.

Thailand

In July 1980, the Petroleum Authority of Thailand signed two major contracts: one called for laying a 32-in., 42-km underwater gas pipeline in the Gulf of Thailand; the other, an onshore pipeline contract for a 28-in., 172-km gas pipeline from the point of connection with the offshore line to the Bangkok South power plant. With a combined total of $92.8 million, the two contracts emphatically proved that Thailand is serious about its gas potential and plans to give it top priority.

The country's optimism about energy reliance was buoyed by Union Oil's gas discovery with an output capacity of 28.6 MMcfd of natural gas and 708 b/d of condensate in a previously unexplored prospect in the Gulf of Thailand. The additional reserves were discovered at three levels in the Satun 1 field northeast of its Erawan field in Block 12. The authority describes the find as commercial and expects more gas to be found in the structure.

Satun 1 was the first well Union drilled since it resumed exploration in its concessions in the Gulf of Thailand. The company's exploration activities were curtailed at the request of the Thai government late in 1979 because they wanted Union to concentrate on the development of gas fields in Blocks 10, 12, and 13.

The company is working on a second well in the field, Satun 2, and already has announced plans for many more.

FAR EAST

Although Japan received a well-deserved nod for its earnest exploratory efforts, the Peoples' Republic of China drew all the attention in 1981 as preparations were made for exploring the potential Chinese fortune reportedly hiding in the Gulf of Bohai, Yellow Sea, and East China Sea.

Japan

Backed by a supportive government, Japan continues to strengthen its offshore efforts based on their 5-year oil resources development program.

The plan hopes for discovery of 90 million kl of offshore oil and gas during the 5 years leading to production in 1985 of 8 million kl, more than double the domestic production of 1978.

Under Japan's exploration and development plan, a total of 75 exploratory wells will be drilled at a rate of 15/year. The government has agreed to pay $129 million to sink eight of these holes in water depths of 200–500 m. The other wells, to be drilled in around 200 m depths, will be covered by private firms to the tune of over $550 million.

But even if the 5 year plan is completely successful, the most optismistic estimates say Japan's offshore recoverable reserves (about 1.3 billion) will never provide more than 2% of the nation's huge demand. Drilling has continued at a moderate pace in recent years, thanks mainly to generous government assistance. Seven exploratory wells were spudded on the continental shelf in 1981, indicating the 5-year plan already is behind schedule. Of these wildcats, the most promising was drilling by Japex Offshore and Niigata Oil Exploration off the mouth of the Shinano River.

Other recent offshore activity includes a Nippon Offshore Oil well, which was sunk off the coast of Shiretoko, Hokkaido, at a water depth of 81 m. Test drilling also began off Miyako Island in Okinawa where seven more wells are slated. New Japan Sea Exploration spudded a wildcat in the Aga field off Niigata, reporting it had struck large quantities of oil. Japex will drill a wildcat in the sea off Akita with the cooperation of Idemitsu Oil Development.

China

The Gulf of Bohai is only the beginning of the vast Chinese potential to be found offshore. By opening its doors to interested foreign firms, the Peoples' Republic of China hopes to boost production as well as its technical know-how by working together.

The flow of oil from fields in the Gulf of Bohai, the only offshore producing area today in the PRC, is small but important. Estimated at 2,000 b/d, the actual production from Bohai wells may be a few hundred barrels more than that, which is hardly a significant contribution to their overall output of 2 million b/d. Therefore, it is China's potential, due more to its vast unexplored territory than to any pure geologic reasoning, that has the world's top oil companies pleading for a chance to drill.

And the potential, both onshore and offshore, is believed to be great. Western oil analysts have estimated it to be 72 billion bbl, of which 12 billion are onshore. Natural gas reserves are tagged at 19,000 billion cu m. Recoverable offshore reserves are given as 30 billion bbl.

All the offshore oil from China today comes from two producing areas in the Gulf of Bohai. Close to the western shore is Hai, which was the first discovery in the gulf coming only 6 years after the oil search began. Platform 4 has nine wells, but only six are producing. These turn out from 30–40 ton/well, although some do as much as 50 ton.

East of that field is Changbei, discovered in 1973 and the best producer in the gulf. Platform 6 at Changbei has nine wells, and eight are producing. Platform 8 has 12 wells, and 10 have oil flowing. These 18 wells are producing about 2,000 b/d, for an average of only 111 b/d each.

From the end of 1966 to 1980, the Chinese drilled 114 wells in the Gulf of Bohai. Total footage of these wells is 320,000 m, for an average depth of 2,807 m, or just over 9,122 ft. After testing, the Chinese reported they have 27 wells producing.

Since China's offshore reserves may be extensive and since these resources are concentrated in areas such as the Bohai, the Yellow Sea, and the East China Sea, oil companies have determined offshore exploration and production operations are no more expensive than frontier areas onshore.

As it is well aware of its limited technical capabilities, China is pursuing an open-door policy for petroleum development both onshore and offshore. Foreign investment will necessarily involve technological, economic, and legal condiserations.

One of the primary aims of the PRC's petroleum law will be to ensure adequate legal protection to foreign oil investors who apparently would not invest in such multibillion dollar projects without first being assured of full protection of their interests by law.

REFERENCES

1. H.D. Klemme, *Oil & Gas Journal, Petroleum 2000* (August 1970), p. 108.

2. *Oil & Gas Journal* (April 12, 1982), p. 57.

3. *Oil & Gas Journal* (August 27, 1981), p. 133.

Index